Alan Hodgkin's experiments led to the now widely accepted ionic theory of nerve conduction, but he believes that chance often plays as large a role as design in scientific discovery. His book charts the balance of the two in his own life, beginning with his childhood in an extended Quaker family. A Cambridge undergraduate in the thirties, he describes the climate of opinion among his contemporaries and the experiments which led to a Trinity fellowship but had to be abandoned on the outbreak of war. He then worked briefly in aviation medicine but mainly on airborne radar, and his account throws light on an important but little known chapter of military history. A final section chronicles the domestic and social detail of the postwar years in Cambridge and the resumption of experiments that led ultimately to a Nobel Prize and the Presidency of the Royal Society.

CHANCE AND DESIGN

CHANCE & DESIGN

Reminiscences of Science in Peace and War

ALAN HODGKIN

CAMBRIDGE
UNIVERSITY PRESS

Published by the Press Syndicate of the University of Cambridge
The Pitt Building, Trumpington Street, Cambridge CB2 1RP
40 West 20th Street, New York, NY 10011–4211, USA
10 Stamford Road, Oakleigh, Melbourne 3166, Australia

First published 1992
Reprinted 1993

Printed in Great Britain at the University Press, Cambridge

A catalogue record for this book is available from the British Library

Library of Congress cataloguing in publication data
Hodgkin, A. L. (Alan L.)
Chance and design: reminiscences of science in peace and war /
p.   cm.
Includes bibliographical references and index.
1. Hodgkin, A. L. (Alan L.)   2. Neurophysiologists–
Great Britain–Biography.   3. World War. 1939–1945–
Technology.   4. Military research–Great Britain–History–
20th century.   I. Title.
QP355.2.H63   1992
591.1′88′092–dc20   91–27737   CIP
[B]

ISBN 0 521 40099 6 hardback

CONTENTS

PREFACE

In 1971 I gave a lecture on nerve conduction at one of the centenary meetings of the Institution of Electrical Engineers. There I met G. W. Edwards of the General Electric Company's Research Laboratory, whom I hadn't seen since the war when we were working together on microwave radar. A few weeks later he sent me a photocopy of the log-book that he had kept of the early flight tests of airborne centimetric radar (AI). This was a fairly massive document as it covered some 170 flights made between March 1941 and the spring of 1943 on Blenheim, Beaufighter and Mosquito aircraft. The main aim of these flights was to explore a new system for intercepting enemy bombers at night, but, as this was the first time that a microwave radar had been installed in an aeroplane, tests were also made on submarines and ships of various size.

The early flight tests were made from Christchurch, a small civilian airfield without a runway, which had been hastily converted into an RAF Station for trying out experimental radar equipment developed by the Telecommunications Research Establishment (TRE) at Swanage, some thirty miles away.

The object of the flight trials described in Edwards's log was to see whether an experimental set, designed jointly by TRE and GEC, could pick up and intercept enemy raiders or minelayers at a range greater than the height above ground. If so, then an engineered version of the equipment would be fitted and tested in suitable night-fighters such as the Beaufighter or Mosquito. The flight trials were usually carried out with an RAF pilot and two civilian observers, one of whom watched the radar set. This last task, together with maintenance and repair of the equipment, was initially carried out by a small team, consisting of G. W. Edwards (GEC) and A. E. Downing and myself from TRE.

ix

After reading Edwards's log, I realized that with my own notebooks and letters from the war period, it provided a basis for a readable account of an interesting but little-known piece of military history. I wrote an informal account along these lines between 1985 and 1988 and then revised it in the light of comments made by a number of wartime colleagues, namely R. J. Clayton and G. W. Edwards of GEC; W. E. Burcham, E. J. Denton, R. Hanbury Brown, A. C. B. Lovell and R. Willmer, all formerly of TRE, and J. F. Coales of the Admiralty Signals Establishment, to all of whom I wish to express my warm thanks. W. E. Burcham generously lent me his wartime diaries together with two notebooks kept by A. E. Downing which came to light many years after his tragic death at the end of 1942. I have also been greatly helped by the Air Historical Branch of the Ministry of Defence who lent me the photographs reproduced in Figure 17.3 as well as the excellent monograph on which the account of the operational performance of centimetric AI in Chapter 21 is based. I am also indebted to Janet Dudley, Librarian at the Royal Radar and Signals Establishment.

Having briefly described the origin of the wartime part of this book, which I have called *Flight Trials and Tribulations*, I must turn to the rest of the book which is concerned with the much longer periods in peace. To my regret I never kept a regular daily diary, but until the war, and to some extent during it, I did the next best thing which was to write weekly letters to my mother, which she kept and returned to me toward the end of her life. These form the skeleton of my account of life in Cambridge and New York before the war, and partly take the place of a diary.

For the period after 1944 I have relied on a different set of letters, both for checking events and for a picture of the general background to our lives. In March of that year I married Marni Rous whom I had first met in New York in 1937. Before leaving that city seven years later, she promised her parents, Dr and Mrs Peyton Rous, that she would write to them every week whenever possible; Peyton died in 1970 but Marni continued the weekly letter until her mother's death in 1985. Her mother kept all the letters and gave them to Marni shortly before she died. They have proved an invaluable source of information for me, and Chapter 36, about the award of the Nobel Prize for 1963, is largely either a direct quote or a paraphrase of her letters.

Something must be said about the title of this book. In 1976 I gave a lecture, entitled *Chance and Design in Electrophysiology*, to the Physiological Society as part of their centenary celebrations, and, with essays by other authors, this was published by the Cambridge University Press in *The Pursuit of Nature*. The theme of my lecture was that the record of published work conveys an impression of directness and planning which does not at all coincide with the actual sequence of events. 'In writing papers, authors are encouraged to be logical, and, even if they wished to admit that some experiment which turned out in a logical way was done for a perfectly dotty reason, they would not be encouraged to "clutter up" the literature with irrelevant personal reminiscences.' After I had decided that I wanted to expand *Flight Trials* into an autobiography, it occurred to me that *Chance and Design* was a good description of my early life as a whole, as well as of the experiments that determined the nature of my research. I certainly did not plan a career in neurophysiology, any more than I knew what I was doing when I moved to radar work in 1940.

Some of *Chance and Design in Electrophysiology* has found its way into the present book as have parts of another brief autobiographical account, *Beginning*, that I wrote in 1983 for the Annual Review of Physiology. I am indebted to the Physiological Society and to Annual Reviews Inc. for encouraging me to write these accounts which helped me to start an autobiography of my early life. I also wish to thank the Physiological Society for permission to reproduce many figures from the *Journal of Physiology*, which appear in Parts I and III. In this connection it is a pleasure to thank Bill Smith of the Physiological Laboratory, Cambridge, who helped me with several of the figures in Part II.

I am grateful to the Royal Society for permission to quote from Christopher Hartley's part of the Biographical Memoir on D. A. Jackson, from Bernard Lovell's memoir on Lord Blackett and from the paper which David Martin, Kingsley Dunham and I wrote about our 1972 visit to China.

Finally, I must express my thanks for many helpful suggestions made by readers of the typescript; in addition to those mentioned above, these are: Marni, Jonathan and Rachel Hodgkin, Bob Robson, Tess Rothschild, Andrew Huxley and Ursula Tokle of the Royal Society.

January 1991

I

BEGINNING

Chapter 1

Childhood. My father's death in 1918

I BELIEVE that I had a happy childhood, though I would not want to have my life over again if it involved going away to school, as I did in my ninth year. My father, George, died in Baghdad in 1918, leaving my mother with three boys, aged four (me), two (Robin), and a few weeks old (Keith). My first clear memory is of going into Mother's darkened bedroom and being told by her of Father's death. She seemed calm, and thenceforward for many years was supported by the feeling that George's spirit was still with us. And indeed in a sense it was, because before we said our prayers at night we always had a 'Father Talk' in which she would speak of some episode, usually quite a cheerful one, in George's life. I think George knew that he might easily not return from his last journey because he wrote eight long letters to Robin and me to be given to us on our fifth, tenth, fifteenth and twenty-first birthdays as Mother did later.

The Father Talks came to an end in a rather comic way. The mental picture which Robin and I had of our father was based largely on a 1917 photograph of George – a rather solemn-looking man in a high stiff collar of a sort which by 1920 was utterly unfamiliar to small children (Figure 1.1). It took me many years to realize that George was not solemn at all and could be the most wonderful company. He was such a good mimic that my mother's younger sisters found it impossible to keep a straight face when they met someone whom they had heard George mimic. But at the time our picture of Father was so dominated by the Edwardian collar that one day Robin and I solemnly dressed ourselves up in very high collars improvised from the stiffened curtain sashes, put on particularly holy expressions and appeared before our mother as Father. I must emphasize that this dressing-up was not naughtiness or

3

Figure 1.1. George and Mary Hodgkin, with Alan and Robin, 1917.

play-acting but a wish to give reality to a solemn occasion. Naturally this episode shook my mother and after that the Father Talks tapered off, though he was certainly never forgotten.

Saying prayers with Mother lasted much longer and led to the episode of the Prayer Strike in which I took no part as I was away at prep school by the time. Shortly after we had moved from Banbury to Oxford, Mother decided to abolish the nursery and make one room into a place where we children could play and get on with our own business. However, she made a firm condition that this room, known as the ARK from our three initials, should always be kept tidy. Of course it wasn't, and after repeated warnings, Mother said, 'One more mess and I'll close the ARK for a week.' Inevitably, this occurred. Robin was deeply indignant and retaliated by refusing to say his prayers in my mother's presence. 'I am quite sure', he said, 'that God wouldn't wish me to say anything to Him when I am in the same room as someone who could be so cruel to her children as you have been in shutting them out of their own room.' I cannot vouch for the exact words, but that, or something very like it, was what I heard when I got home from school. Like many strikes this one ended in a draw. After a week, mother reopened the ARK and Robin resumed saying his prayers with her – so neither side won.

The reader may wonder what my father was doing in Baghdad in 1918. Other people's relations are apt to be as tedious as their dreams, but I shall need to write something about George's life as it affected my own in more ways than one.

In late Victorian times, undergraduates often took university examinations even more seriously than they do now and it was not uncommon for young men to break down in health through overworking for their university finals. Something like this seems to have happened to both George and his Trinity friend, Keith Lucas, when they both obtained firsts in the Natural Sciences Tripos. Keith Lucas will come in again later so all I need say now is that he became a very distinguished electrophysiologist and was killed in an aeroplane crash in 1917. On coming down from Cambridge, the two young men were advised to travel and take life easily for a while. However, Lucas could not bear to do nothing, so he accepted an invitation from the New Zealand government to make a bathymetrical and zoological survey of the deep lakes in both the North and South Islands. He asked George to join him, which my father was very happy to do, spending six months in 1902 on Lakes Taupo, Wakatipu and Waikare Moana. An account by George of this expedition appeared long after his death, together with chapters by E. D. Adrian and W. M. Fletcher, in a little book on Keith Lucas published in 1934.

After returning from the New Zealand visit, George's health, particularly his eyesight, was considered to be still too poor to allow him to go on to study medicine and become a doctor, like 'Uncle Doctor' of Hodgkin's disease, as he originally intended. He drifted along for six or seven years, learning Civil Engineering in Sunderland and applying it to piers and harbours in the Isle of Man. He also spent much time on various Quaker concerns and in acting as courier for his father and mother on a long trip to Australia, New Zealand and countries on the way. Much later my mother told me that she thought that George, who was the youngest of the six Hodgkin children and by that time the only one unmarried, had perhaps been a little bit exploited by his ever-loving family and that he really should not have been deflected from settling down to a regular career for so many years. Eventually, in 1913, when he married my mother, Mary Wilson, he got a fairly humble job in Gillett's Bank (later Barclay's) in the small market town of Banbury. Here they lived in great happiness until life was darkened by the war in a way that needed all their ardent Quaker faith to withstand.

The Society of Friends, to which my family had belonged since the seventeenth century, was deeply divided by the war and in particular by the Military Service Act of 1916. Some Quakers enlisted enthusiastically at the beginning of the war or followed later when large numbers of their contemporaries volunteered. Others were clear that they should not take life but were prepared to help with the Red Cross or the Friends Ambulance Unit, still others took up agricultural work, either voluntarily or on the instructions of a tribunal. The 'absolutists', of whom George was one, felt they could not take up alternative work prescribed by a tribunal under a Military Service Act. They felt that to do so would merely be to force other people to fight instead of themselves. The position of such people was not easy. Some had to serve sentences of hard labour in prisons like Princetown and Wakefield. A few were even condemned to death, though the sentence was never actually carried out.

Partly because George was married and partly because his opinion was so well known, he was left alone for a while, though he and my mother had to put up with a great deal of abuse and unpleasantness.

In March 1916, George's great friend, Harold Buxton, asked him to join in an investigation of distress in Armenia with the object of arranging the allocation of money collected by the Lord Mayor's Fund. This George could do with a good conscience because it was in no sense part of the national war effort. The party went as far as Lake Van, via Petrograd and Tiflis (now Tbilisi) where, to their surprise, they found the tercentenary of Shakespeare's death being celebrated with enormous enthusiasm. However, this note of gentle comedy stands out against a very grim background of the real misery and suffering of the many Armenians driven from their homes by the Turks. This trip led George to decide that he would return to help Armenian refugees – providing that this could be done voluntarily and not as war service prescribed by the tribunal.

The Buxton expedition to Armenia lasted only for three months for which George was given leave by the Bank. On his return, he was faced by eighteen months of harassment by tribunals who offered him the choice of the Army or alternative service like the Friends Ambulance Unit. Matters became more acute in the summer of 1917. This was one of those periods when Zeppelin raids were frequent, and in Banbury as in other British towns, there were meetings to discuss reprisals against German cities. George opposed the whole idea of reprisals at a public meeting in June 1917. The Mayor refused to allow him to speak, but some of those present became so enraged at his making the attempt that they started to drag him towards the canal with the intention of ducking him. This was prevented by the police but George and my mother were followed home by an angry crowd, booing, jeering and threatening further violence as they went. A long letter from Mary to her mother conveys the misery occasioned by this episode and its aftermath.

61 Broughton Road, Banbury.
21.vi.17.

Dearest Mother,
I have a lot to tell you since last Sunday, which proved a very eventful day for us.

On Sunday morning we heard that a town's meeting had been called by the Mayor for that afternoon, to demand reprisals. This was done at rather short notice, but there were bills up in a good many shop windows and posters.

At meeting, Friends were wondering whether they should do anything, but nobody seemed very decided about it. By the time the afternoon came, George and I felt clear that we ought to go down. George wrote an amendment which he thought might be able to get someone to propose or failing that, propose it himself. I enclose a copy.

It was a roasting afternoon, and the meeting was being held down in the recreation grounds near the station, at 3 o'c. When we got there, we found the two Miss Havelocks, but no other Friends, and very few people whom we knew. I think most of the people who disapproved had showed their disapproval by staying away. George approached one man, who admitted that he disapproved of reprisals, so George asked him if he would propose the amendment, as he was a well-known man in the town. He said he would ask to have it proposed. (In the end, he didn't even vote against the reprisals resolution). Well, the meeting began and of course we heard very bloodthirsty speeches from the Mayor and the other members of the Town Council (most of them are also on the Local Tribunal). Then, just before the Mayor put it to the meeting, George stepped up to the cart from which they were speaking, and asked if he might propose an amendment. The Mayor said 'No, we can't allow that – you can vote against the resolution if you like but we don't allow an amendment.' Then they took the vote for and against. There were, I think, 2 other people besides ourselves who voted against. As soon as the crowd saw that George voted against, they began closing in round him. I was further away, and couldn't see what was happening – as I was talking to some women who were very much worked up. Eventually 2 policemen came up, and led George off the field followed by a booing crowd and cries of 'Duck him in the canal'. He got to the Blue Bird, where we had left our bikes and then I came on and joined him there. The policemen departed, and we waited there, for about a quarter of an hour. The crowd seemed mostly to have subsided, and we thought we had better get home. As soon as we got out into the road, everyone seemed to spring up again from somewhere. They got hold of George and the bicycle, and began knocking him about, and then started dragging him off down a side street towards the canal. One girl was trying to hold off the man she was with, and there was a young man on the outskirts with a troubled face, but otherwise I think the crowd was fairly unanimous. Eventually two policemen came up and we got away.

There was only one thing I minded. I shouldn't have minded if it had been a *rich* mob. But these were all people from the poorest part of Banbury. There was a wounded soldier, a munition worker, and a discharged soldier who had just been called up again (these were the ones who had hold of George); they were all people who have a sense of injustice and bitterness constantly in their lives, and they were people with whom [one] was really in great sympathy. I tried to say something to them, and managed to get hold of one man's hand but the other said 'No I'll never shake hands with you, you pro-Germ-woman'. They had just been worked upon and their bitterness just happened to burst out at the first person who came along. The crowd couldn't have known, at the beginning, that George was a C.O. [conscientious objector] though of course the men on the platform did. I believe almost anyone who opposed would have got 'booed' – though I expect the ducking in the canal was reserved for a C.O. And you see George had not said one single word. Some of the men called out 'Let him speak, let's hear what he's got to say' – but I don't suppose if he had spoken that it would have made much difference.

On the Monday, everyone seemed to have heard about it. George had rather a bad time at the Bank. Mr Braithwaite was quite nice, but of course thought it a very unwise thing to do; the manager talked to George like a schoolmaster – though he's really a nice man, and stuck up for George to another violent Jingo man, saying 'Of course we all disagree with Mr Hodgkin, but we all like him – he has such a charming manner.'(!) After this ordeal, George came into the outer office and there was a man withdrawing his account! George turned very red and went back again! The manager went round afterwards and found that he *had* withdrawn it because of George. As George was coming home, two men passed him and one said to the other 'Just knock that gentleman down, will you?'

One does have that sort of feeling; when I went out with George yesterday, there were men who looked and nudged each other and muttered things. And the people who don't think that way, mostly think George very 'unwise' or 'ill-advised' or rather a fool.

One or two people have been very nice. Specially the Penroses. We went up there to supper last night. They hadn't known of the meeting, but as soon as she heard, Mrs Penrose wrote a letter to the Vicar, asking him to protest. He said he didn't feel clear to. He's a very kind, broad-minded man, but he is rather *too* nice to everyone. Someone else asked the Nonconformist ministers to do the same. One was anxious to protest, but the others 'didn't see their way to proceed with the matter'. I think the Penroses asked us up partly to show that they were in sympathy. We had a lovely time and I felt much better afterwards.

I would never have believed that feeling would run so high over a thing like this.

One good thing has come out of it. We see in the paper today that the resolution has been sent up to the government with the words 'representing the Borough of Banbury' omitted. They just say 'this meeting calls upon the government' etc. which is a good thing, as we know that the meeting did not really represent the town.

George says he has felt great comfort in thinking of Daddy at that Chamber of Commerce meeting early in the war.

I wish I could have talked all this to you. I should have given you a much better idea of it. If you think the rest of the family would care to see it, do send it round.

Dear love to everyone,

Your loving Mary

In the autumn of 1917, George, who was now 37, found that Gillett's had a single 'reserved' place left which he was told he could occupy if he wished, but he realized that if he did so, another married man would have to go to the front. He therefore resigned from the Bank. He fully expected to be put in prison, and to avoid the appearance of running away, he decided that he must wait at least three months before attempting a further visit to Armenia. However, he was not summoned, perhaps because the Pelham Committee, a sort of high-level tribunal before which he had appeared earlier in the year, had been so impressed by his personality.

George had the support of Aneurin Williams MP who was Chairman of the Lord Mayor's Relief Fund. He was anxious for George to start for Armenia as soon as possible, but by late 1917 the question of getting there had changed very drastically from what it had been eighteen months earlier, since the northern route was closed by the Bolshevik Revolution and subsequent events in Russia. Perhaps through his influence, George managed to get permission to attempt a route through the Persian Gulf and Mesopotamia. Later, the family learnt through George's brother, Robin, in the War Office that the plan was for him to travel north from Baghdad to Baku in the wake of a military expedition under General Dunsterville and then to go on alone over a mountain route to Armenia.

No-one in the family, except George and Mary and perhaps my Uncle Robin, knew how dangerous this proposal was, but there was very little fuss over George's departure. It was probably a great relief to be going on a definite mission rather than facing a blank future with the continuing possibility of a long prison sentence. He left at the end of April 1918, reaching the Persian Gulf by sea and then travelling up the Tigris by river boat, from Basra to Baghdad. On 30 May, Mary had a brief cable from Basra, and a few hours later a cable was sent back to George to say that his third son, Keith, had been born that same day. This news, however did not reach him until he arrived in Baghdad a fortnight later. He had written many letters on his long journey, including the eight to Robin and me, but as would be expected in 1918, nothing reached home for many weeks. After the cable from Basra, the next news was another telegram, dated 20 June, to his brother Edward, saying IN HOSPITAL BAGHDAD NOTHING SERIOUS GEORGE. This did not reach England until 25 June and was followed almost at once by a wire from General Hawker (a cousin of Harold Buxton) who was then Governor of Baghdad: REGRET INFORM YOU YOUR BROTHER DIED YESTERDAY [24 June] IN HOSPITAL BAGHDAD DYSENTERY.

This is all that anyone knew. Much later there were kind letters from the Matron and Sisters of the Officers' Hospital where George died. From a shaky pencilled postscript at the end of a letter it is clear that George had heard of Keith's birth and sent a message to him. But otherwise nothing.

Figure 1.2. Mary Hodgkin, with Alan, Robin and Keith, *c.* 1920.

One might have expected that my father's death would have made my mother, who was then only 26, unduly protective of her young family as we all grew up (Figure 1.2). It seemed, however, to have had the opposite effect, either because she was buoyed up by some inner faith or because she made a deliberate effort to avoid molly-coddling. At all events, when we were old enough we were allowed to wander about the pleasant country round Banbury or, after we had moved to Oxford, to spend whole days walking, bicycling or canoeing, often going miles from home. Later, when we had learned to use map and compass, we were even permitted to walk long distances in the snow-covered hills of the Lake District where we occasionally spent the Christmas holidays.

From an early age I was interested in Natural History and was greatly encouraged by Aunt Katie, a talented but somewhat eccentric relation with whom we used to stay in a primitive holiday cottage in Northumberland. She taught me to record my observations in a bird diary, and although her approach to natural history was thoroughly scientific, she managed to endow the subject with an exciting quality that had a special appeal for a small boy. When I was about fifteen I was recruited by a professional ornithologist,

11

Wilfred Alexander, to help him with the surveys of rookeries and heronries that he had helped to initiate. And at my second school, Gresham's, near the Norfolk coast, I overlapped with the ornithologist David Lack and spent many hours with him or another friend, looking at rare birds on the salt marshes or hunting for nightjar's nests on which Lack was making a behavioural study of some importance. All this got me interested in biology and helped to blur the distinction between learning and research.

Chapter 2

Staying with relations

Treworgan Granny

Holidays played such an important part in my life that I need make no apology for devoting a chapter to the people and places that made them possible. To avoid giving too grand an impression, I should explain that one of the few luxuries – apart from servants and foreign travel – that Quakers felt they could allow themselves, was the creation and maintenance of very large gardens. So although we were not at all well off, a large chunk of my childhood seems to have been spent in wandering about enormous gardens. This was true even at home in Banbury, where the small plot of 61 Broughton Road was allowed to open into a spinney connecting it with the larger garden of Woodgreen where the Gilletts encouraged us to explore several acres of park and even build tree-houses in out-of-the-way places.

As a girl, my paternal grandmother, Lucy Anna Fox, spent much of her life at Glendurgan, in Cornwall, where her father had planted a wonderful collection of semitropical trees on both sides of the steep ravine that runs down to the Helford River. In 1861, when she married my grandfather, Thomas Hodgkin, the historian, he took her to far-away Northumberland where they lived first in New-castle, then in the Keep of Bamburgh Castle (then unrestored), and finally at Barmoor Castle which they rented from the Sitwell family. But although she became a loyal Northumbrian, she always hankered after the milder air and soft warmth of south-west Cornwall. And indeed, her influence induced the family to make an annual pilgrimage from Northumberland, where they lived for fifty years, to stay in one of the family houses in Falmouth. Of these annual journeys, my Aunt Nelly wrote:

13

. . . I can never remember a time when March and April did not bring a family migration from the chills and damps of Benwell to the beginning of spring at Falmouth. It was a migration of the first magnitude, comprising parents, six children, Mum, Littlefair, and enough lesser staff to run the small house at Falmouth: the other maids were left behind to 'do the spring clean'.

. . . After a winter in Newcastle, the arrival at Falmouth was like coming out of a tunnel into sunny landscape. I remember kissing the polished mahogany banisters of the commonplace little staircase, just to show how much I loved coming back to them. And then, almost before we got our satchels unpacked, came the glowing impact of uncles, aunts, cousins – lots and lots of cousins – and our own formal visit to Grandmama Fox.

She was the central figure, imposing, rather formidable, but truly loved by us all . . . She was a frank old lady, and never hid her opinions; but I think on the whole we children were less frightened of her than we were of some of the grown ups. Once, when Cousin Hannah Mary offered to read aloud to grandmama, she was met with the straight reply: 'Why do thee suggest it, dear, when thee knows I cannot bear the sound of thee's voice!' To a young relative informing her of his new job the reply was: 'Well, Howard, I always thought it needed a person of some intelligence to take a post of that sort. But perhaps I was wrong.' To Blanche Pease who ventured to play the violin: 'Thank thee Blanche I used to think I liked the violin.' . . . But though merciless to the older ones, she was tolerant to children . . . I often picture her, a young woman, as she has been described to me by her contemporaries – the beautiful Sarah Lloyd with her long neck and upright carriage, her sharp eyes and quick tongue, coming among the gentle nature-loving Fox circle. I have never heard if there were any incidents before she settled down, but knowing her and remembering some of her neighbours, I can think that the assimilation was not easy.

In the end, Lucy Anna's dearest wish was fulfilled when, in 1913, she and Thomas settled in Cornwall at Treworgan, another house with a thickly wooded garden that runs steeply down to the sea. Thomas died in 1913, but Lucy Anna lived until 1934 and regularly invited us to spend several weeks of our summer holiday there.

The garden was not as large and beautiful as the one at Glendur-gan but it could keep small children occupied for days on end and I resented leaving it to visit Foxes and other relatives in the neigh-bourhood. An exception was Glendurgan which we loved in spite of having to kiss Great-aunt Rachel – who lay on a divan for twenty years until, at the age of ninety, she rose 'and walked smartly around the house'.

Children see the world from a lower level than adults and, if observant, will often find nests, caterpillars and other treasures that grown-ups miss altogether. One summer, on the morning of our arrival, Granny said: 'There is a terrible plague of wasps and I am offering a reward of half a crown [roughly £4 in today's money] to anyone who finds a wasps nest. Cornwall [the head gardener, misnamed as he was a Northumbrian] has orders to poison any that are found.' At lunch I returned saying, 'Look, Granny, I have found three wasps nests and I claim seven and six!' She found this hard to believe but I was vindicated by Cornwall when I showed him the nests that I had found by the simple method of following the wasps as they returned to their hollow tree or hole in the ground. I could have found many more, but Mother thought I should not press home my financial advantage.

The garden sloped down to a little bay where we collected cowries and learned to swim at a suitable interval after meals. The beach was midway between the Helford and Fal estuaries. There wasn't much shipping out of Falmouth, but it was still possible to see great three-masted sailing ships returning from long voyages. My visual memory has them always returning, which is possible but perhaps not very likely.

On sunny days, I would wait for hours by a buddleia, with a butterfly net in which I hoped to catch a silver-washed fritillary, or would slowly scan the trumpet vines for one of the many humming-bird hawk moths that regularly migrated from France to England in those days. A more exciting kind of entomology was 'treacling' when we coated trees with a mixture of treacle and some strong alcoholic drink, very hard to find in a Quaker household. I was then allowed to get up near midnight and visit the trees where there might be some tipsy beauty like a poplar hawk moth or red underwing greedily sucking up the alcoholic treacle through its long proboscis. There was also a chance that I might see a badger in the moonlight. I cannot remember seeing one clearly, but there were many badger holes in the garden and we used to hear them making a terrific racket in the spring.

As we got bigger, we grew more ambitious, and built a raft of pine logs on which we hoped to sail or paddle to Maenporth, the beautiful sandy beach round the headland to the east. The first year we made the classic Robinson Crusoe mistake of constructing a raft that was too heavy for us to move. Next year we did better, and built a raft made buoyant with petrol cans, on which we regularly paddled to Maenporth, sometimes in quite rough weather.

Treworgan Granny was a sad, gentle person who seemed to spend most of her days grieving for the loss of her husband, Thomas, and two of her sons, George and Edward. She wished to join them and was confident that she would but yet had a great fear of death. This imparted a certain melancholy dignity to her character which otherwise was a perfect match for her soft Cornish background. Her principal occupation was arranging flowers which was indeed a major task since every room was filled with fresh blooms arranged with an extravagance that seemed alien to her simple Quaker nature.

Granny was something of a hypochondriac and took her alimentary canal very seriously indeed. I guess that she must have taken some rather powerful laxative every day because I have never met anyone whose insides made such loud and long-lasting gurgles. We children had the utmost difficulty in keeping straight faces after some particularly loud rumble had echoed round the dining room. All this, of course, was good training for Quaker Meetings where one is sometimes subjected to the same overpowering wish to laugh.

Like all my family, Granny was a great reader-aloud, but she had no hesitation in skipping if she thought the material unsuitable. After her death I inherited a pocket Shakespeare that was obviously intended for family reading because the bad bits (though not the worst) are marked with brackets. My mother, too, was a talented skipper but in her case it was mostly removing dull rather than bawdy bits. For this there is some justification, particularly with writers who have clearly padded out chapters to fulfil the obligations of serial publication. But my mother's most remarkable trick was to give books a happy ending if she felt they were getting too sad. It was not until I read *The story of a red deer* to our own children that I realized the book had a completely different ending from the one that I remembered.

Occasionally Granny would reminisce, particularly if you were alone with her. In her long life, she had clearly met a great range of people, mostly good but some bad. However, all were treated in the same gentle, slightly wistful manner: of Hatry, the famous financial swindler: 'Such a nice man, so sad he made a mistake and had to go to prison for so long.' Or when I asked about the ex-Emperor Napoleon III, whom she had met in Surrey some years after Sedan: 'He was a most beautiful carver; he cut ham as thin as paper.' Like all Quakers, she had been strongly anti-slavery and therefore never took sugar in her coffee although she much preferred it sweet. On wet days we played non-gambling games like Parchesi or Halma and sometimes tackled a big jigsaw puzzle. One day we started on a large map of Europe in which Germany had originally been painted black. Granny explained that she did not think this right and had 'washed Germany white'. This led me to expect pieces glistening with that dazzling white which comes for some strange physiological reason, after sinners have been washed in the Blood of the Lamb. It was a great disappointment to find that what we were looking for were pieces of bare plywood-colour, a rather nasty pinkish grey.

The house was full of texts. I remember one in the upstairs loo which said '*With God all things are possible*' and another in Granny's bedroom, given to her by Aunt Vi, which annoyed her younger sister, Aunt Nelly, because it said '*Leave it all to Me*'.

Although Granny had left her own gentle imprint on Treworgan, the house was really a mausoleum to my grandfather, Thomas. Everywhere there were traces of his attractive and scholarly personality, as well as rooms full of the great library that he accumulated when writing *Italy and her Invaders*. Most of these books were too stiff for me, but everywhere also were many books of the *Heroes of the Nations* sort which are really suitable for any age. Nowadays we disapprove of historians who moralize, but it does help things along if people are clearly divided into 'goodies' and 'baddies'. There was also a fair amount of light literature by authors like Stanley Weyman or H. A. Vachell. I particularly remember a book by a Quaker author, I think Henry Wallis, about prehistoric men and women, where the heroine rescues the man who is trying to capture her for his wife and has broken his leg in the attempt. Of course physical relations were never mentioned, but nevertheless, like *Jane Eyre*, it managed to be extremely sexy.

My mother did not like Treworgan and was often ill there. I think she found the atmosphere too cloying, both physically and mentally, and much preferred holidays in the north of England with her own side of the family.

Elmfield Granny, Theodora née Harris, and Grandfather, Harry Wilson

Every Christmas we used to stay with my mother's parents, usually in Birmingham, but occasionally in the Lake District. Aunt Nelly once described the Birmingham 'cousinhood' as 'a vast spreading network of entwining roots, like growth under thick turf, of Lloyds, Albrights, Wilsons, Staceys, Gibbins . . .' She adds Foxes and Howards, but these are really much closer to the Hodgkin side. Many birthright Quakers are interrelated, but the genetic effects are not too bad because Quakers strongly discouraged marriage between first cousins.

My grandparents spent many years at Selly Wood, but my memories are mostly of Elmfield, a smaller house though still big enough for a large family and several servants. To begin with we were slightly afraid of Grandfather Wilson, who was a magistrate, had a gray moustache and a stern expression. Probably he was tired after a long day of business and hadn't time for chattering children. But he wasn't really hard to get on with as I found after a long walk in the snow when he showed me the tracks of different animals and birds. He also gave me hints about taxidermy on which he had once been expert as we could see from the stuffed nightjar, polecat and other creatures, each with a proper surround in separate cases in his study. I am not sure how these were acquired as he never used a gun, either in war or for sport – like most Quakers.

As a fourth son, Harry couldn't go into the family firm, Albright and Wilson, so he entered what one of his sisters, Aunt Katie, used to call 'that wretched screw business', a shaky concern which foundered through no fault of his own. Eventually he became a senior partner in J. & E. Sturge a small chemical company that specialized in the production of citric acid, at first extracted from citrus fruit and later made by a fermentation process.

I remember Theodora Wilson, or Elmfield Granny as we called her, as a delightful person, who combined running a large house with an energetic public life as a Labour member of the Birmingham City Council. She was brought up in Leighton Buzzard, where her father Theodore Harris was a partner in the local bank. She had three elder brothers, Fletcher, Tyndall and Henry and one younger half-brother, Alderson, whose mother died at his birth. We saw a good deal of Uncle Alderson, who was always present at Sunday lunch. I remember him as a tubby, benevolent bachelor with a gleaming bald head, an appalling stammer and an absolutely uncontrollable laugh. Granny, who entertained us with endless stories about her wicked elder brothers, such a pleasant contrast to the good Hodgkins, told us that Tyndall, who sat in Quaker Meeting between his father and Alderson, had perfected the art of keeping one side of his face in perfect repose while making a hideous grimace on the other. This set off Alderson's extraordinary choking laugh for which he would be severely reprimanded, though not beaten, as he was extremely delicate. (So far as I can make out Quakers rarely inflicted corporal punishment on their children though Treworgan Granny once told me about a Quaker child who had the tip of her tongue snipped off for telling a lie.)

Tyndall's ability to make faces got him into trouble soon after he and his brother Henry joined the family bank of Bassett and Harris in Leighton Buzzard. One day his father left the two brothers in charge. Henry had to go out for a moment and when Tyndall thought he heard him returning he tried out a new act which involved turning up his eyes until only the whites showed and stumping around on his knees, like Jose Ferrar in the film about Toulouse Lautrec. As the door opened, Tyndall heard a gasp, and on turning back his eyes was just in time to see the bank's most important customer leaving the office in a hurry. There were many other such escapades in which Granny who ran the house until Theodore Harris remarried, had to act as intermediary between a stern father and his wild sons. I think that both Fletcher and Tyndall died young, perhaps from alcohol, after lives of dissipation involving drink, unlawful games and, in Tyndall's case expulsion from Oxford and marriage to a barmaid (who turned out a good sort). Details are lost to history, for Quakers do not keep annals of wickedness any more than they record the many people and good causes they have helped. I did meet Uncle Henry and was surprised to encounter a handsome, fussily dressed Edwardian character who smelled strongly of scent and cigars and spent an idle but pleasant life drifting from one hotel to another.

We spent two Christmas holidays at Wood House in the Buttermere valley of the Lake District, where Michael Wilson (mother's youngest brother) taught us about rock climbing. There was a firm rule that we must not do this on our own, but we were allowed to climb the snow-covered hills after we had learnt the obvious mountaineering rules. Since then I have done a good deal of hill walking, but think that my early rock climbing was mostly bravado as I have a bad head for heights.

The holidays in Cumberland had something of the quality of a family pilgrimage. Apart from the Harris line who came from Hampshire, Theodora's ancestors were all from the north of England, with Allasons and Fletchers from the Lake District and Tindalls from Scarborough. The Wilson connection with Cumbria was even stronger and a large web of cousins can be traced back to Isaac and Rachel Wilson of High Wray – a small village to the west of Lake Windermere. You can form a vivid picture of life in rural England and America, which Rachel Wilson visited on a mission in 1768–9, from the diary and letters which she wrote on her extensive travels (Somervell 1924). In the seventeenth and eighteenth centuries, Quakers were widely scattered by persecution and by the Civil War, and many ended up in remote communities in England and America where they were able to get on with their quiet lives in peace. Their faith and cultural traditions, especially their belief in truth, was held together by 'preachers' who might be of either sex. After bearing eight children Rachel went to America, with the full support of her husband, who remained at home, on a mission lasting eighteen months, on which she visited many towns between Rhode Island and North Carolina. She comes out of her letters and diaries as a most attractive person, quite unlike the bossy female Friends, whose testimony caused the meteorologist Luke Howard to leave Quakers and join the Plymouth Brethren in 1836.

Aunt Katie

Although we looked forward keenly to the holidays at Treworgan or Elmfield, they did not have such a strong effect as the fortnight that we spent each Easter with Aunt Katie in her cottages on Budle Bay near Bamburgh in Northumberland. You may feel that I have said enough about relations, but I can't omit Aunt Katie who encouraged my interests in Natural History and insisted that I record my ornithological observations in a bird diary, which should preferably be illustrated. She also occupies a central position in our family tree; having been born a Wilson, my grandfather's sister, she married Edward Hodgkin, my father's elder brother, and was thus both an aunt and a great-aunt.

I barely remember Edward Hodgkin, who died in 1921 of one of those mysterious illnesses that were not discussed in the presence of children and in his case was sometimes attributed to overwork during the war. I knew that he had been to Trinity (Cambridge), where he had got firsts in both parts of the Natural Science Tripos. Years later I found that he really wanted to stay on at Trinity to do original work, but to avoid disappointing his father he gave up this ambition in order to become a partner in what was still a private bank. This regret seems to have stayed with him all his life for he continued to read scientific periodicals, put up a small observatory in his garden and lectured in Newcastle on *Light as a repulsive force, Sidereal astronomy, The interior of the earth* and *Animal life and instinct* to name some of the subjects listed by his sister Nelly. A more transient but keenly felt sorrow was the abrupt ending of his rowing career at Cambridge. 'He had good hopes of getting his Blue, but his mother saw him in the May Week races, she consulted his doctor and that finished it. The disappointment was bitter. He accepted it but never spoke of it afterwards.'

As a banker in Newcastle Edward was successful if not contented and continued his father's philanthropic activities and interest in the university. My brothers and I have every reason to be grateful to him for he left us a legacy which helped with our education and brought in about £300 per annum in the 1930s. But for this I doubt if I would have had the courage to attempt a research career in physiology without being medically qualified, a course of action of which most of my teachers disapproved.

Edward and Katie lived at Old Ridley, another house with a large garden which tumbled down through steep woods to the Tyne, near Hexham, some twenty miles from Newcastle. Aunt Katie continued to live at Old Ridley after Edward's death in 1921 although it was much too large for a single person. But she filled the house with a fantastic number of things, some useful and beautiful, others not, but none needed in the vast quantities in which my mother found them when she undertook the 'unpicking of Old Ridley'. She was also intensely hospitable – to animals and birds as well as the friends and relatives who shared her many hobbies. When old and rheumaticky she shared her breakfast in bed with red squirrels from the garden who would introduce the next generation to her when they arrived; and she had fixed up a mirror beside her writing desk so as not to miss the greater spotted woodpeckers

21

when they arrived at the nut-board outside the drawing room window.

I did not go to Old Ridley until much later. At Easter we joined her in one of the very primitive cottages on Budle Bay looking out across the mudflats and tidal pools to Ross Links and Holy Island. It was a nice time of year with gorse and primroses on the headlands, sandwich and arctic terns arriving, but still with great flocks of wild geese on the mudflats, and the Cheviot Hills white with winter snow. Aunt Katie spent much of the day at her desk, writing letters or painting another gifted study of the strange shifting colours of the mudflats. But if it was fine she would sally out to study the diving birds in the bay or to look for the neat nest and beautiful blue eggs of a stonechat amongst the bents or gorse bushes.

Nest hunting was something Aunt Katie took very seriously; in her hands it required careful planning and had all the excitement of a military campaign. 'Alan, you must watch from the pier, and I will take up the washhouse position.' The nest most prized was that of the golden plover. For that, Katie's ex-coachman, Nichol, had to be mobilized and the family was driven some six miles to Belford moor where one or two pairs nested. Nichol had found it difficult to make the transition from carriages to cars and was in fact a very poor driver so the journey was puctuated by cries of 'Nichol, Nichol, you are in the ditch'. Occasionally we would stop and Nichol would be asked to destroy the nest of a carrion crow, at the top of a tall tree. These birds, though nasty predators, usually build completely inaccessible nests so that poor Nichol was repeatedly set an impossible task, which he carried out to the extent of putting a gaitered leg on the tree and then sadly shaking his head.

My bird diary and an essay that I wrote for an RSPB competition record that we found our first golden plover's nest in April 1928 having hunted without success in the two preceding years. The search followed a well-established pattern and was not unlike a minor piece of scientific research. You collected evidence as to where the nest might be and confirmed your hypothesis by finding the four beautifully marked but well-camouflaged eggs. But sometimes your initial hypothesis was wrong and there were no eggs, either because it was too early in the year or because you had been watching the male instead of the female bird (the two are superfic-ially identical) and it had been sitting on a 'scrape' or dummy nest. To obviate the latter possibility it was necessary to leave two watchers behind when the rest of the party left the area. But this didn't always work and I remember my deep indignation when Aunt Katie said, 'The hen must have seen us, Alan's pink face was rather conspicuous.' 'Not to mention,' I thought, 'the vast grey bulk of Aunt K. sitting on the heather.'

For me, watching birds with her was the most absorbing thing but every aspect of the holiday was filled with the same life and excitement. You tried to sketch and she would tell you exactly what paints to use in order to catch the colour of the light on the bay, although there was no hope that anything would come of your efforts. She spoke of the danger of the channel in the bay and it became a sinister eddying current to be avoided at all costs. The headland was being eroded by the wind, so you saw it crumble away in no time and were relieved to find it much the same next year. And after the bright days in the Easter wind, there were the evenings in the lamplit room when she talked of every part of the world and strange birds and butterflies she had seen: a greenshank's nest at Altnaharra, an expedition to Palestine, a lappet moth in the porch at somebody's wedding, safely captured because 'I alway carry a pill box with me.' It all had a charm and excitement which remained with us for the rest of our lives.

Of course there was a good deal of repetition and I am afraid that we three boys played a mean game in which each chose a story and then tried to work the conversation round to the point at which it would appear, for example, bags I the dragoman (a story of a visit to Palestine) or the wigeon's nest in Sutherland ('So I marked where the duck plummeted to the heather and sprinted up the hill') or the hyena and cousin P. ('But Aunt Katie, do you really mean that cousin P. had all his face scraped off?' 'Yes I am afraid so. He was always overconfident.')

Chapter 3

Mainly schools, 1923–32

The Downs School, Colwall

Perhaps because there was no very suitable school at Banbury where we lived I went to a preparatory school, The Downs, near Malvern when I was nine years old. I think the parting must have been more painful for my mother than it was for me, but she consoled herself with the picture of the charming wife of the headmaster reading *Treasure Island* to a group of small boys in front of a glowing fire. There was nothing wrong with the picture, and bedtime reading was certainly a consolation; but small boys can behave devilishly to one another, and going away to school is something I would wish to avoid if I were offered my life again. Still there was much that was enjoyable about school. We were allowed plenty of time to follow our own pursuits, which in my case involved bringing up a pet owl or hunting for birds and flowers on the hillsides that look out across the wooded Herefordshire plain to the distant Welsh mountains. Another more sociable activity consisted of helping the headmaster build a model railway with an engine and truck large enough to carry six or seven boys a distance of several hundred yards. I do not think we can have learned a great deal because I found myself in the bottom form of my next school, Gresham's, where I went at the age of thirteen and a half.

After we had moved to Oxford my younger brothers went as day boys to the Dragon School, a large and famous preparatory school where the clever boys got Winchester scholarships and learnt Greek and Calculus at the age of twelve. The Downs was certainly not in that category at all for which I am not particularly sorry as I could no more have got a Winchester or Eton scholarship than I could fly, and it is better not to be discouraged at the beginning of life. But I

24

think The Downs did all right by its pupils just the same. My contemporaries included Fred Sanger and Alex Bangham on the scientific side, neither outstandingly brilliant at school, and the painters Lawrence Gowing and Kenneth Rowntree. The youngish headmaster, Geoffrey Hoyland, had a series of boyish enthusiasms which he communicated almost automatically to his pupils. In this he was greatly helped by his marriage to Dorothea, one of George Cadbury's daughters, who was keenly interested in the school and helped Geoffrey to make it the stimulating place that it undoubtedly was. Without the Cadbury wealth it seems unlikely that the school would have been able to acquire Brock Hill Wood, in which different packs of boys were encouraged to build log huts each surrounded by a palisade. Nor would Geoffrey have been able to develop his large model railway or his collection of modern painting without Dorothea's help. Both Dorothea and Geoffrey were Birmingham Quakers and at one time my mother had shared a governess with Dorothea. It was probably for this reason that I went to The Downs rather than to some more conventional preparatory school. As she had noticed, Dorothea read aloud to the smaller boys and Geoffrey to the older ones. He particularly liked reading M. R. James in a darkened study and insisted that there was one story that was too frightening to read, much worse than *Whistle and I'll come to you* for example. Of course although secretly terrified we begged him to read the really frightening one but he never would. Since then, although in no way believing in the occult, I still admire M. R. James and retain a slight fear of the supernatural. I am also very doubtful if the especially terrifying story actually existed.

One of Geoffrey's many schemes was an elaborate system of merits and demerits in work and general behaviour. If the school had reached a certain target and the weather was fine a shilling holiday was declared; on such occasions each boy was given a shilling and after buying food for the day little groups sallied into the countryside, often going a long way from home. I remember bicycling some forty miles to Ross on Wye and back and another long day in the Eastnor woods picking wild daffodils.

With all this extracurricular activity there was relatively little time for regular work and it is not surprising that the school had few star scholarships to its name. None of this did any harm but there was a darker side to Geoffrey's character which led to occasional break-downs in health and on one occasion to his leaving the school for a year. My guess is that although apparently happily married he had strong homosexual feelings and was subject to sadistic impulses that he could barely control. I never suffered in any way but I remember the horror expressed by a nurse as she examined the savage wounds that Geoffrey had inflicted when caning one of my contemporaries for stealing.

Towards the end of my last term the school broke up in some disarray. It appeared that Geoffrey Redmayne, a boy with whom I was friendly and had shared the upkeep of the young tawny owl and magpie, had been found to have hanged himself in an out of bounds woodshed. Redmayne was in trouble with the headmaster for a minor crime which he had aggravated by persuading the head-master's nephew to own up for him. At the same time Redmayne was made a general butt and bullied by his contemporaries, something that had happened to me earlier in the year. This hardly seemed enough to account for suicide, particularly as Redmayne was due to leave the school for good in a fortnight's time anyway. Later we heard that at the inquest the coroner had brought in a verdict of accidental death; at the same time Geoffrey Hoyland wrote to me that our owl had been released and had been seen hunting successfully in the gardens. My mother asked me if I believed the verdict. I said that I didn't and was surprised to find that she agreed with me. I suppose there must always be an element of doubt in such cases and accidental death was certainly the kindest verdict. I felt no guilt as perhaps I should have done. It is true that sympathy from a friend might conceivably have prevented the tragedy but I felt that I had done enough in not joining in the persecution of someone who had bullied me earlier in the year.

Gresham's School, Holt

Most boys from The Downs went on to one of the Quaker public schools, principally Leighton Park or Bootham. The reason my mother chose Gresham's, a small public school in Norfolk, is interesting in view of the way that my scientific life subsequently developed. As I mentioned earlier, my father's friend the physiolo-gist Keith Lucas was killed flight-testing military equipment in 1917. Like my mother, Mrs Keith Lucas was left to bring up three boys, in her case with very little money. However, she made an outstanding success of running a day school in Tunbridge Wells where she lived. All three of her boys were at Gresham's and she spoke highly of its scientific teaching and modern approach to education. One thing that appealed strongly to my mother's pacifist views was that the Officers' Training Corps was not compulsory and I went on condition that I did not join the Corps. This, though in no way her fault, was probably a mistake. Gresham's with only 200 boys was near the lower limit at which it could send a contingent to the annual OTC camp and my housemaster, Colonel Foster, was desperate to get all the recruits that he could. If you did not join you were regarded as a sort of pariah and this, though I stuck up for my family's views, was something that I did not like at all. If she wanted me to be a pacifist my mother would have done better to let me join the OTC and find out for myself how tiresome the military establishment can be. Those of us who did not join were known as non-recruits and on Tuesdays when the others were square-bashing in uniform we went for a gentle run. Our stock was raised when we were joined by the quiet personality of Benjamin Britten. We may not have recognized his genius as a composer, but realized at once that he was a wonderful performer on both organ and piano and his organ voluntaries made chapel more entertaining than it otherwise would have been. Britten was always delicate and it is sometimes said that he was unhappy at school. This may well be true but I think Gresham's behaved sensibly about him, for they made no difficulty about his going once a week to London for music lessons, and Greatorex the senior music master made no attempt to teach Britten but simply asked him to play whatever he felt like. Rather surprisingly, Britten had a good eye and looked like becoming an effective cricketer. However, he was allowed to give up games when the danger to his hands was pointed out. And if the school did nothing else the deep Norfolk countryside in which it was set clearly contributed to Britten's feeling for East Anglia.

In some ways Gresham's was a rather odd school. It was founded as a free Grammar school in 1546 by Sir John Gresham, who endowed it with property in Norfolk and London and entrusted its management to the Fishmongers Company. It remained a small day school in Holt until the end of the nineteenth century when some long leases in London fell in, giving the Fishmongers a large sum of money to be used for the school's benefit. They decided to found a new, modern public school with an attached junior school, at the same time maintaining the Holt scholars as a day school element.

27

The first headmaster, Howson, and his successor Eccles were very successful in attracting able pupils and staff and both devoted their lives to the welfare of the school in a single-minded fashion. All this was to the good but Holt is a tiny town, miles from anywhere and the school tended to become static and ingrown. Roughly speaking everything was brand new in 1900, and was then retained exactly as Howson ordained it for ever afterwards. This applied even to the school uniform which looked remarkably like something designed thirty years earlier. Discipline and conduct were supposed to be regulated by a Howsonian invention known as the Honour system. When you arrived you had to make three 'minor' promises to your housemaster and three 'major' ones to the headmaster. In my time the three minor promises were 'Not to eat sweets', 'Not to use bad language' and 'Not to personally interfere'. This last sounds more like a major promise, but it was explained that all that was meant was that you should not have pillow fights in the dormitory. At all events if you broke a minor promise you were supposed to own up to your house master; breakage of major promises had to go to the headmaster. I am afraid that I have forgotten the major promises but they included 'Not to smoke' as one and 'being pure in thought word and deed' as another. Promise-breaking was in fact rarely reported except during the run up to confirmation when housemasters' lives were made a misery by boys reporting that they had said 'damn'. When I was there the promise about sweets was being eroded by the chocolate digestive biscuit and the stuffed date. I think we had reached the point that all sides had to be covered with chocolate before the biscuit became a sweet. But you could see that the end was near.

Fairly soon after arriving at Gresham's I found that it was a wonderful place for watching birds. On Sunday we had a long afternoon free, from 1.30 to 5.00 p.m. I soon discovered that if you walked fast, or ran if you had lingered over an ornithological rarity, you could make an excellent round: across fields and heath to Cley, along the bank across the salt marshes to the shingle beach and then home by the pretty village of Salthouse. The round had to be made on foot because bicycling was not allowed on Sundays (probably not Sabbatarianism but to prevent boys straying too far afield) and we were all in our Sunday best with straw hats which had to be held between your legs when using binoculars on the windswept shingle bank.

Ornithology was not the fashionable subject that it became later and at Cley there were no lookout shelters, equipped with telescopes and bird pictures, as there are now. But there was an ancient watcher called Bishop, probably an ancestor of the present watcher, who saw to it that wildfowlers did not shoot on the small protected part of the marsh and might tell you if there was any great rarity about. To begin with I did the walk alone, but I soon made ornithological friends, though not from my own house which was considered slightly bad form. The companions that I remember were David Lack and his younger brother Charles and a talented boy called Maury Meiklejohn, with whom I later spent several holidays, sometimes at our holiday house in Islay and once in Florence and Siena. Meiklejohn, who died in 1974, ended up as Professor of Italian in Glasgow, but ornithology was his real passion and he might have been happier if he had been able to take up the subject as a career. David Lack who was three years older than I, eventually became the first Director of the Edward Grey Institute in Oxford and was already making distinguished studies of bird behaviour and ecology. One of my first contacts with biological research was in helping David to complete an elegant study of nightjars. He showed that these birds, which like to nest at the edge of woods on sandy heaths, manage to cram two broods into their brief summer visit. After the first brood has hatched the male takes over the maintenance of the young and the female lays a second pair of eggs. Lack proved his theory by ringing adults captured on the nest, a procedure which also enabled him to determine sex from a characteristic difference in the feet of the male bird. Nightjars sit very tight so that their capture is easy if you know exactly where the nest is, but impossible otherwise as the birds are beautifully camouflaged. Lack needed to know the location of some dozen nests and enlisted me in the search which consisted of dragging a rope held between us across bits of heath in the hope of disturbing a brooding nightjar; if successful the bird would flutter across the bracken, trying to draw us off in a broken wing act.

Like The Downs School, Gresham's can claim distinguished alumni with Auden as well as Britten on the Arts side, and several scientists including William Rushton, whom I first met when he returned as a distinguished Old Boy to talk about electrophysiological experiments on nerve and muscle. I soon decided that I wanted to go to Cambridge and having started in the bottom form I had to work hard in order to reach the scholarship class. As one might expect in so small a school, teaching was somewhat patchy. Of the subjects that later became important to me, mathematics was well taught but I have never been able to do sums against the clock and more or less gave up the subject during the two years that I worked for a scholarship. In the lower forms physics was taught atrociously by the headmaster, with extreme rigidity, not only as to the wording but also as to the pronunciation of the laws and definitions that he made us learn by heart. Thus you had to distinguish between 'that' (to rhyme with hat) as an adjective and 'th't' as a conjunction. I could not bear this and gave up physics before I reached the sixth form where I might have learnt something from a different master.

For some time I could not decide between history and biology, but in the end natural history won the day and I managed to get a scholarship at Trinity College, Cambridge, in Botany, Zoology and Chemistry. In all these subjects I was well taught by youngish masters. We were encouraged to read widely and to work on our own, and this was perhaps the most important thing that I learned at school. During one summer holiday I spent an enjoyable week investigating the distribution of specialized plants that grow on the sand dunes and salt marshes of Scolt Head Island off the Norfolk coast. This must have helped me in my scholarship examination as I was lucky enough to be set a question covering the sort of ecological work that I had done. It may also have brought home to me the powerful physiological effects of a hostile ionic environment in which only the most thoroughly adapted plants survive.

In retrospect such an education seems much too specialized and it was clearly absurd to give up physics. But I did continue with classes in English, German and French. Sixth form English was fun as it was taught by Denys Thompson, the editor of *Scrutiny* and an ardent Leavisite. I had and have considerable reservations about the writings of Leavis but his approach to the teaching of English was stimulating and encouraged us to read much that we would not otherwise have done. French was taught by another gifted but eccentric master, MacEachern, who eventually quarrelled with the headmaster and moved to Shrewsbury. We were supposed to learn languages by the direct method, which meant that you never spoke a word of English and used Larousse rather than a dictionary. MacEachern's interpretation of the method was to get the class to recite French poetry in perfect unison. He particularly liked the sound of Alexandrines rolling out and we got through much of *Phèdre* and several other plays of Racine. But he didn't stop there and by the time we had finished we had absorbed poems from Villon to Valery, including some like Mallarmé or Rimbaud of which I can have understood very little. But I had been brought up to learn poetry by heart and much of this stuck with me for many years. Perhaps I would have done better to learn French in a more conventional manner with a better knowledge of grammar, irregular verbs and so on. But I have never regretted the poetry, which paid off later when I had to answer a question involving symbolist poetry in a rather highbrow exam that Trinity used to set its fellowship candidates.

Getting a scholarship emboldened me to visit my future Director of Studies in Trinity, Dr Carl Pantin, an experimental zoologist of great charm and distinction. I asked him what I should do in the nine months before I came to Cambridge and he gave me some excellent advice, which I had the sense to follow. The advice was that in my last term at school I should do no more biology but should concentrate on mathematics, physics and German. He also said 'You must continue to learn mathematics' and this I have endeavoured to do during the rest of my life, or at any rate till a year or two ago. For a long time my bibles were Mellor's *Higher Mathematics for Students of Chemistry and Physics* and Piaggio's *Differential Equations*. But I cannot claim to have done all the examples in Piaggio, as another admirer of that work, Freeman Dyson, has evidently done.

I told Pantin that I was going to a German family in June and July but that I would like to spend May of that year at a Biological Research station. I suggested the marine laboratory at Plymouth, where I had once been on a schoolboys' course; but Pantin thought a shy eighteen year old would be lost in a relatively large laboratory like the one at Plymouth. He suggested that I go to the Freshwater Biological Station which had just been set up at Wray Castle on Lake Windermere and was directed by two young men, Philip Ullyott and R. S. A. Beauchamp. I jumped at the idea, not least because it would provide an opportunity (as it happened the only one in my life) of spending May in the Lake District.

Wray Castle, May 1932

This was my first experience of research and a fairly odd one at that. Wray Castle was a large ivy-covered, Gothic-revival castle built in the nineteenth century on the western shore of Lake Windermere. It was large enough to house many scientists but was otherwise utterly unsuitable as a laboratory. There was also something strange about the work that was being done there. Beauchamp was carrying out fairly standard freshwater ecology, but Ullyott was trying to discover how a light-shunning planarian (*Dendrocoelum lacteum*) moved down a non-directional light gradient. This sounds straightforward enough but Ullyott had managed to build up a reputation for black magic in the quiet rural community outside the castle. He worked in a cellar wearing a black cloak and mask in order to make certain that no stray light reached his apparatus, in which all rays were supposed to be normal to the plate on which the planarians crawled. At the time I admired these precautions, but I have wondered since whether they were not partly done for their dramatic effect. Later on when Ullyott moved to Cambridge after being elected to a Trinity Fellowship his colleagues in Zoology tended to laugh at his methods and had no particular respect for the problem he was trying to solve. This is sad because Ullyott's experiments were interesting and his theory of movement in a light gradient was very like that advanced some forty years later by Berg and others to explain bacterial chemotaxis.

Ullyott suggested that I should study the effect of temperature on another planarian, *Polycelis nigra*, and in particular should see if the animals congregated at the cold end of a temperature gradient. I spent some time building an incredibly haywire apparatus but reached no definite conclusion except that the animals did congregate in the cold and that this was only partly explained by the fact that they moved faster in the warm. Six months later I tried to continue the experiments in the spare bathroom at home, but nothing came from this, apart from the disturbance to our guests.

Just before I left Windermere Ullyott told me teasingly that I had rescued his reputation in the local community, first by playing village cricket and second by going to church. As I am almost as bad about going to church as I am at playing cricket I can only attribute these astonishing facts (which I have verified in an old letter) to the

32

persuasive powers of the local vicar, whom I dimly remember as a most engaging old boy

A strange feature of this visit, which I realized only quite recently, is that the tiny village of High Wray in which I lived was the home of Rachel and Isaac Wilson, the eighteenth century ancestors from whom my mother's family and the vast clan of Quakers chronicled in the Wilson pedigree book are descended. It is pleasant to think that they came from such a nice place.

Philip Ullyott's history is somewhat melancholy. After a year or two as a Research Fellow at Trinity he moved to a teaching post at Peterhouse which he found too conventional for his taste. In the late 1930s he took part in a zoological expedition to Lake Ochrid in Albania. When it was time to return to Cambridge he decided that he would rather stay, which having a little private income he was able to do, living in a somewhat irregular homosexual ménage which the Nazis when they invaded would certainly not have tolerated. Philip got out just in time in a local caique and somehow made his way to Istanbul where he got some kind of job as a freshwater biologist. He returned home in the seventies, a serious alcoholic and a thoroughly disappointed man who eventually killed himself. I do not know whether he was aware that his work on planarians had anticipated a fairly important development in biology or whether such knowledge would by then have made any difference to him.

Chapter 4

Summer 1932: Oxford, Frankfurt and Scotland

Oxford

After Wray Castle I returned briefly to Oxford before setting out for Frankfurt. Perhaps I should explain that in those days it was thought highly desirable that anyone intending to read science should have a reasonable knowledge of German. A good way of doing this was to stay with a German family for a few months in the interval between school and university. My mother had arranged my visit with Frau Wirth who had been highly recommended to her by friends in Oxford.

I think it must have been during that week in Oxford that I first met Reader Bullard, who will appear several times in this book and who had a considerable influence on my life. We were soon to acquire a family connection with him as he was married to one of A. L. Smith's seven daughters and my mother would shortly marry their elder brother Lionel, as I knew when I met Bullard.

At that time Bullard, who had served in the Middle-Eastern consular service for many years, and for that reason was universally, if inaccurately, known as Haji, was Consul-General in Leningrad, having previously held the same position in Moscow for a year. I had heard glowing accounts of Soviet Russia from various Quakers who had made short visits there, and knew that others like Bernard Shaw or the Webbs, who had made apparently critical studies, regarded it as a kind of Socialist paradise. So I was taken aback to find that Haji, who spoke Russian fluently and was in a position to know what was going on, took a very black view of Soviet Society. He agreed that it was still possible to see good opera and ballet or performances of the *Cherry Orchard* and *The Government Inspector*. But he gave many horrifying examples of the role of

34

informers and the way in which children were encouraged to denounce the bourgeois origin of their parents who might then disappear without leaving a trace behind. I began to think that the 'dictatorship of the proletariat' was a very unpleasant concept. You can read a detailed account of what Bullard saw in his autobiography, *The camels must go*. His criticisms cut a lot of ice with me because, unlike most members of the consular or diplomatic service who might be expected to take an anti-Soviet line, Haji had a very humble origin and had achieved his present position and later, much more eminent one, by hard work and a remarkable gift for languages. There was therefore no question of his toeing an out-of-date Proconsular line handed down from his ancestors. For many years Haji's father worked as a casual labourer in the London docks and although he eventually achieved a regular position, he rose no higher than wharf foreman earning about £2 per week.

This background, as well as Haji's quiet and convincing account made me very sceptical of the claims that Soviet Russia, if not perfect, was at least a potential paradise. For this I am very grateful.

Another event which influenced my political thinking during that summer had the immediate effect of pushing me to the left rather than the right. About half a mile from our home in North Oxford lived the Arthur Gillett family. Arthur Gillett, a Quaker banker, had been a great friend of my father's and perhaps for that reason kept a fatherly eye on us. At all events we often went on joint expeditions to one of the many attractive woods or bits of downland to be found near Oxford. The eldest boy, Jan, about two years older than I, was a scholar at King's College Cambridge, studying biology. He planned, and I think achieved, a career not unlike the one I planned for myself, as an applied botanist working overseas, in his case mainly Kenya. Tona, my age, was going up to St John's College, Cambridge, in the autumn to read engineering and like many engineers was spending the preceding year in an industrial concern, I think in Manchester.

Unfortunately both Jan and Tona were dedicated communists. I do not know how or why, but would guess for the usual basically humanitarian reasons: dissatisfaction with massive unemployment and the hardships associated with a meagre dole, disapproval of the existing inequalities in wealth, and a feeling that a small country like Britain had no business exploiting and attempting to rule something like one third of the world. In their case their beliefs were made more fanatical by an event which led to Tona being sent to prison.

In the country near his engineering works, access to some pleasant hills was blocked by a grouse moor on which the owners allowed no trespassers. Tona had joined, and perhaps helped to organize, a large demonstration which was to march across the moor establishing the public's right of access to the hills beyond. The owner of the moor responded by attempting to block the demonstration with a party of keepers, beaters and workers on his estate. Tona became involved in a scuffle with a keeper, was arrested and sent to gaol for three months – a pretty severe sentence for a boy of eighteen, which would prevent his ever getting a post in the Civil Service. Fortunately it did not prevent him going to Cambridge in the autumn.

At this distance in time, and without knowing any details, it is impossible to form any opinion as to the justice of the sentence. But it was represented to me as an example of the wickedness of the capitalist state in which we lived, and for a time made me sympathetic to Tona's requests that I should help him distribute Daily Worker type material in the poorer parts of Cambridge – something that for quite apolitical reasons I absolutely hated doing.

Frankfurt

Apart from a summer holiday in the south of Brittany, the visit to Frankfurt was my first trip abroad. I seem to have arrived short of sleep and in a somewhat dazed condition. The Wirths were evidently horrified by my inability to speak more than feeble schoolboy German. In perfect English Frau Wirth, whom I liked at once, said 'For one evening we will all speak English but from tomorrow it will be only German.' My German was in truth very feeble. At Gresham's regular German was offered instead of Latin but my mother, quite rightly, insisted that I continue with Latin for a while and take German as an extra. She also arranged a few lessons in the holidays. But most people need much more than that before they can speak any language, let alone a complicated, inflected one like German. However, the Wirths remained firm by speaking no English themselves and insisting that I struggle with my German. To help matters they arranged for me to have German lessons with Fraulein Burnitz, an elderly unmarried lady with the unusual distinction of being a member of the very small Frankfurt Quaker Meeting. When I got back after my first lesson the Wirths were startled to find that I was on 'Du, deiner' terms with my teacher. In Germany, at that time the second person singular was reserved pretty strictly for your immediate family and it struck them as comic that I should have got on intimate terms so very quickly with this elderly lady. It hadn't seemed so odd to me because I was used to my elderly Quaker aunts 'thee, thou-ing' one another.

At all events Fraulein Burnitz was quite a good teacher and gave me a lot of homework which I did in the Wirths' pleasant garden in Sachsenhausen, a suburb south of the River Main. That year, 1932, was the centenary of Goethe's death and as Frankfurt was his birthplace and he spent much of his life there I learnt a great deal about him during my stay. I didn't mind this because I think I realized vaguely that he stood for a liberal tradition opposed to the Prussian, militaristic philosophy which was so soon to be in the ascendant. The immediate effect so far as my German homework was concerned was that Fraulein Burnitz set me to read Biels-chowski's two-volume life of Goethe and to keep a notebook in which I wrote down every word that I didn't know. This occupied my mornings in the garden for much of the next six weeks.

Dr Wirth, who was very much the head of the family and of whom I was slightly in awe, held a senior administrative position in the big city hospital located nearby. There were two daughters, Marlene aged 16 and Renate aged 11, with whom I spent many pleasant hours without falling in love, though I think Frau Wirth feared I might. A year later, when Marlene was staying in England I invited her to Cambridge, and was somewhat put out to find that her family would not allow her to spend more than a few hours there, let alone stay in a hotel, as I had planned. Marlene was a serious, ambitious girl who impressed me with the hours that she worked, both in school and at home. She planned to study Law at the University and hoped to become a professional lawyer. But I don't think she had much luck with that plan or indeed with life in general. A year later I had a letter from her saying that she doubted whether a 'Deutches Mädchen' would be allowed to study Law at a university and that she had been given '*die Frau und der Kochtopf* [*cooking pot*]' to read. Hitler's Germany was not an easy place for an intellectual feminist, and I gathered from Renate, whom I met in Cambridge some forty years later, that her elder sister had had an unhappy life and was inclined to be difficult. Neither sister married, but Renate had clearly made a success of her career as a music teacher and violinist of professional standing. She earned enough money to share a cottage in the Alps above Lindau and to go on enterprising tours to Sikkim or other remote mountainous parts of the world.

Although I grew weary of Bielschowski's biography I derived considerable benefits from the 'Goethe Jahr', and a letter home tells of seeing *Fidelio* and *The Magic Flute* at an especially low price. I also went to an out-of-door performance of *Goetz von Berlichingen*, more of a pageant than a drama, with five hundred soldiers in armour and a squadron of cavalry charging round in the square in front of the Town Hall. It was a beautiful summer evening and the occasion was marked by the presence of the great airship, the Graf Zeppelin, which flew low above us in the middle of the performance. This filled the audience with patriotic pride of a harmless variety that contrasted sharply with the unpleasant jingoism of the Nazis which was then stalking the streets and was soon to become official doctrine.

Having been brought up as a supporter of the British Labour Party my sympathies were naturally with the Social Democrats who were then losing ground to the Nazis on the one hand and the Communist Party on the other. To judge from what happened later in France, one might have expected that Hitler's rise to power would have been resisted by an alliance of left-wing parties such as the French Popular Front. I have since been told that an approach of this kind was very much in the mind of the German socialists, but that the communists under Herr Thaelmann had been told firmly by Stalin that the Social Democratic Party was the greater enemy. I am uncertain of the evidence for this assertion but it fits well with the later and better known policies of Stalin, as well as with the line that was dished out by the British Communist Party at about that time.

Although not Roman Catholics, the Wirths voted Zentrum, a predominantly Catholic Party headed by Herr Brunning, who was Premier for a short time before giving place to Schleicher and then von Papen as the government edged uneasily towards the right. What Frau Wirth cared passionately about was the question of reparations and still more of war guilt (*Krieggeschultigkeit*).

I went along with her views on reparations and in a letter home wrote that I felt ashamed of being a member of the Allied Nations who had imposed the Versailles Treaty. There had recently been some amelioration of the reparations position, but the Wirths were bitterly disappointed that Germany should still carry all the blame for the war. I expect I went along with their views at the time, but when I got home I read as much recent history as I could and came to the conclusion that although the Allies were not blameless, the Central Powers did have a great deal to answer for.

In the summer of 1932 the effects of the economic depression in Germany were very evident and one could not escape the influence of six million unemployed living on a pitifully small dole. If you walked in the state forest you came across large groups of dejected looking men in threadbare clothes apparently camping out for the summer. They were considered somewhat dangerous and the Wirths would not allow their daughters into the Stadtwald under any conditions and tried to prevent my going there by myself, which I thought over-fussy, but in retrospect seems only reasonable. It was also necessary to be careful about theft, as I learnt to my cost at the Stadion, the splendid 100 metre swimming pool outside Frankfurt where I went on most sunny afternoons. Here I had all my clothes stolen but luckily had enough money to telephone and initiate a rescue mission by one of the Wirth daughters.

This was a time of great demonstrations that sometimes led to violence and multiple deaths as in the *Altonaschlacht* in Hamburg which happened a little later that summer. Herr Dünner, to whom I had an introduction from an Oxford friend, suggested that I watch one of the large Communist Party demonstrations. In my letter home I reported that those on the demonstration looked decent people, like British working class, and that the unpleasant characters in the crowd opposing the march were probably Nazis. However, I retained a less favourable impression of the people I met at the Communist Party headquarters where Dünner took me later on. I do not know what happened to any of these people but I would guess that their chances of survival were not high, either in Germany or Russia where Dünner told me that he planned to go if the Nazis got into power.

I naturally kept quiet about these doings, but something must have got back to the Wirths for I received a long lecture from Dr Wirth about the bloodthirsty nature of the Communist Party and the danger of mixing with that movement.

39

All this sounds a bit grim, but much of my time was spent in an almost idyllic way, picking a vast crop of strawberries in the garden, expeditions to the Taunus mountains, or the Rhine, where I found a colony of breeding kites and saw many storks nesting in the traditional way on village chimneys. But all too soon it was time to return home and, after a brief visit to Hamburg, where the Wirth sisters were staying, I sailed from Bremen and arrived back in England with so little money that I had to borrow half a crown in order to get home. I would in fact have liked to spend longer in Germany as my German had become fluent but was in no way ingrained. However, I had a strong reason for coming home as my mother was very shortly to marry again and we were to move from Oxford to Edinburgh, where my future stepfather had recently been made Rector (headmaster) of Edinburgh Academy.

My mother's marriage to Lionel Smith

I first got to know Lionel in 1929 when we spent our summer holiday in The Captain's Lodgings, a restored part of Bamburgh Castle that we shared with my Uncle Robin Hodgkin and his family. They went there every year, but my mother considered it rather too grand for us to be an annual event and this was a special treat. It *was* grand, too. You could sit in a window embrasure in ancient walls and look down on people working in their gardens as though you were a mediaeval Baron inspecting his villeins. And there were morning prayers attended by many servants at which Mrs A. L. Smith played the hymn. In the evening those over seventeen changed for a late-ish dinner, whereas the rest of us 'children' ate a high tea that included fish or crabs we had caught or, I am sorry to say, an occasional snipe or plover shot by Christopher Hartley, the oldest member of the junior party.

Lionel Smith was home on leave from a government post in Iraq (then mandated to Britain), which he was spending with his sister Dorothy, her husband Robin Hodgkin and a bevy of nephews and nieces (like us) as well as her own sons Thomas and Edward.

Lionel, a somewhat legendary figure to us, was the eldest son of A. L. Smith, Master of Balliol from 1916 to 1924 and the redoubtable Mrs A. L., née Baird, a familiar figure in Oxford for 25 years after A. L.'s death in 1924. Lionel's birth in 1880 was followed at roughly two-year intervals by that of seven girls – Gertrude, Molly,

Dorothy, Biddy, Maggie, Rosalind, Barbara – and finally in 1899 another boy, Hubert. In varying degree the daughters were tall, handsome, intelligent, witty and, like their mother, possessed of strong personalities; one of their principal recreations was gossip, preferably about other members of the family. They made happy and successful marriages, all to Balliol men, except for Biddy who married Reader Bullard of Queens' College, Cambridge. Lionel was deeply attached to his mother and over a period of some sixty years wrote to her practically every day, though whether out of affection or because he was under her thumb it is hard for me to say. He saw less of his father, who until he became Master of Balliol had to work very hard indeed to maintain his large family.

After exposure to this intensely feminine, gossipy family background it is not surprising that Lionel should have been reticent about himself, and also that, although extraordinarily handsome and attractive to women, he should much prefer male to female company. It is true that he did have some close female friends but these tended to have an adventurous masculine streak in their character: Gertrude Bell, Freya Stark and Vita Sackville-West, whom he met in the Middle-East, are examples of women whom he evidently liked and who wrote or spoke of him with respect and affection. Of Gertrude Bell, who was eleven years older than Lionel, E. C. Hodgkin writes that 'Lionel found the company of this superbly vital and intelligent being a relief and a stimulus, while she called him "the most lovable person in the world and one of the most distinguished minds I have ever known". After her death in Baghdad in 1926 when Lionel had been one of those who carried her body to the grave, Baghdad was never quite the same for him again.'

A convenient, though perhaps not very kind, way of summarizing Lionel's varied life and accomplishments is to look at the *Who's Who* entry for 1972, the year of his death, which is revealing as much for the omissions as for what it contains.

SMITH, **Arthur Lionel Forster**, CBE 1927; MVO 1914; Iraqi Order of Al Rafidhain; Hon LL.D Edin and St Andrews ; late 9th Batt. Hants Regt; *b* 19 Aug, 1880; son of late Arthur Lionel Smith, Master of Balliol. *Educ.* Rugby; Balliol College, Oxford. Fellow and Tutor of Magdalen College, Oxford 1908–20; Fellow of All Souls 1904–8. Director of Education Mesopotamia 1920–21; Adviser of Education Iraq 1921–31; Rector of Edinburgh Academy 1931–45. Address 84 Inverleith Place, Edinburgh.

The first omission one notices is that he does not mention his mother, and then, even odder, he omits his wife although he had been married to her and lived in the same house with her for forty years.

Other more intelligible omissions, which can be attributed to Lionel's extreme modesty and reticence, are his failure to mention any of his remarkable athletic achievements, such as the fact that he frequently played hockey for England in the key position of centre-forward, and the absence of anything about his difficult political service in 1918 in the Holy Shia cities of Kufah and Najaf, where one of his predecessors Captain W. M. Marshall had recently been murdered.

I do not remember exactly when my mother told me that she was going to marry Lionel, but I would guess in 1931 when I was seventeen and a very young seventeen too. I didn't mind at all and on the whole was pleased. Almost everyone liked Lionel, particularly the young, and I had found that he was as enthusiastic an ornithologist as I was. Bamburgh was an excellent place for bird-watching so we had been able to enjoy our hobby together. Another bond was the cheetah. For much of his time in Baghdad, Lionel shared a house in the Alwiyah, about four miles from his office (which he always walked both ways), with three other British officials, one of whom kept a cheetah as a pet. After a while the responsibility for looking after the animal, which was shared between the trio, grew too much for them and they arranged for it to be presented to the London Zoo, where I used to visit it from time to time. A report on the cheetah's welfare together with my ornithological news formed the basis of a correspondence with Lionel in Baghdad.

To begin with my only reservation about the forthcoming marriage was a faint feeling that I thought Lionel was more my friend than my mother's. During the early summer of 1932 various things my mother said made me think that Lionel had had second thoughts about the marriage and would like to get out of it. I also did not understand her remark to me that 'One can't put the clock back' which in the context of the conversation might have meant either that they had decided not to have children, or that she was prepared to accept a *marriage blanche*. My mother was only forty-one at the time and almost certainly could have had more children if they had wanted, and as Mrs A. L. ardently hoped. But I stifled such misgivings as I might have had and in any case there was nothing I could do about it.

I think now that the marriage was never consummated and that it brought my mother much unhappiness. However, they managed to present a united front to the world and together gave much happiness to many people during their forty years of married life.

When I got back from Germany at the beginning of August 1932 events were rushing on towards a great party at Old Ridley in Northumberland after a quiet wedding at Bywell Church, looking out over the Tyne. Whatever forebodings may have been felt were thoroughly suppressed and there is no hint of them in photographs of the occasion. However, there was one last-minute snag which had to be dealt with urgently.

It had been arranged that while Lionel and Mary had a 'honey-moon' at Rodel in the Outer Hebridean island of Harris, we three boys should have separate holidays on our own. Robin and I would take part in a climbing expedition to Norway and Keith would join cousins in a canoeing trip. After a fortnight we would all meet in Skye where my mother and Lionel had taken a house for six weeks. However, on the day before the wedding, Robin managed to break his leg in a fall on the steep river bank in the Old Ridley garden. The fracture was not very serious, but bad enough to keep him in plaster for six weeks and an absolute bar to his joining the Norwegian expedition. This didn't worry me, as I had never been keen on joining what was primarily a school expedition, but it was hard for Robin who was looking forward to it very much. And there was the tricky question of what was to happen to us during the Hebridean honeymoon. After much deliberation it was decided that I would drive Robin to Bamburgh where Lionel's sister Barbara had kindly asked us to stay for a fortnight; after that I would take Robin in the family car on the long journey to Skye.

The elder and more cautious member of the family insisted on two precautionary measures that I thought unnecessary. The first was that, as I had only recently learnt to drive, my Uncle Robin should give me a few lessons and make me pass a sort of driving test. This caused roars of laughter amongst my cousins who considered Uncle Robin to be one of the worst and most absent-minded drivers in the world. However, I passed the test all right and was quite grateful for my uncle's advice.

The other precaution was that in Edinburgh we should pick up a male nurse cum masseur, who would help look after Robin on the journey and in Skye. This we did, but there really wasn't much for Mr Glass to do, as you can't massage a leg in plaster, and poor Robin, who wasn't really in need of his attention, was driven mad by his continuous attempts to be helpful. However, it turned out that Mr Glass was determined to shoot a grouse, for which purpose he had brought a gun. Finally, after a careful stalk of a covey that could barely fly he returned in triumph with one of the smallest cheepers I have ever seen. With that ambition satisfied, he was willing to depart and we were able to enjoy the rest of the holiday before settling into our new home in Edinburgh.

Chapter 5

Cambridge, 1932–5

Life in Trinity

When I arrived as a freshman at Trinity I felt distinctly nervous – not quite as nervous as when taking a scholarship exam the year before, but quite nervous enough. I hadn't been unhappy at Gresham's but I much preferred the holidays. It came as a very pleasant surprise to find that I positively enjoyed Cambridge and looked forward to the beginning of term instead of dreading it as previously.

I have tried since to decide why this was. Of course there were many pleasant things about Cambridge, and Trinity in particular: the diversity of people, the beautiful buildings, the historical tradition and the knowledge that exciting things were being discovered in the Laboratories where one was taught. All these became important later on, but were not so obvious at the beginning. What I noticed at once was that provided you were not positively anti-social, no-one minded what you did or thought. You could play games – or not, as you pleased. You could work all day – or do virtually no work at all if you read for an ordinary degree. You could be religious or agnostic. You could be intensely political or shut yourself in an ivory tower like the one described in Christopher Isherwood's novel, *Lions and Shadows*.

Another feature of life in a large college that I liked particularly was that the concepts of popularity and unpopularity which had been so terribly important at school no longer seemed to matter very much.

Long afterwards I found that the same feeling of exhilaration at a newly found freedom had been expressed more exuberantly than I could ever do by one of Trinity's galaxy of poets. After

45

Byron had been at Trinity a few days he wrote to his sister Augusta saying:

> As might be supposed I like a College Life extremely . . .

and then:

> I am allowed 500 a year, a Servant and Horse, so feel independent as a German Prince who coins his own Cash, or a Cherokee Chief who coins no cash at all, but enjoys what is more precious, Liberty.

Today if there is a German Prince alive he certainly coins no cash. And if a Cherokee Chief still exists he certainly enjoys little liberty. But Oxbridge colleges which have had their ups and downs are still going strong, and the tutors can tell the undergraduates that they enjoy as much freedom as was going in Byron's day, even though they cannot do everything that Byron was able to do.

A large college like Trinity has the drawback that it can be a lonely place, particularly for someone living in lodgings a fair way from the College, as many of our freshmen did then. I was lucky in that, being a scholar, I acquired a room in college: in Whewell's Court, not the most beautiful site, but a great deal better than all but the most expensive of lodgings.

I was also fortunate in that the Trinity Chaplain, Michael Gresford Jones, who was my first cousin, introduced me to a large number of undergraduates and some dons. This had the possible disadvantage of classifying me as a religious rather than an agnostic undergraduate, but that was not something that bothered me for a year or two. In any case Michael was a tolerant worldly person who tried to get to know as many undergraduates as possible and I soon found that I was meeting a wide variety of new people.

With the first-year scholars in college and all other freshmen living out, it was natural that the former should constitute a well-defined social group, in which you knew and were on potentially friendly terms with nearly all the other scholars. These feelings were strengthened by various formal ceremonies, like admission, involving an ancient ritual which applied only to the scholars.

Soon after I arrived at Trinity a group of new scholars was invited to dine at the Master's Lodge. In 1932, and until his death in 1940, the Master was Sir Joseph Thomson (J.J.), who contributed to many branches of Physics but is best known for his measurement of the mass and charge of the electron. He was then aged 76 and full of life, though becoming increasingly eccentric. One of the Head Porters at Trinity, Mr Prior, who died in 1986, remembered that one of his first duties as a junior porter was to rescue the Master, who had been misled by the newfangled open display of goods in a well-known chain store, from an accusation of shop-lifting. The shop in question couldn't believe that anyone so shabbily dressed could be Master of Trinity, and Mr Prior looked so young that he had difficulty convincing them of J.J.'s position. However, all ended well and the shop realized that J.J. thought that the HELP YOURSELF counter meant what it said.

In my letter home I reported that this awe-inspiring evening was relieved by a comic episode. After dinner the Master turned on the wireless to some 'rather vulgar' dance music and went around chuckling to himself, but Lady Thomson who was extremely conventional, was shocked. She transported as many scholars as possible to another part of the great drawing room.

The only other time that I can remember dining with J.J. was just after I had become a research fellow in 1936. One of the guests was Enoch Powell, also a research fellow but two years my senior. At that time he had no interest in politics but worked at classics with a single-minded ferocity that I have never seen equalled. One of his heroes was Bentley, the famous classical scholar and friend of Newton, who had managed to remain Master of Trinity in spite of the strenuous efforts of the Fellows to have him impeached and ejected. At dinner Powell insisted that Bentley was the best Master that Trinity had ever had. Lady Thomson did not like this at all, partly because she disapproved of Bentley's scandalous conduct and partly because she had no doubt that J.J. was better than any former Master. As we left, she said severely to me, 'Really Mr Powell, I do think Mr Hodgkin's views on Bentley are too distressing for words.' I was never asked to dine again but one dinner as a research fellow was the ration so I had no cause for complaint.

Soon after our dinner with the Master, I and several of his pupils went to lunch with Andrew Gow, my tutor, of whom I was rather afraid as he had an unwelcoming manner and, very often, an acid tongue. Later, I got to know him well and he had a considerable influence on my life, particularly in interesting me in Italian painting, of which he had a detailed knowledge. But it took a year or two for our relationship to reach that stage.

47

Gow was attracted to good-looking young men and had a wide circle of male friends, but he was not a misogynist like the philosopher C. D. Broad, another Trinity Fellow. All Gow's pupils noticed the photograph on his mantelpiece of the pretty girl resting in an Alpine meadow, and at the age of 75 he electrified the college by striking up a great friendship with the actress Dorothy Tutin. Apparently Gow, though not a theatregoer, had a passion for *The Beggar's Opera* and greatly admired her performance as Polly Peachum in the 1961 production at the Aldwych Theatre. Through the good offices of Raymond Leppard, then Director of Music at Trinity, who was responsible for the Aldwych version of *The Beggar's Opera*, Gow got to know Miss Tutin and they continued to meet fairly regularly until, and to some extent after, her marriage.

On the right occasion, Gow could be a very good raconteur, but in general he had little small talk. He liked certain favoured undergraduates to call for an hour around 10.00 p.m., and if you did so, it was not uncommon to find Gow and another favourite sitting perfectly silent on either side of the fireplace, apparently waiting for some conversational inspiration. My letter of August 1933 (p.53) indicates that I had begun to acquire favour but was still finding the conversation hard going.

Things were much stickier to begin with. After that first lunch, which was on a Sunday, I wrote, 'Yesterday I went to lunch with my tutor, which was an awfully grim show but a very good meal.' To my embarrassment I found that conversation consisted of Gow asking each of us how we had spent the morning. I replied that I had been for a walk. On being asked where, I replied, 'Between the Newmarket Road and East Road'. Fortunately, Gow did not inquire why I was visiting this rather slummy part of Cambridge but confined himself to saying, 'Not a very salubrious part of the town'. I was relieved that I did not have to say that my communist friend Tona Gillett had asked me to accompany him to this area, where he and others collected grievances, for example about evictions, and if possible tried to help by stating the case to the town council and getting up demonstrations.

My particular friend during my first year was a sweet-natured scholar, John Bolton, from Blundell's School, who had got a major scholarship in mathematics at the early age of 16, and then, perhaps because he had been pushed too hard too early, had given up mathematics for biology and was reading the premedical subjects.

Our friendship terminated abruptly the following summer when he got a particularly painful attack of the Oxford Group movement. In his case not only did this new brand of Christianity require him to confess his own sins in public, but it also led him to try to persuade every one else to do the same. Whoever was John Bolton's leader must have been a bit of a sadist because poor John's list included many senior Fellows. I heard about this later from Andrew Gow who said that Bolton came to see him on an evening when he was entertaining the explorer Peter Fleming, who became convulsed with suppressed laughter as Bolton ploughed on through his list of highly personal questions. After several years Bolton abandoned the Group Movement and led a happy and useful life as a general practitioner.

Some other early friendships which have lasted all my life were those with John Morrison, John Raven and Michael Grant, all classicists, and Charles Fletcher, David Hill, Dick Synge and John Humphrey, broadly from the biomedical field.

Although Trinity was very large, there were many opportunities of meeting undergraduates and dons in subjects different from your own. One that I much enjoyed, and which still flourishes, was a vacation party, known as the Trinity Lake Hunt which takes the party of dons and undergraduates to the Lake District for five or six days at the end of the May term. This was an offshoot of the Trevelyan Man Hunt, a game invented by George Trevelyan and Geoffrey Winthrop Young well before the First World War, which involved chasing people at high speed over the hills and valleys of the Lake District. There I first met E. D. Adrian, and got to know several of the younger dons and research fellows, Patrick Duff, Outram Evennett and Walter Hamilton in particular. I also made friends with a research scholar, Wilfred Candler, then working for a Fellowship in theoretical astronomy, who often helped me with the mathematics that I was trying to teach myself.

In June 1933 when I first went on the Lake Hunt it was very much less energetic than the original Trevelyan Man Hunt, and, unless you volunteered to be a hare, you could get by with a gentle hill walk perhaps meeting others for tea at Gatesgarth at the head of the Buttermere valley. Otherwise I do not think it would have been possible for dons as unenergetic as Evennett to be regular members. In any case the number of senior and junior members of the college that you met on the Lake Hunt was bound to be a small fraction of the total. But there were many other ways of meeting the bachelor Fellows who lived in college, as it seems to have been traditional for them to do a mild amount of entertaining, whether or not they were tutors. Some of those that I remember especially were the economist, Dennis Robertson, the Vice-Master and Cambridge historian, Winstanley and the steward E. F. Collingwood. But the person who influenced me most initially was my Director of Studies, Carl Pantin, who had already given me such good advice before I came up.

Mainly Science, 1932–5

Just before the beginning of my first term Pantin advised me to take Physiology with Chemistry and Zoology for the first two years. He suggested that I give up Botany, which I was reluctant to do as it had been my best subject in the scholarship examination. However, after going to one or two lectures, I concluded that Pantin was right and gave up the subject quite happily. Physiology, then combined with Biochemistry, was new and exciting, particularly in the experimental classroom. The regular lectures, except for some of Barcroft's and Winton's, were not brilliant, but I remember superb lectures by Krebs on the ornithine cycle and by Adrian on referred pain. I enjoyed Zoology supervisions (tutorials) with Pantin in the evening in Trinity, and it was a great disappointment to me when a serious attack of tuberculosis took him away from Cambridge for nearly two years. In Physiology, Roughton was less conscientious as a teacher, but I learnt a great deal by sticking firmly to carbonic anhydrase and haemoglobin, which were the subjects he liked to talk about.

The scholarship examination that I had taken a year before was a very stiff hurdle and I soon realized that I had already done much of the Part I syllabus in Chemistry and Zoology. I was too diffident of my own abilities to cut lectures, but found that I had a reasonable amount of time for library reading, either in mathematics and physics or in general physiology and cytology. Some of the books that influenced me particularly were those by A. V. Hill, E. D. Adrian, James Gray and J. B. S. Haldane. Like William Rushton, I was much influenced by Keith Lucas's collected papers, which I read all through as a student. Here my choice of author must have been partly determined by my father's friendship with Lucas.

At Easter I went on a vacation course to Plymouth and learned more invertebrate zoology there than in a whole year at Cambridge. This was partly because I have always disliked seeing the delicate beauty of marine creatures transformed into the formalin-pickled relics of the museum or zoology dissecting room. It was also a great thrill to watch the fertilization and division of a sea-urchin egg.

After a few months at Cambridge, I attended, and eventually became a member of, the Natural Science Club, a small élitist organization which was founded in 1872 and by 1981 was celebrating its 2,500th meeting. The active members of this society were all junior members of the university (either undergraduates or research students). The club took itself pretty seriously and had elaborate rules and weekly meetings at which someone read a paper for an hour or so. Some of the members in my day were Edward Bullard, John Pringle, Dick Synge, Maurice Wilkins and Andrew Huxley – all subsequently became FRSs and four of us Nobel prize winners. I have never been a clubbable person, but I thoroughly enjoyed this club and belonging to it did much to counter the narrowness associated with studying one branch of science in depth at a university. The printed records of the club's proceedings make interesting reading. One finds that the subjects chosen for student papers are often those that scientists take up many years later, probably without remembering their early interest. These are some examples: Synge, *protein structures, techniques in biochemistry, cystine, distillation;* Wilkins, *mirrors, watches and clocks, seeing structures*; Huxley, *the ear and its functions, the conduction of nervous impulses, the use of the microscope, experiments on single cells;* Hodgkin, *nature of cell surface, membrane theory of nervous action, the behaviour of sense organs, light-reception, nerve-muscle function.*

When the Physiological Society met in Cambridge, anyone keen enough was allowed to sneak into the audience and I remember a splendid debate on humoral transmission with Henry Dale, G. L. Brown and Feldberg on one side and Jack Eccles on the other. My scientific sympathies were wholly on the side of acetylcholine but I thought Eccles put up a good fight. In May 1934 I was lucky enough to be present on the famous occasion when Adrian and Matthews demonstrated the effect of opening the eyes or of mental arithmetic on the Berger rhythm, using Adrian as a subject (his rhythm was unusually responsive). A year earlier a letter home describes another of Adrian's demonstrations to the Physiological

Society – this time recording auditory impulses from the brain of an anaesthetized cat. I quote parts of this and a later letter below as they give a good idea of my extreme youth as well of the way in which my time was spent after the May examinations were over.

Trinity College
Cambridge

I was quite relieved to hear Lionel [my stepfather] had turned down Eton. [He had been strongly pressed to accept the offer of the Headmastership but continued to decline to the distress of many of his friends and relations] . . . a good walk today with Thompson [probably 15 miles] . . . to breakfast with Mrs Gillett and Jan and Tona.

Yesterday, a meeting of the Physiological Society was held in Cambridge . . . Professor Adrian was demonstrating some of his experiments, the most striking of which is one in which he whispers into the ear of a cat (dead) and, from a couple of wires stuck in the brain, the electrical changes are amplified and connected with a loud-speaker, so that one can hear what he is saying, I.E. one listens through the brain of a cat. There were several other very interesting demonstrations which are too technical to describe.

Last Saturday Mellanby (of Sheffield) gave an interesting public lecture on Vitamin A and the nervous system. I also went to a meeting of a small society to hear a man called Lundsgaard talk on the chemistry of muscular contraction. This is a subject which is in such a state of flux that a theory, which was held to be quite definite till two years ago, has now been almost completely overthrown.

We had an amusing (ordinary) lecture on an aspect of Respiration, which was the scene of a great battle between two rival theories, one worked out by the Oxford and one by the Cambridge school. At the moment the Cambridge theory is completely victorious.

I had a very good game of squash this last week.

I have been reading . . . *Kleiner Mann was nun* by Hans Fallada . . . very good . . . the chief characters are a German working class man and his wife. The fear of unemployment always hangs over them; and the novel ends with the man being unemployed – but it's not altogether a tragedy as the wife more or less keeps things together. It is more or less socialist in its outlook (like all the best modern German art and science), but not at all propagandist.

The Biochemistry Lab. here is very anti-Nazi and sticks up anti-Hitler cartoons – but then it has a reputation for having Communist tendencies – and of course with one exception [probably Warburg] the great German Biochemists are either Jews or Communists and have been turned out. Some have migrated and probably more will migrate to the Cambridge biochemistry lab.

Early August 1933,
Trinity College.

I find that a person whom I know a little is the son of Philby of Arabia [known well, and disapproved of, by my stepfather]. He is a very charming sort of person, has done Economics Part II and is a keen Socialist.

I had an invitation to spend the evening (10–11.30!) with Gow, who likes to have undergraduates in at that hour. We went to the Fellows Garden, which was nice as it relieved the conversation difficulty; I think he is about the most difficult person to talk with [that I know].

On August 4th I distributed (with other people) Anti-War pamphlets. I don't know if you have come across the Anti-War movement. Its aims are organisation for practical measures against war, either in the event of a war, or when war is imminent – laying stress on the effective though unpleasant methods of strikes etc. It is fairly left in outlook [and] other parties such as the LNU [League of Nations Union] and peace movements have refused to cooperate [with it] but has fairly broad individual support. In Cambridge it is largely run by Communists and hence has frightened off many SCM [Student Christian Movement] and LNU people.

That late summer holiday was spent at the Bridgend Hotel, Islay, and was the first of many that my family spent in that delectable island. It was Lionel's idea that we should go there, and it did not take long to persuade us that Islay really was infinitely preferable to Skye, to which my brothers and I had become strongly attached, in spite of the lack of real mountains, like the Cuillins, in Islay. But, as my mother said, rock climbing does not really go with family holidays, and if we wanted hill-walking, we could easily take a boat across the narrow Sound and climb one of the Paps of Jura in that large, but empty, island. And this is something we frequently did.

The advantages of Islay rest in its pleasant, prosperous looking farms and picturesque towns like Bowmore, unusual in the west of Scotland, in the attractive mixture of wooded valleys and farmland in the middle of the island, but most of all in the hundred miles of wonderful coastline interrupted by sandy bays and beaches like those in Connemara or the outer Hebrides. The island is on one of the main bird migration routes and altogether is an ornithologist's paradise.

To begin with, we stayed at the Bridgend Hotel, which was all right except at the time of the Agricultural Show when you were apt to find whisky-sodden farmers asleep in every bathroom. But after a year or two my parents rented an attractive eighteenth century house, Newton, from the Islay House Estate, for the absurdly small sum of £25 a year. Here my family and friends spent many weeks, at Easter and the late summer, or best of all in June: an arrangement that continued very happily until the Second World War brought it to an end.

There were so many things to do in Cambridge that I always came away with a huge reading list for the vacation. In Islay, I liked to work in the morning and then bicycle out to join my family for a picnic lunch on a hill loch or a sandy bay, to spend the rest of the day fishing or water-colour sketching. I was never much good at either of these activities, but they are pleasant ways of passing the time and they have the merit of embedding some bit of the countryside deeply in your visual memory.

In Edinburgh where we now lived, the Physiological Department let me use their Library and I used to work there, or in my bedroom at home in Inverleith Place, where I had a beautiful view looking across the new town to the Castle, with Arthur's Seat and the Pentland Hills in the background.

In spite of its beauty and romance, I never really felt at home in Edinburgh and thought that its society was rather stuffy compared to that of Oxford. I got some relief from walks with my wild and amusing step-cousin Mary Jameson (later Cowan) who was in a much more open state of revolt against Edinburgh society than I was. Her father Sheriff Jameson, who was known to all as Johnny, was a wonderful looking, white-haired lawyer, who was often to be found preaching the gospel most eloquently and sincerely in Princes Street, an activity which no one seemed to think peculiar or even eccentric and which didn't prevent his advancing to the position of a Scottish Law Lord.

There were of course many things to do and many interesting people to meet in Edinburgh, but for some reason I regarded it primarily as a place to work and was glad to be returning to Cambridge at the end of the Long Vacation (October 1933).

Like most people I found the second year of the Natural Science

Tripos rather hard going. This is mainly because after eight months you have to take an examination in which you are questioned on what you have learnt in the previous *two* years. In the event I did all right, but I was very pleased when I could forget about remembering three or four memory-intensive subjects and concentrate on dealing critically with one.

I had some difficulty deciding whether to read Physiology or Zoology in Part II. My original intention had been to read Zoology and subsequently become an applied zoologist – the sort of person who decides on the right way of controlling agricultural diseases in Africa. I was now less keen on this and in any case preferred Physiology to Zoology. However, I was advised that there was little prospect of my getting a job unless I was medically qualified. There was a good deal of force in this. In the depression, academic jobs were hard to find and in most universities a medical degree was an essential requirement for a physiologist. Cambridge was exceptional in this respect, but the department found it essential to retain a reasonable number of medically qualified physiologists on its staff, as Rushton found when Barcroft offered him a lectureship on condition that he spend three or four years getting a medical degree. Nevertheless, my new Director of Studies, Roughton, strongly advised me to take up Physiology and this is what I did in the end, though I am not sure that I would have made this decision without Uncle Edward's legacy. Three hundred pounds a year does not seem much now, but it was enough to live on in those days.

Although I was becoming increasingly involved with cell physiology and neurobiology, I retained my interest in zoology and even played cricket for that laboratory. One consequence of that interest was that I reluctantly became secretary of the undergraduate Natural History Society and had to organize the large evening exhibition of Cambridge Biology that the Society puts on every year.

The expedition to the Atlas Mountains, summer 1934

Another consequence of my interest in zoology was that in the spring of 1934 John Pringle asked me to join an undergraduate expedition to the middle Atlas in Morocco for which he was collecting funds. He was successful in this as in many other ambitious projects that he started. The expedition was called the

Cambridge University Biological Expedition to the Atlas Mountains and was supported by the Natural History Museum (London), the Royal Geographical Society, the Percy Sladen Trust and grants from Cambridge University and Colleges. I was rather torn about going. I longed to see this little-known part of world, but I was also trying to start the nerve experiments described in a later chapter and felt I should not spend the entire Long Vacation in Africa. So I compromised by saying that I would pay my fare and go for about half the time.

We had decided to start by making a thorough ecological study of a lake situated at an altitude of about 6,500 feet in the cedar covered hills of the middle-Atlas. Here the hope was that we might discover glacial relicts like the free-living flat-worm, *Planaria alpina*, which had been found in high-mountain lakes, as far south as Spain, apparently living on there since the Ice Age. But it was pushing the notion to expect such relicts as far south as the Atlas Mountains and the idea proved illusory. However, the general collecting that the Natural History Museum encouraged us to do proved successful and we apparently discovered several new species, one being a large flightless grasshopper that infested our camp and ate any food that was left lying around. We also hoped to get to the High Atlas and climb one of the major peaks in that range. The following extract from an article in the Morning Post (24 September, 1934) which was probably based on an interview with John Pringle, just after his and Chapman's return, gives the impression that all this was accomplished with ease and efficiency.

> Giant toads, large land crabs and giant wingless grasshoppers are among the specimens brought back from southern Morocco by the Cambridge University Biological Expedition to the Atlas Mountains . . .
>
> The members of the party, which has just returned to England were . . . Chapman, Cotton, Hodgkin and Pringle.
>
> Very little was previously known of most of the forms of life to be found in these mountain districts . . .
>
> Two months were spent in Morocco, mainly in the middle Atlas Mountains, the camps being at various heights from 2000 to 6500 feet above sea level. One camp, was however pitched in the Great Atlas range and two members of the party climbed a 14500 foot peak. All the camps were within the official 'zone of insecurity' and, by the courtesy of the French military authorities, the expedition was accompanied by an escort from a native regiment.

The specimens collected, which include about a thousand insects as well as fish, snakes and lizards, are now being examined at the Natural History Museum.

This is accurate zoologically and was probably the right thing to say to the newspapers as Pringle would not want to alarm Cotton's family unnecessarily, but it leaves a good deal unsaid, particularly that Cotton and I ended up in the French Military Hospital in Meknes (the base hospital for the Foreign Legion) and, as far as I remember, that Cotton was still there when Pringle and Chapman got back, having had a perforated appendix and subsequent complications.

Although it is right off the main stream of my life, I have pieced together an account of this Moroccan adventure from vivid memories and letters to my mother, which, to avoid alarming her, make out that a somewhat daunting experience was little more than a rest cure.

13th July 1934
S.S.Strathnaver

A satisfactory set-off this morning with no hitches. The Thames and Medway attractive in grey, misty weather.

Near Tilbury there were a number of the nice red-sailed barges which were going a good pace before the wind. This is a 22,000 ton boat and goes eventually to Australia. I was quite excited by the Lascar crew in costume and am now getting generally excited about the journey . . .

I am really quite glad to have gone away from Cambridge because although the Long [vacation term] is very pleasant it would have been a grind to go on with my experiment which was giving negative results.

I have taken *Madame Bovary* and Dante (*Inferno*) as literature and a mathematical book [Piaggio] in case I feel like doing any work. [I did, which so alarmed my fellow passengers that some one was put on to jolly me along.]

16th July . . . I have enjoyed the last two days enormously. This morning I saw a whale which blew about 50 yards from the ship and swam away, blowing frequently. I don't know what kind it was – it seemed about 20–30 feet long. Yesterday two porpoises followed the ship for some way – jumping out of the water and altogether behaving differently from the more leisurely ones that I have seen from the shore . . .

The evenings and nights are delicious with a hot wind blowing off the land. Last night the wind was faintly scented with a smell reminiscent of pine and cedar woods. I could not think what it was due to, but tonight it is much stronger and I think it must come from the cedar woods in Spain or Algeria – by the land breezes which set up in the evening. (Land and sea breezes are one of the few things I remember of Eccles physics.) . . .

From Lake d'Aguelman, Azrou, Morocco.
c. 20 July 1934.

I am writing this after supper by candlelight so I expect it will be rather incoherent, I will try and give things in their chronological order, or otherwise I shall leave out a great deal. I was landed at Gibraltar, as it was too rough to land at Tangier . . . Started by bus at 4.30 am to Meknes via Rabat. The roads were very good and for 50 miles round Tangier were lined with eucalyptus and run through pleasant undulating country – a background of brown grass with broom and other shrubs and occasional fields of maize. The last 100 miles to Rabat were across flat rather dull country, but I saw many birds – egrets, stilts and masses of little white herons, feeding among cattle.

The old part of Rabat is built of a deep red sandstone and there is a magnificent mosque. Some of the walls are covered with a lovely blue convolvulus. My bus stopped for two hours so I went to see Haji [Reader Bullard, then Consul-General in Rabat]. I had lunch with him before going on. We went for some way through cork-oak forests and then began to climb up into mountainous country. The roads are magnificent and the buses go very fast.

I was very impressed with Meknes . . . There are two sets of walls about eight miles in circumference enclosing the gardens and an inner wall enclosing the old town. The gates are magnificent – blue-green patterns on the yellow sandstone wall.

The old town is a dense mass of yellow sandstone houses, and occasional mosques, tunnelled by narrow streets which are only now and again open to the sky.

I met an English chemist [pharmacist] at Meknes who is a most enthusiastic entomologist. He took up his job at Meknes in order to be able to collect. He said he used to collect up here (at the lake) before the Rif war, but that his collecting was handicapped because he had to look out for a stab in the back or a rifle shot: just as an enthusiastic fisherman might object to midges on the grounds that they interfered with good casting . . .

I started from Meknes at six on a lovely morning. The telegraph

58

wires were sentinelled by rollers, and occasionally bee eaters, peregrine falcons, eagles and kites . . . After Azrou the road climbed steeply through thick cedar woods which were full of butterflies, and then through some rather barren stony hills.

I got out of the bus and met Pringle on a horse escorted by two Arab 'soldiers' who watch over us, and a donkey, which was soon piled up with my baggage and the two soldiers on top. Our camp [about five miles away] has a superb view down the lake on which there are ducks and a magnificent goose with white and blue green wings . . . I have seen an antelope, rather like a roe buck and a couple of fawns. Kenneth Chapman saw a cat rather like a lynx and a kitten . . . The air is extraordinarily dry and ones lips and throat get quite sore. There are a few mosquitos – grasshoppers are almost the worst insect.

24 July. Our days seem very full. Everything is going well . . . Our camp is at the top of the lake near a good spring – this has attracted a camp of nomadic Moroccans, from whom we rather irregularly get milk, eggs and occasionally bread and cheese. At times the stream is full of Arab women washing babies which interferes with our attempts to collect the fauna. Near to our tent is one put up by the Arab soldiers, who are supposed to keep a guard on us. This they do with varying efficiency – one lot escorted us wherever we went, but those here now are more sensible (in this respect) and only come out if we are going some distance. One soldier, Rahu, is amusing though at times irritating. He knows hardly any French except 'la haut' and 'la bas', so we mostly use the few Moroccan words that we have learnt. He asked for cigarettes, yesterday, excusing himself by saying 'Fatima Maroc' meaning that he was entertaining some Arab woman and was very pleased with a pair of gaiters (several sizes too large) which we had given him and had invited the woman in to admire them.

We get a good deal of our food sent up each week from Meknes, and get bread twice a week from the café restaurant on the road about five miles away (we also collect letters there). We try and supplement this with food that we shoot but so far we have only got pigeons. The fish in the lake are supposed to be good, but we have not been able to catch any . . .

We generally go out in the woods from 5–8 pm which is the best part of the day as there are shadows on the hills. On one side the hill slopes steeply into the lake and there are fine outcrops of red limestone, eroded into pinnacles and arches. There are great cedars everywhere, and higher up, evergreen oaks which hum with cicadas. Rollers are quite common and I succeeded in finding a pair feeding young in a woodpecker hole at the top of a cedar tree. The rollers brought beetles, lizards and eggs – they make a noise half way between a fulmar and a jay; this morning I saw a young one peeping out of the hole. There are many choughs, both alpine and ordinary, a pair of golden eagles and several pairs of kites nesting round the lake.

. . . One night two Arabs, the energetic efficient ones, got very excited over the trail of this lynx-cat thing. They tracked it to a large cliff where they fired two shots and insisted that they had killed it and that it had got down a hole. However, I saw the beast slinking away a little later.

There are very good sunsets and it gets dark quickly so we generally come back in the dark. At night the noise of cicadas and grasshoppers changes to the more musical purring of nightjars, but in the evenings the Arabs make a great noise, with an interminable song which appeared to have no bars or stopping places. We countered with 'London's burning' but could not touch them for persistence (or monotony).

I had two health scares: one which came to nothing and the other more serious. I came back to our living tent one day with a raging thirst and quickly drank a mug of water which I thought I had filled from the bucket of spring water that we kept filled for that purpose. At the time Paul Cotton was skinning birds and preserving their skins with white arsenic for the museum. This was done by applying an arsenical paste with a stiff brush which he dipped in water from time to time. After a while he said, 'Hey Alan, I hope you haven't been drinking water from my mug.' We looked at the two mugs which were identical and he decided that his, potentially arsenical, mug had gone down in level. After talking with the others we decided that I must make myself vomit by drinking concentrated salt water. But this was easier said than done on an empty stomach, as I am hardly ever sick, and the only effect of massive amounts of brine, followed by half a pound of solid salt, was to give me the purge of all time, with no symptoms of arsenical poisoning whatever.

To understand the other trouble you need to remember that in 1934 sulphanilimide was still in the laboratory and penicillin even further away. I think too that we were probably not as careful as we might have been about disinfecting insect bites and minor scratches. At all events at the end of July one of my feet which had been mildly sore for a few days swelled up a lot round the ankle and I began to run a high temperature; this was followed by massive swelling of

60

the whole leg which prevented me walking more than a step or two.

At the same time Paul Cotton was afflicted with vomiting and incapacitating abdominal pains. With considerable difficulty John Pringle succeeded in getting a car to within about a mile of our tent and Paul and I managed the intervening distance on horseback. We got into Azrou that night. 'The doctor there did not think either of our complaints serious – as a matter of fact Paul Cotton's was – so we spent the night in a hotel. The next day [1 August] we moved to the civilian part of the military hospital in Meknes and Paul had his appendix out the next day.' The appendix had already perforated and he had adhesions and other complications which kept him in hospital for many weeks.

I got better quite quickly, but there was one nasty moment when I was wheeled into an operating theatre and examined on a surgical table by a Moroccan surgeon in a fez (with a tassel) – surely not the right headgear for operating. For a moment I feared they were going to amputate my leg, so I summoned up my remaining French, never up to much, and said something like 'Il n'est pas necessaire de couper ma jambe.' At which the surgeon shook his head and smiled. Another detail which sticks in my mind was that when thermometers were first brought round we naturally tried to put them in our mouths. 'Non, non,' shrieked the rather blowsy nurse, 'Il faut rentrer dans l'anus!' demonstrating something quite unfamiliar to us.

My memories of the hospital, for which we paid nothing, are somewhat grim, for instance the unpleasant smell which we attributed to the septic wounds of someone who had had a bad car smash. But none of this appears in the slightly querulous description that I gave my mother on 6 August. 'We are in the civilian part of the hospital, which is not bad though nobody seems to worry much about things like washing and they give you little attention generally. There are two very nice French matrons, who are sisters; both speak English, one has climbed in Skye and knows Edinburgh and Oxford quite well. She told us a lot about the Grande Atlas in which she has also climbed.'

I got out of the hospital in a week and went to stay with Haji Bullard in Rabat for a few days, while I arranged for a somewhat earlier return passage than I had planned. I was still not feeling at all well and it would have been crazy to go off to a very remote camp with Pringle and Chapman. So, though disappointed, I was quite happy to sit about reading Haji's books and eating delicious meals until it was time to go home. Before leaving I combined a trip to Fez, where I spent the night, with a call on Paul Cotton in the hospital at Meknes. I was distressed to find that he was still running a temperature and very low because he thought, rightly, that this illness would probably dish his chances of getting his Blue in the Cambridge Rugger fifteen.

Chapter 6

Cambridge, 1934–7: starting research

I RECOVERED from my leg infection fairly soon after getting home from Morocco in late August 1934, although for a while I suffered from boils and minor health worries. However, towards the end of September, about a fortnight before the beginning of the autumn term, I had recovered sufficient energy to return to Cambridge and have another go at the nerve block problem that I had started in early July.

During my first two years at Cambridge I had become interested in cell membranes, mainly through reading James Gray's *Experimental Cytology* and A. V. Hill's *Chemical Wave Transmission in Nerve*, both of which were then relatively new. I had also read the excellent review by Osterhout on *Physiological Studies of Large Plant Cells*, and was impressed with the evidence obtained by Blinks (1930) for an increase of membrane conductivity during the action potential of the large cells of *Nitella*.

It seemed to me that evidence on this crucial point was lacking and I tried to test it by the method illustrated in Figure 6.1*A*. Using class apparatus, much of which had been designed by Keith Lucas some twenty years earlier, I arranged to block a nerve locally by freezing it, and applied two appropriately timed electric shocks at the position shown in the figure. I argued that if the ionic permeability and conductivity of the membrane increased during activity, then arrival of a nerve impulse at the block should increase the fraction of current which penetrated the nerve and hence lower the electrical threshold at the distal pair of electrodes.

I set up the equipment using a silver rod and a tin of ice and salt to cool the nerve, a Keith Lucas spring contact-breaker to time the two shocks and a smoked drum to measure the size of the muscular contraction, and hence to determine the number of nerve impulses

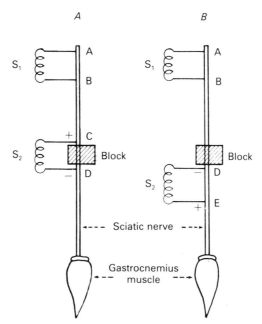

Figure 6.1. Diagram of method of testing the effect on excitability of a blocked nerve impulse, using sciatic gastrocnemius preparation.

set up by the shocks. The experiment consisted of establishing a block, and then alternating between shock 2 by itself and shock 2 preceded by shock 1 with an interval of a few milliseconds between them. If there was any facilitating effect then the muscle twitch evoked by the two shocks was greater than that produced by one alone.

When I tried the experiment in early July I got a negative result, but on trying again in October the experiment worked well and I was very pleased indeed. Later that year I made a note to the effect that 'Freezing must be light and reversible. In July non–reversible freezing or ligaturing was used and no effect could be detected.' The threshold near the block was very variable and it was hard to obtain quantitative results. However several controls established the genuine nature of the effect; for example it was abolished by crushing between B and C, was unaffected by reversing A and B, and developed with a shock separation consistent with the time taken for the impulse to travel from B to C.

Then after five or six weeks I had a horrid surprise. I switched the anode from just above the block to a position beyond it, i.e. from C to E as in Figure 6.1B, and found that the effect persisted. It

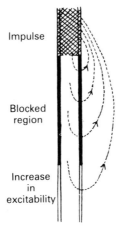

Figure 6.2. Diagram illustrating local electric circuits spreading through block and increasing excitability beyond it. (From Hodgkin 1936, 1937a, b).

therefore had nothing to do with an increase in membrane conductivity and was most simply explained by assuming that local electric currents were spreading through the block and raising excitability beyond it, as shown in Figure 6.2. More generally, the effect might be attributed to whatever agent is responsible for the conduction of nerve impulses. The existence of the effect did not provide any evidence for electrical transmission, but it offered a nice way of testing the theory, and it was this subject that I chose when I started whole-time research in the following year. At the time, I was disappointed that I had not obtained any evidence for an increase in membrane conductivity during the impulse and I gave up the experiments for the undergraduate Part II work that I should have been doing all along. This was no great sacrifice. A good deal of the course was taught jointly with Biochemistry and I enjoyed learning about the way in which phosphate bonds were handed around in glycolysis and elsewhere. At the end of term we spent a week doing practicals in the Biochemistry Lab, which again I found rewarding, though I never acquired the real laboratory skill that a good biochemist has.

At the beginning of the autumn term (1934) I visited Carl Pantin, who was back in Cambridge, though still far from well. 'We had a long talk and I was nearly persuaded to do Zoology. So I spent Tuesday rushing round seeing people and trying to make up my mind. I decided to stick to Physiology and am now glad I have done so' (14 October 1934). One of the things that decided me was a rather unfair remark made by my Director of Studies, Jack Roughton, who said, 'All experimental zoologists do is to apply to many animals the conclusions which physiologists have reached by working on one particular animal; if you want to find out anything really new you must join us in Physiology.' As the Physiological Laboratory in Cambridge then contained people like J. Barcroft, E. D. Adrian, B. H. C. Matthews, Grey Walter, F. J. W. Roughton, G. S. Adair, and E. N. Willmer, there was something to be said on Roughton's side. At all events I was converted by the remark, although it cannot be defended against the examples of J. Gray, D. Keilin, H. W. Lissmann, P. B. Medawar, C. F. A. Pantin, J. W. S. Pringle, V. B. Wigglesworth and J. Z. Young, to name only a few of the distinguished British scientists who have approached biology from the zoological side.

A fairer and perhaps more interesting way of distinguishing between the two subjects is that the zoologist is delighted by the differences between animals, whereas the physiologist would like all animals to work in fundamentally the same way.

During my Part II year I read nearly all the papers of the St Louis school. This made it clear to me that the leading 'axonologists' were thoroughly sceptical both of the membrane theory in general and of the local circuit theory in particular. I came to the conclusion that it would be well worth while to see whether the transient increase of excitability beyond a localized block was an electrical effect, and decided that I would take this up as a research project after I had finished Part II in July.

Several interesting things happened during the intervening period, including holidays in Austria (skiing), Florence and Siena (pictures) and Norway (mountaineering), but as this chapter is primarily about research I shall move on to July when I started life as a research scholar of Trinity and a research student in Physiology.

In those days laboratory life was rather informal, at any rate in Cambridge. I never worked for a PhD and didn't have a research supervisor. You might easily start in a bare room and have to build most of your equipment yourself, apart from a few standard bits like smoked drums, Palmer stands and kymographs. This sounds depressing but actually it wasn't. Nowadays scientists are apt to become neurotic and give up work if they know that their equipment is markedly inferior to other people's. Indeed it is regarded as somewhat unscientific to carry out experiments with any but the best equipment. This certainly wasn't my feeling when I started research and, to begin with, all that mattered was that I should have enough equipment to do something new. Perhaps I was rather extreme in this respect, but the general attitude to equipment was very different from that existing today. In his comments on the prewar Cavendish Laboratory in Cambridge, the distinguished physicist J. A. Ratcliffe (1975) explains that in the 1920s and 1930s an elegant experiment meant one that could be carried out very cheaply. He says:

> There was, I think, a feeling that the best science was done in the simplest way. In experimental work, as in mathematics, there was 'style' and a result obtained with simple equipment was more elegant than one obtained with complicated apparatus, just as a mathematical proof derived neatly was better than one involving laborious calculations. Rutherford's first disintegration experiment, and Chadwick's discovery of the neutron had a 'style' that is different from that of experiments made with giant accelerators.

Ratcliffe illustrates his point with this anecdote:

> . . . as a young research student I wished to try out a radiocircuit, in the way that was then common, by screwing some components to a wooden 'bread board'. When I went to get a piece of wood for the purpose Lincoln [the head of the workshop] pointed to a pile of scrap wood in the corner and invited me to take a piece, but as I was leaving the room he ran after me and said 'Here, Mr Ratcliffe, do you really need mahogany?'

I was lucky because I inherited a Matthews oscilloscope and other electrical equipment from Grey Walter. In those days it wasn't considered proper to use an amplifier built by someone else. So I constructed a condenser coupled triode amplifier in a series of biscuit tins which I painted bright blue. At that time there were no electrical soldering irons, no resin-cored solder and the valves, which were usually microphonic, needed anti-vibration mountings; so building an amplifier took longer than it would today. But I was helped in this and other things by Martin Wright, who already showed signs of the mechanical ingenuity for which he later became well known. I also received much help from Charles Fletcher who worked next door to me on *Mytilus* muscle.

The building we occupied was put up in 1912 with £20,000 given by the Drapers Company. Though not a thing of beauty, the old building has proved immensely serviceable and, with additions in every direction, is still the home of Cambridge Physiology. A lecture theatre and a new wing were added in the mid-1930s, but I have no clear memory of the disturbance the additions must have caused.

The head of the laboratory was Joseph Barcroft, who had succeeded Langley as Professor in 1925. I learned later that he was nicknamed Soapy Joe by some and that there were those who regarded him as a tricky character. But I saw nothing of this and he treated me with great kindness and friendliness right up to the day he died when running to catch a bus after working late in the laboratory at the age of 75. Barcroft's enthusiasm was infectious and I admired *The Respiratory Function of the Blood* (1914) as well as his splendid review on *La fixité du milieu intérieur* (1932) which provided me with the kind of philosophical background to physiology that I felt I needed.

When I finished writing my first paper in 1937 I took the manuscript to Barcroft and asked if it needed his approval before I sent it to a journal. He was quite taken aback and explained first that we did not do anything like that in Cambridge and, second, that anything I wrote was entirely my own affair. The only time I remember his getting cross was when I rather tactlessly asked him if he wanted me to light the way to the door late one evening. He said, indignantly, that after thirty years he could find his way around the lab with his eyes shut. My concern was understandable, because the basement where we were was pitch-black and unless you were careful you were liable to fall into a large bath containing bloody saline and a dead sheep on which a caesarian operation had been performed earlier in the day. But they were Barcroft's sheep and I suppose he knew where they were. This was after the war, perhaps in 1946–7.

I do not remember any laboratory secretaries or typists in the prewar period, though I have a mental picture of the chief assistant operating with one finger on an old-fashioned typewriter in the office next door to Barcroft's lab. His name was Secker and both his son and grandson have worked in the same laboratory. Old Secker was an ex-sergeant major and had a fine military moustache. He looked irascible but wasn't, except when struggling with the lab accounts. If you were sensible you did not ask for anything when the auditors were around. ('They have me sweating,' he used to say.)

There were no centralized laboratory stores and if you wanted a valve or an electrical component you went to a local wireless shop that provided components for several laboratories. You might be able to scrounge nuts and bolts or sheet metal from the workshop but it was generally better to go to one of the local ironmongers. Mr Hall, the machinist in the workshop, was often busy with class equipment, but if you could persuade him to make something for you it worked pretty well and his ideas about design were generally better than yours. I think we were allowed to spend about £30 per annum, which was quite a lot in those days. If you wanted anything more, or needed your manuscript typed in the town, you paid for it yourself. It was some time before I realized that one might be able to get a grant from the Royal Society or from a Research Council.

I first met E. D. Adrian (later Lord Adrian) at Seatoller on the Lake Hunt, and have a clear memory of him as a figure in a mackintosh cape running swiftly downhill, emerging briefly from the mist and then disappearing again as another cloud swirled up the Ennerdale valley. Three years later I had the alarming experience of being driven by him from Cambridge to the Lake District at what then seemed a terrifying speed. I am afraid that by modern standards Adrian would be considered a dangerous driver because he relied to a great extent on the quickness of his reflexes. Indeed this did not always work, as I found on another occasion when he ran into a taxi in Trafalgar Square. As a young man driven by his Professor I kept my mouth shut, but I was surprised to find that Adrian's command of language was quite equal to that of the taxi driver.

Adrian regarded the Physiological Laboratory as a place to work and we rarely talked there for more than a few minutes. After I had given a lab seminar about my nerve-block experiments in 1936 I remember him saying, 'Why not work on crab nerve?' which was advice I took. On another occasion he did not actually say anything but managed to show his disapproval of the way I had soldered a joint by rushing out of the workshop with a muttered curse.

Although Adrian was not given to obvious enthusiasms you could see that he was really pleased if you told him about some technical advance such as isolating a single nerve fibre or getting a microelectrode into a cell. This was very encouraging. In addition I owe Adrian a great debt for all he did to promote my scientific career. I am sure that he was largely responsible for my becoming a

University Demonstrator and Teaching Fellow at Trinity in 1938, as well as for a great deal of help after the war. But it was not possible to thank him for such things and I know that if I had tried he would have choked me off at once.

As a research student I did not have a formal supervisor, but there were plenty of people one could talk to about technical matters, including Bryan Matthews and Grey Walter in Physiology and A. F. Rawdon-Smith in Psychology. On the theoretical side I learned a lot from William Rushton, with whom I did a joint piece of work in 1939. Victor Rothschild, who was then working in Zoology on fertilization, lent me apparatus from time to time and had a considerable influence on my life by softening the strongly puritanical streak I had acquired from my Quaker upbringing. The Rothschilds were infinitely hospitable and gave splendid parties, sometimes illuminated with fireworks, at their beautiful house on the Backs in Cambridge. This provided a welcome change from political discussion or gloomy contemplation of the international scene – these being two of the principal occupations of most of young Cambridge in the late 1930s.

Another person who helped me in many ways was A. V. Hill. Polly and David, his two elder children, were close friends of mine and I sometimes stayed with the Hills at Three Corners, their house near Plymouth, when working there in the summer.

One of the advantages that Plymouth shares with Woods Hole is that you meet people from other laboratories and disciplines. Besides A. V. Hill, I got to know J. Z. Young and Bernard Katz at Plymouth, with both of whom I have kept in close touch ever since. In prewar days, scientific meetings meant little to me. I went to the Physiological Society from time to time, but never attended a large international congress until the Oxford Congress in 1947. For this reason informal contacts such as those I made at Plymouth or later at Woods Hole were particularly important.

Apart from learning how to dissect crab nerve, nothing much came of the laboratory work that I did at Plymouth in the few weeks that I spent there in September 1934 before starting work as a research student at Cambridge.

After a few weeks I managed to record the electrical potential spreading in front of the nerve impulse through a blocked region. However, the Matthews oscilloscope wasn't really fast enough – although excellent for recording the presence of impulses – and the whole set-up was terribly cumbersome, with an arc lamp, rotating mirrors and a cylindrical paper screen all arranged to give the same effect as a cathode ray oscilloscope. So I bought a cathode ray tube and accessories from Cossor, and a second-hand film camera from Wardour Street in London. I also persuaded our workshop mechanic, Mr Hall, to build a rotary contact breaker for starting the time base and timing two shocks. This did what I wanted, but made a terrible noise as a series of huge cams smacked into three magneto contact breakers ten times a second. As I had to pay for all this equipment myself, I bought the cheapest kind of cathode ray tube; this was a 'soft' tube in which electrons are kept in a column by positively charged gas molecules, rather than by focusing electrodes.

Bryan Matthews was away in the Andes that year, but I received much help and advice from A. F. Rawdon-Smith who had a first-class knowledge of electronics and worked next door in Psychology. At that time there was a sort of mystical idea that the noisiness of an amplifier varied inversely with the skill of the man who built it, and amplifier noise was regarded as a sort of moral penalty for bad workmanship. As I have always been rotten at making things I naturally attributed my noisy base line to poor workmanship and went over the whole amplifier resoldering joints in critical places. In fact, as I eventually discovered, the base line in my set-up was relatively noisy because the frequency response of my cathode ray tube was much higher than that of the Matthews oscilloscope used in the basement. Then the whole business of shock-artifacts was shrouded in mystery and I didn't learn to think rationally on this subject until I went to the Rockefeller Institute in 1937. There I met Dr Toennies who looked after the electronics in Gasser's group; he told me to forget about radiation fields and other irrelevant ideas that I had been struggling with, and to think only in terms of electrical leaks, stray capacities, and actual spread of current in the tissue.

By mid-July 1936 I had been through the main experiments and the results came out pretty strongly in favour of the idea that nerve impulses are propagated electrically by electric currents spreading passively in a local circuit ahead of the active region. My results showed that when an impulse arrived at a local block it might decrease the electrical threshold by 80–90 per cent at a distance of 2 or 3 millimetres ahead of the active region. This transient increase in excitability was accompanied by an electrical change which had the characteristics expected of an electric current spreading passively along the cable-like structure of the nerve fibre. Quantitatively, the results indicated that there was a large safety factor in nerve conduction: in other words that the action potential was much larger than it need be in order to excite the next region of nerve. One of the implications of this result is that a nerve impulse should be able to skip over 1 or 2 mm of inactive nerve, a prediction later verified by Tasaki and others.

At the time I did not know whether or not conduction was saltatory, and therefore drew the local electric currents as spreading continuously along the nerve fibre (Figure 6.2) instead of being concentrated at the node as I might have done. But perhaps this does not matter very much as conduction is continuous in many types of excitable tissue, and in any case the question does not affect the argument as to the electrical nature of nerve conduction.

By mid-July 1936 I had been through the main experiments and wrote up the results in a Fellowship thesis for Trinity College, where I had been living happily for four years. I was surprised and very pleased to be successful, but also a little alarmed to be joining a society which included people like J. J. Thomson (the Master), Rutherford, Aston, Eddington, Gowland Hopkins, Adrian, Wittgenstein, Hardy and Littlewood.

I gradually got to know Trinity's great men of whom Rutherford was the most dominating. He divided science into physics and stamp collecting but made an honorary exception of physiology, and was very nice to me. He had a loud voice which could be heard all over the College dining hall, and hated anything that interfered with his weekend golf. I remember the indignation with which he spoke at a College meeting called for a Saturday afternoon to decide whether or not to give £1,000 for a squash court at a women's college. 'Of course we must give it,' he said, 'and not spend a beautiful summer afternoon debating this topic like a bunch of schoolgirls' – or something of the kind. This did not go down well with some of the elderly bachelor dons, but Rutherford won the day as he usually did.

It was generally believed that the Master and Rutherford didn't get on and that J. J. was responsible for keeping out some of Rutherford's star research students at the annual election, which is supposed to be decided mainly on the promise shown by the thesis.

The failure of Mark Oliphant to get a Trinity research fellowship has sometimes been attributed to J.J.'s antagonism, but in his defence it should be said that it is very difficult for someone to show early promise if, as in Oliphant's case, his research depends on the design and construction of large and expensive machinery.

There were many stories about J.J. and Rutherford. Gow told me that on one occasion, F. W. Aston, another Nobel prize-winning physicist, but with a quiet retiring disposition quite unlike Rutherford's, came in some distress to Rutherford saying that he had discovered a new isotope, but that the Master had refused to believe him. 'Then my boy,' said Rutherford, 'you can go down on your bended knees and thank Almighty God that the blighter hasn't swiped it.' This no doubt was a reference to the widespread rumour that in describing his famous determination of the mass/charge ratio of the electron, J.J. had been less than generous to his research student J. S. E. Townsend. Townsend, later Sir John, became Professor of Experimental Physics at Oxford, where it used to be said that 'doing a J.J.' was a standard term of abuse in his family.

In his later years J.J. had an extraordinary way of talking through his false teeth, which made him a gift to the mimic and tended to perpetuate, and perhaps exaggerate, anecdotes about him. But there is one story for which there is documentary evidence. One of the Master's duties as Chairman of the College Council is to write to the special lecturers whom the Council selects each year. Of these the Clark Lecturer in English Literature is one of the oldest and most distinguished. So it was with some consternation that the Council heard J.J. announce that he had written to R. W. Chambers instead of E. K. Chambers, whom Council had chosen, and that the former had promptly accepted. I know about this because I was on the Council ten years later when an attempt was made to put matters right by an invitation to E. K. Chambers, who, however, was unable to accept.

Another literary anecdote was that, on seeing that W. B. Yeats was to be sitting next to him in Hall, J.J. opened the conversation with the remark, 'Been writing much poetry lately, Mr Keats?'

After my election to a research Fellowship I looked forward to the Fellowship Admission dinner with some apprehension because J.J. was reputed to mix up his welcome with his own private thoughts. For instance when Anthony Blunt was elected, he said, 'This is the first time we have elected a Fellow in Art-History – and

I very much hope it will be the last'; or with Victor Rothschild he felt it important to stress that we looked only for intellectual merit and paid no attention to race, creed, social position or great wealth. However, I got off lightly, because as far as I remember J. J. confined his remarks to the observation that there was nothing special to say about any of the new Fellows this year.

After getting the Fellowship I spent several months repeating and tidying up experiments and eventually two papers were published in the *Journal of Physiology* (1937) almost exactly three years after the beginning of the experiments.

It is natural now to ask whether it was worth while spending so long in proving something so obvious as the Local Circuit Theory. But the theory really was not regarded as probable or even plausible by many people, even though it had been batting around since the time of Du Bois Reymond and Matteucci nearly a hundred years before. An example of this scepticism is provided by the very pleasant letter that I received from the distinguished physiologist, Joseph Erlanger, who shared the Nobel Prize with H. S. Gasser in 1944:

Washington University, School of Medicine,
Saint Louis
6 January 1937

Dear Mr Hodgkin:

I am returning to you herewith the manuscript of your thesis which was sent to me by Professor A. V. Hill via Dr. H. S. Gasser. The material was read with the keenest of interest. It is a very direct and well executed attack on a fundamental problem. It is to be expected that the number of questions raised by an investigation will be in direct proportion to its importance and a number of questions have suggested themselves to me as I perused your manuscript. I am going to mention some of these as they occur to me.

In general I find it hard to believe that nerve impulses are propagated by currents eddying outside of the conducting structure. Teleologically such a mechanism of propagation seems queer. The fact that we have found it impossible to demonstrate in an intact nerve any alteration in the excitability of inactive fibers lying parallel to active fibers lends support to our skepticism.

Your curve of the excitability changes set up by a blocked impulse beyond the block does not have the temporal dimension of the curve of the temporal summation of impulses across a nerve block , in that your curve falls to zero in a period that exceeds slightly the duration of the recorded action potential, whereas temporal summation of a blocked impulse is demonstrable sometimes out to 100 msec. If both of the effects are electrical, and we believe they are, the latter duration must be determined by the negative after-potential. This would not be impossible granting your assumptions relative to the effectiveness of the action potential as a stimulus. If one-twentieth to one-fiftieth of the spike suffices to stimulate, then an after-potential having one-hundredth of the spike height might exert an appreciable effect on the excitability beyond the block.

There are two further pages discussing various possible sources of error and the danger of depending on one record. These are not easily intelligible without consulting my thesis and published papers.

Letter ends:

In any reply you might care to make please do not hesitate to let me have the benefit of your experience relative to the questions I have raised.

Very sincerely yours,
Joseph Erlanger

I replied to the first part of Erlanger's letter in the following way:

Trinity College, Cambridge
1 February 1937

Dear Dr. Erlanger,

Thank you for returning my thesis which arrived here a few days ago. I was extremely interested in your letter and I will try to answer the various questions you raised. Naturally I can't do this at all fully and you will forgive me if the necessity for being brief makes my opinions sound dogmatic.

Your first point is that a mechanism involving current eddying outside axons is curious teleologically. In answer to this I can only reply that this concept seems to me to fit in with the general implications of cell physiology. For many reasons it is plausible to suppose that excitation depends upon a change at the cell surface. Now if the surface is to be activated electrically one would imagine that current (ohmic or capacity current) must flow through it. And this could only happen if there were local circuits, with current flowing in one direction inside and in the other direction outside the surface membrane. There is no histological evidence for a gap between plasma membrane and myelin sheath so that one would expect the local circuits to be completed by current flowing through the interstitial fluid between the axons.

75

I had read your paper on the extrinsic effects of nervous impulses, but I was glad to have the reprint for reference. It does not seem to me that there is necessarily any contradiction between your results and the view that transmission depends upon electrotonic currents. The crucial point is whether electrotonus can spread along one fibre without affecting others. This would be possible if the resistance of the interstitial fluid were low compared to that of the axon cores. For in this case only a small fraction of the current flowing in the external circuit of the active fibres would penetrate the resting axons, since these would be shunted by the interstitial fluid. In order to try and make up my own mind I did some calculations for a case where one would expect the effect of extrinsic impulses to be fairly large. The assumptions were as follows:

1. All the axons in a nerve trunk except one active. I assumed the resting axon had a cable-like structure and was exposed to the field produced by the others. The shape of the potential field was obtained from one of your records of an action potential at about 10 mm from the cathode. Conduction velocity assumed to be 30 m.p.s.
2. The space constant of the resting axon was assumed to be such that an electrotonic potential would fall to 1/e of its value in 2.3 mm in the absence of an applied potential.
3. Resistance of interstitial fluid is equal to the resistance of all the axon cores in parallel.
4. The potential for rheobasic excitation is equal to 1/20 of the action potential developed by that axon.
5. Hill's k (Chronaxie) assumed to be 0.3 msec.

I had to solve the differential equations involved by a rather rough graphical method. With these assumptions the result was that the excitability in the resting fibre would have increased by at most 25%. I don't know whether you would regard this result as in your favour or in mine. At any rate the calculation shows that the extrinsic effects of an action potential are small compared to the intrinsic ones. Until more is known about current distribution in nerve it seems precarious to use this kind of argument at all.

At the time of writing my thesis I had only done a few experiments with pressure blocks. Since then I have found that with pressure but not with cold blocks there is usually a long tail to the increase in the excitability curve. This corresponds very closely with the unusually large negative after-potential shown by the [pressure] blocked nerve. Of course the agreement between potential and excitability can only be obtained if the electrotonic spread from blocked impulses is large compared to action potentials propagating through the block . . .

Thank you for sending me your reprints and for the two other letters. And may I again say how much I appreciated your interest in my thesis.

Yours very sincerely,
[Alan Hodgkin]

We continued this and other arguments, in the same polite and friendly way, when I stayed with the Erlangers in St Louis for a few days in the spring of the following year, 1938.

When I wrote up my first two papers I called them *Evidence for Electrical Transmission in Nerve*, and explained that the starting point for the investigation had been the observation that a nerve impulse arriving at a localized block increased the excitability in the nerve beyond the block. This was perfectly true and I did not feel it necessary to add that the blocked impulse effect was discovered by accident when trying to test something quite different. Indeed, had I included anything of the kind I am pretty sure that, even in those more spacious days, as I explained much later in the original article, *Chance and Design in Electrophysiology* (1976), the editors would have made me take it out.

Chance and good fortune, combined with some theory, were equally important in my next piece of research. I had grown interested in cable theory and had come to the conclusion that it should be necessary to excite a finite length of nerve in order to start an action potential. The argument, which was developed independently and much further by Rushton (1937), led to the idea of a subthreshold response which might explain the unexpected results obtained by Katz (1937) and Rushton (1932) in their studies of excitability. But I didn't make any deliberate attempt to test these ideas and instead followed a suggestion of Professor Adrian that I should work on crab nerve. One of the things that I hoped to do with this nerve, which has no outer sheath (perineurium), was to test the idea that accommodation might be due to the polarization of some structure in series with the nerve fibres. I was also still interested in the possibility of measuring an increase of membrane conductivity during the impulse and thought this might be more obvious in crab nerve. At all events I found it very easy to split crab nerve into fine strands and one of the strands I picked up turned out to be a single nerve fibre – to judge from its enormous all-or-nothing action potential. This really was a great piece of luck as I had no dissecting microscope and the chances of picking up one of the half-dozen or so 30 μm fibres in a nerve trunk a millimetre thick are not high. The next day I borrowed a dissecting microscope and from that day to this I have hardly ever worked on a multifibre preparation again.

Soon after I got the preparation going I noticed that a shock which was just below threshold produced something like a small, graded action potential, which grew rapidly in size as the stimulus approached threshold. This clearly was what was needed to explain Bernard Katz's results and I was very pleased to be getting evidence of something as unorthodox as a graded response in a single nerve fibre. My electrical technique was not really up to recording from single nerve fibres as can see be seen from the illustration in the preliminary note describing the Cambridge experiments (Hodgkin 1937c). However, help was at hand because Herbert Gasser, who was then Director of the Rockefeller Institute in New York, had invited me to spend a year in his group, and I had been awarded a travelling fellowship by the Rockefeller Foundation. Soon after I arrived, Dr Toennies, the electronics expert in Gasser's group, pointed out that it was essential to use a cathode follower if one wished to make accurate recordings of rapid electrical changes from a high resistance preparation like a single crab axon. He provided the necessary equipment and I learnt a great deal about electronics and electrical recording from Gasser and his group which included Lorente de Nó, Grundfest and Hursch. This aspect will be taken up again in Chapter 8.

Chapter 7

Cambridge, 1932–7: mainly politics

IN READING the letters to my mother written over a five year
period it soon becomes clear, both from the handwriting and
from the style, as well as from the thoughts expressed or suppressed, that the writer changed radically in character and maturity
during his first five years in Trinity (Figure 7.1). These changes are
closely connected in my mind with the progressive improvement in
the location of my Trinity rooms: these being as follows:

1932–3: ground floor Whewell's Court.
1933–4: ground floor New Court.
1934–5: attic, south side Great Court (Trinity's largest and grandest
court, mainly Tudor).
1935–7: second floor room in Essex Building with splendid view of
Great Court.

And finally to complete the list, after a year in the USA:

1938–9: second floor Nevile's Court with views of that court one
way and the river the other.

I do not mean that my character was determined by the beauty or
otherwise of my surroundings, though there may be something in
that, but simply that I associate different feelings and types of
behaviour with these different places.

In my first year in Whewell's Court I saw a good deal of James
Klugmann who occupied rooms nearby. Later on he became an
open and enthusiastic member of the Communist Party, and is
widely thought to have been the person who converted Anthony
Blunt and Guy Burgess. However, when I came up he was living
the life of a recluse among a great library of books that he had
accumulated. His sister Kitty and brother-in-law Maurice Corn-

Figure 7.1. From the collection of Fellows' portraits, Trinity
College Library, October 1936.

forth were open communists, but none of the three made any
serious effort to convert me, either then or later. At that time I was
most anxious to keep up my German, and Klugmann lent me
novels by Arthur Schnitzler, Hans Fallada and other Austrian or
German writers of the Weimar Republic period. We also spent a fair
amount of time discussing the political situation in Germany which
was clearly developing in a most unpleasant way.

80

In my first year, 1932–3, Communism was not fashionable. Apart from James Klugmann and Philby, the only serious communist that I remember in Trinity was David Haden Guest, who was later killed in Spain. He was then a scholar reading Moral Sciences who expressed his disapproval of society by being permanently half-shaven, yet never growing a beard. Still I was subjected to a good deal of left-wing pressure as may be seen from the following extracts from letters written to my mother in March 1933:

> This weekend Jack Hoyland is down and had lunch with me today; we had an interesting talk about various things: Russia to Oxford Groups being the limits. I met David Buxton last week; although he got quite a good second in Zoology Part II he can't get any kind of colonial biology job; in fact with this depression they are axeing people. It's a depressing prospect; present day society seems unable to make any good use of science at all – witness poison gas, chemical warfare, bad influences of mechanisation etc. And an economic biology job which would probably only result in more profit going into some capitalistic hands instead of to the good of society as a whole doesn't seem worth while. However at the moment things are quite the reverse of depressing – I get great pleasure out of the informal side of science – things like the Science Club etc. Last week's paper was on colour change in animals. There was an exhibition in the Experimental Zoology Lab. Some of the research apparatus was very exciting. I also exhibited the parasites I had found which behaved quite well.
>
> On Saturday there was an Antiwar demonstration in Cambridge – chiefly demonstrating against the increased expenditure on the Navy and various other militaristic activities. I dislike demonstrations etc very much but I think it was quite a good thing.

In the light of subsequent events the last extract seems illogical for someone as deeply opposed to Hitler and the Nazis as I knew myself to be. But at that time many people so feared and hated the prospect of war that they felt it right to oppose rearmament or any sort of militaristic activity, no matter what the consequences of their action might be.

Jack Hoyland, who is mentioned in the first letter, was keen that I should join one of the allotment schemes by which the Quakers were trying to help people in areas of massive unemployment. The idea was that relief funds should be used to purchase derelict land which volunteers and miners would convert into allotments for the use of the latter. He wanted me to join a party going to a South Wales mining area in the Easter Vacation. I could not manage this as I was going to the Marine Biology course at Plymouth at the end of March but said that I would attempt something of the kind in early September (1933).

So I found myself staying for a week in Wigan, then a mining town, with an unemployed miner, Joe Banks, and his wife and small daughter. They lived in one of those long rows of industrial dwellings with no bathroom and an external earth-closet type lavatory at the back of the house, cleared every few days by a refuse cart. The food was simple but adequate, with a great deal of bread and butter and tea at every meal. However, I paid for my keep (and I hope a bit more) so perhaps they were doing better than normal – also they had only one child. But I didn't get the impression that conditions were quite as grim as those I had seen in Germany in the preceding year.

Joe Banks was very nice, but I found it hard to like his wife who was one of those devout puritanical Nonconformists whose religion takes the form of disapproving of practically everything and everybody. Joe had injured his shoulder while mining and so could no longer work at the coal-face, but he was obviously a capable organizer and would not have much difficulty getting a job if mining ever started up again in Wigan. He spoke with regret of the pre-Depression days, when he was a bachelor and could earn enough money in eight or nine months to keep himself at seaside resorts for the summer, which he and his friends would do, knowing they could get re-employed whenever they wanted. He was deeply attached to Wigan, which he thought much maligned by music-hall jokes about Wigan Pier and insisted on showing me the local beauty spot – there is one – and taking me to a football match. This wasn't exactly what I had planned but it was impossible to refuse and in fact fitted in quite well. For I found when I arrived that I was the only volunteer and though land for the allotments had been obtained and some had been started, a long dry summer had made the ground iron-hard and digging with a spade was virtually impossible. But I had come to dig and struggled on, which the unemployed miners thought ridiculous as they sat around and watched my amateur efforts. So I am afraid that my contribution to the Quaker allotment scheme, of which I thoroughly approved, was more or less negligible. I do not know what happened to the

Wigan allotments, but allotments became important during the war and have persisted ever since, so I hope the Wigan ones survived. But I found the Wigan experience discouraging and, after various rather feeble efforts that I made to help a boys' club in Edinburgh, I began to think that I was not cut out for social work.

My communist friends in Cambridge disapproved of the Wigan visit for very different reasons. They argued that Quaker relief was a bourgeois trick, and set about removing my religious beliefs. This was not very difficult, because I had begun to have serious doubts about most of the central aspects of religion which I found incompatible with my scientific knowledge. But it was a tactical mistake on the part of those who wished to make me a communist or fellow traveller, because the reason that I distributed their pacifist leaflets was basically a religious feeling that I ought to do so. With religion gone, I was much less use to them. Of course, like half the college, I was quite prepared to join the Hunger Marchers out of general sympathy with the unemployed. But I never came any-where near joining the Communist Party, and after a while my left-wing friends gave up trying to convert me.

In the mid-1930s the Trinity College Magazine contained an excellent Who's Who which consisted of brief but perceptive comments about senior and junior members of the College. For example, against one of Anthony Blunt's great loves there was the single entry: *Cleopatra*. And for Patrick Duff who devoted his life to the College the entry was: *Like fine and luxuriant ivy, could not live without clutching on the walls of some institution*. I am rather proud of my entry in 1933 which reads: *Can't quite believe in the Marxian interpretation of science*.

John Cornford and the left-wing explosion, 1933–7

John Cornford occupies a central position in this chapter, partly because he had a great influence on his contemporaries, and partly because his arrival in Cambridge coincided with a change in the tactics of the Communist Party. Instead of concentrating almost exclusively on the working class, Party members felt it would be right to recruit the intelligentsia and particularly students to the revolutionary cause.

John was the eldest son of the Professor of Ancient Philosophy Francis Cornford and Frances, a minor poet, who was the granddaughter of Charles Darwin on her father's side and related to the poet Wordsworth on her mother's.

Apart from his writings on Plato, which I understand are still important to scholars, Francis is best known for his skit on university politics, *Microcosmographia Academica*, which has gone through several editions since it was first published in 1908, and which still has a certain relevance to the conduct of university business when the *Principles of the Wedge, Of Unripe Time* or *The Dangerous Precedent* all appear if the conservatively-minded wish to block some much-needed reform.

Frances published several volumes of pleasant Georgian poetry and was awarded the Queen's Medal for Poetry in 1959, but is best known for her 'curiously memorable though undistinguished lines' in the poem *To a fat lady seen from the train* which runs

> O why do you walk through the field in gloves,
>   Missing so much and so much?
> O fat white woman whom nobody loves,
> Why do you walk through the field in gloves,
> When the grass is soft as the breast of doves
>   And shivering sweet to the touch?
>   O why do you walk through the fields in gloves,
> Missing so much and so much?

It seems strange that two such gentle, cultured intellectuals as the Cornford parents should have produced such a fanatical revolutionary as John Cornford undoubtedly was. His death in Spain at the age of 21 also seems a terrible waste of a brilliant intellect.

After early schooling in Cambridge, John, who was born on 27 December 1915, was educated at Stowe where, at the age of sixteen, he obtained a Major Scholarship in History at Trinity College, Cambridge. After spending two terms at the London School of Economics he came up to Cambridge in October 1933, took first-class honours in Part I of the Historical Tripos and a starred first in Part II in June 1936. He was then awarded the Earl of Derby Research Studentship, which he resigned on going out to Spain, almost immediately after the beginning of the Civil War, in early August 1936. After a month's service in Aragon, he returned briefly to England to form the company which became the nucleus of the British part of the International Brigade.

After five weeks of severe fighting and bombardment in Madrid,

Cornford was made commander of the machine-gun unit in which he served; this was done at the wish of his men after Cornford had at first refused the position. In University City he was wounded in the head, but returned to the front after twenty-four hours in hospital, to face heavy shelling from German troops at Boadilla del Monte. In late December his unit was sent to Cordova and there, on the 28th December, one day after his twenty-first birthday, he was killed in action at the head of his men. News of his death reached Britain about a month later and until the Nazi-Soviet pact of August 1939 he was to remain a heroic figure to many with left-wing sympathies, and his poetry was regarded as an important contribution to revolutionary literature.

In his years in Cambridge, Cornford brought the same fierce energy to student politics as he did to fighting Franco. To begin with he shared a double set over the eastern gateway into New Court with a friend from Stowe, Philip Gell, whom I liked very much. As well as being a good scientist, Gell was a talented painter and gradually covered the walls of their enormously long room with Orozco-type frescoes. Otherwise there were practically no decorations or furniture and Cornford and the people who came to meetings mostly sat on the floor.

Cornford's talent for organization and revolutionary fervour made him joint leader of the Cambridge University Communist Party and later joint secretary of the Socialist Club. In addition to fostering links with working people in the city, his main aims, like those of the Party, were to build a massive anti-Fascist movement among students and to form a strong revolutionary Socialist organization.

His work and that of his colleagues was partly propaganda and partly intensive Marxist study and education. At that time there was relatively little left-wing literature apart from the Marxist classics and one or two books by Strachey and Burns. The 'line on students' and the 'line on culture' had yet to be worked out.

The events of Armistice week in November 1933 were the first illustration of the new spirit in Cambridge. They involved a demonstration and counter-demonstration over the film *Our fighting Navy*, then showing at the Tivoli Cinema, which ended in a free-for-all fight between left-wing students protesting against militarist propaganda and anti-intellectual, 'patriotic' undergraduates. This led on to a large-scale, anti-War demonstration on 11

November organized jointly by the Socialist Society and the Student Christian Movement, which was to march through the centre of Cambridge to the war memorial, where it would lay a wreath with the inscription 'To the victims of the Great War, from those who are determined to prevent similar crimes of imperialism.' The Socialists, including John, were anxious to include the reference to imperialism even though it kept out the League of Nations Union, but eventually the police insisted on the inscription being removed as likely to lead to a breach of the peace. Nevertheless, there was a scuffle if not a fight, and attempts were made to seize the Socialist Club banner.

From 1934 to 1936 the annual Hunger Marches of the unemployed were supported by many students, at least a hundred from Trinity I would think. Of course only a handful walked all the way from Cambridge to London. The majority were taken in buses which brought them to the outskirts of central London from where they walked to Westminster. I joined on one occasion that was massively popular for the paradoxical reason that Claud Cockburn's magazine, *The Week*, had said that the authorities were moving machine guns into the parts of London that the demonstration would cross. Of course we saw nothing of the sort, but I do remember the sight of Guy Burgess in a zipped-up cardigan which he could zip down to reveal an old-Etonian tie. He said this would come in useful if he were arrested by the police.

John Cornford's attitude to university life is summarized by the following poem which does have a certain power, however much one may dislike the sentiments and the language in which they were expressed:

> Keep culture out of Cambridge
>
> Wind from the dead land, hollow men
> Webster's skull and Eliot's pen,
> The important words that come between
> The unhappy eye and the difficult scene.
> All the obscene important names
> For silly griefs and silly shames,
> All the tricks we once thought smart.
> The Kestrel joy and the change of heart,
> The dark, mysterious urge of the blood,

86

The donkeys shitting on Dali's food,
There's none of these fashions have come to stay,
And there's nobody here got time to play.
All we've brought are our party cards
Which are no bloody good for your bloody charades.

Subsequent events have not dealt kindly with John Cornford's articles on communist theory. It now seems ridiculous to write of Stalin's empire of the 1930s in such glowing terms, as he does in several articles; or to speak of 'the consistent and unwavering fight of the USSR to preserve the peace at almost any cost, its willingness to make any concession for peace and its complete freedom from any aggressive aims . . .'; or to counter the charge that Communism like Fascism is undemocratic with the following argument: 'But even here the fog is dispersing. Publicists like Shaw, social investigators like the Webbs, have a considerable influence on the middle classes. And when both proclaim that the Soviet system is in certain respects the highest form of democracy yet seen, those students and intellectuals who are not too prejudiced to face reality at all begin slowly to revise their opinions' [*Communism in the Universities 1936* in John Cornford 1938, ed. Pat Sloan. Jonathan Cape].

The Cambridge Apostles

In late 1935 I was asked by Victor Rothschild and the physiologist Grey Walter if I would join the Cambridge Conversazione Society. They explained that this society met once a week in term-time, when one of the members read a paper which would lead to a discussion of some question of general intellectual interest in politics or the arts and sciences. I gathered that this society, which was sometimes known as the Apostles and sometimes as The Society, was of considerable antiquity and distinction, having been founded in 1820, and including the poet Tennyson and physicist Clerk Maxwell among its former members. My initial reaction was to refuse on the grounds that the time of the weekly meetings clashed with that of the Natural Science Club to which I was firmly attached. This caused great offence and my excuse unfortunately did not work as The Society was just about to change its time of meeting from Friday to Saturday night.

The custom was that the active members, who were mainly undergraduates and research students, attended regularly and senior members, known as angels, came occasionally but were not expected to read papers. When I was first elected there were six or seven active members of whom Alister Watson (mathematical philosophy), Hugh Sykes Davies (English literature) and Guy Burgess (history) tended to lead the discussion. Senior members who came from time to time were George Trevelyan, G. E. Moore, Dennis Robertson and Maynard Keynes. Anthony Blunt, whom I had met in Trinity, and liked, had ceased to be active and rarely came to meetings when I was a member. Although the Society contained several communists, I did not get the impression that it was used as a breeding ground for party members, let alone moles or traitors, of whose existence in peacetime most of us had no idea at all.

In the autumn of 1936, just after I had become a Research Fellow of Trinity, I found that nearly all the active members had either resigned or left Cambridge, the only ones remaining being the economist David Champernowne and myself. This deeply upset Maynard Keynes, who summoned me in the spring of 1937 and asked what should be done. I fear that I was not enthusiastic about recruitment as I was deeply immersed in my nerve research, in love with a very pretty girl and shortly going to New York. However, Keynes and perhaps Champernowne must have got busy because the Society was rapidly restocked with new members, starting with Michael Straight in 1937, and was in a flourishing state when I returned from America a year later. I wanted to resign from active membership at once but was forced to read one more paper before being allowed to do so.

Looking back, the aspect of the Society that I enjoyed the most was the annual dinner which was held at the Ivy Restaurant and was attended by many of the Bloomsbury luminaries like Leonard Woolf, Desmond MacCarthy and E. M. Forster.

Chapter 8

New York, 1937–8: The Rockefeller Institute

TOWARDS THE end of September I travelled from Southampton to New York on the USS *Manhattan* with my
beautiful friend Phyllis Gill, who was on her way to do a year's
research in Montreal with Hans Selye. We were exact contemporaries and had read similar subjects, she taking Biochemistry and I
Physiology in our respective Part II years. After that she worked in
Physiology on tissue culture, under E. N. Willmer, for her PhD.
She had many admirers in Trinity, sometimes known as the
Phyllistines, but would have none of us and married a Dubliner, Dr
O'Donovan, to whom she got engaged in Montreal, to my great
distress. We had seen a great deal of one another during the previous
year, spending part of the vacations together, either with my family
in Islay or with hers on a visit to a superb performance of *Figaro* at
Glyndebourne. Phyllis was a talented violinist and a lasting effect of
this first romance of mine was to make me much more interested in
music than I otherwise would have been. We had planned to spend
two days together seeing New York. But, as I wrote to my mother,

> A horrid thing has happened to Phyllis. On Saturday morning she felt
> very unwell – pulse going very fast and irregular, very tired and very
> depressed. The doctor said there wasn't anything the matter beyond a
> rather severe 'period' and it seems to be the same kind of thing that she
> had after doing Part II.
>
> She was better this morning and I put her into a train for Montreal;
> she will go straight through and a scientist we know is travelling with
> her; she will be met by her University Women's Society at the other
> end, and they will look after her. This sounds rather hard for her, but
> the doctor advised it and it is better than struggling on in hotels in New
> York. I am rather depressed about all this . . . very behindhand in
> consequence – the Rockefeller expects quick starting.

The next letter to my mother is more cheerful, and gives my first impressions of New York:

Beekman Tower, 1st Av. and 49th street, New York
6 October 1937

The boat arrived very early on Friday morning and we dragged ourselves up from below at about 6.00 a.m. to look at the Statue of Liberty. This isn't very exciting – its legs are too short and freedom is the last thing that it suggests. But our first view of New York was very impressive – like the postcards only much grander.

. . . Off the ship about 12.00, having spent hours in customs and passport offices, and drove to our hotel [the New Weston, not as up-market as it is now]. Some of the streets are magnificent and altogether New York is a much more beautiful and spacious town than I had expected. I had imagined that it would be something like a tall forest and that you would feel hemmed in by skyscrapers. But it is really much more like a mountainous district with ranges of high buildings in between. [Another early letter says that 'All the main streets are incredibly clean and tidy'.]

The most exciting moment was when I went to the Rockefeller office at the prosaic address of 49 West 49th Street. The office is in the middle of a magnificent group of buildings known as Radio City. I went inside, asked for the office and was told to go to the 55th floor. I didn't appreciate what this meant until I was shown into a waiting room and looked out of the window. The room was about 800 feet above the street and had a magnificent view of the northern part of New York. I hadn't expected that there would be so much colour. The skyscrapers are built of yellow or grey brick, or white concrete and they make a fascinating pattern of yellow ochre and grey colours with deep violet shadows in the streets. The first few days were sunny so that the yellow and grey of the town were set off by the blue of the rivers on either side of Manhattan.

A great many of the skyscrapers are spoilt by over-decoration and some have spires and pimples on top to make them a few feet higher. The Empire State building has a horrid pawn-like object on its summit which makes it slightly higher than the Radio City.

As you are told, everything moves at a tremendous pace. Characteristically there are no amber lights at cross roads and the traffic either moves full speed when the lights are green or stops dead for the red. In the skyscrapers there are express and local lifts and the expresses move at 20 miles an hour.

There are practically no digs in New York, so I decided to stay in this hotel [the Beekman Tower, on 1st Avenue and 49th Street; more like a block of furnished flats than a hotel]. I have a room on the 19th floor and have a fine view which includes a bit of the East River and hills in the distance. The hotel is quite a comfortable place, but it is a bit too automatic for me, and in 2 or 3 months I shall be sick of lifts, telephones and cafeteria meals. The 'cafeteria' principal is almost universal and it is really very convenient. You get your meals on a tray from a hot counter and then eat them in relative peace and quiet in the restaurant.

Three weeks later, (26 October 1937), I continued to my mother:

We are having beautiful 'fall' weather just now, everyday is clear and sunny. It is nice to live in such a big town and to have so little smoke and dirt. I should think there must be as much or more poverty in New York as in London, but the principle of housing people in high blocks of flats conceals and concentrates the dirt. One of the most extraordinary things is the quick change between poor and rich districts. At one point there is a group of residential apartments belonging only to the extremely rich, and fifty yards further on where the road goes underneath an 'elevated' you find blocks of tenements containing only the extremely poor. The abrupt changes in the skyline of New York reflect one aspect of this very quick transition from poor to rich districts.

A letter to Charles Fletcher from about the same date (5 November 1937) gives an impression of the Rockefeller Institute and my immediate colleagues:

The Rockefeller Institute is a very nice place to work. It is a large but relatively low building by the East River, surrounded by tennis courts, and gardens, in which I have already seen various new and exciting kinds of birds. I have an 'office' room to write in and a lab with superb equipment which looks much more professional than anything you see in Cambridge. I shall run to seed because so far I have had extremely little 'gadgeteering' to do. Everything was put into working order when I arrived and if anything goes wrong I ask the advice of Toennies, the engineer–physiologist who works next door. He designs the amplifiers for Gasser's department and these are made in a private shop by trained assistants. I think it is a pity that there aren't more highly trained assistants in Cambridge. Of course it is much easier to have them in a lab which is entirely devoted to research than in one which has a great deal of teaching to do. But still, as we have often agreed, there is room for a great deal of improvement in Cambridge.

One of the things which has surprised me most was to find how national research is. For instance, in England, the acetyl choline hypothesis is at least considered seriously everywhere even if not universally accepted, but here they look upon it rather as we might look upon isochronism.

The people who work in Gasser's department are all quite young and very nice and friendly. They are completely different from one another. Toennies is a typical German – quiet, methodical and incredibly efficient. Lorente de Nó is a naturalised Spaniard; he talks about 6 languages and has written papers on mathematical, anatomical and a wide range of physiological subjects. Grundfest is an American (Jewish?). He is quiet and friendly and I should think is the best physiologist here, apart from Gasser (Lorente de Nó is rather too dashing for my taste). Gasser is a queer person. He has a treble voice and looks about twenty except for his eyes which are rather old and sad. The odd thing is that after a short time, you don't notice anything abnormal about him. I suppose this is because he is completely unselfconscious about himself. [Gasser, later a Nobel prize winner, was Director of the entire Rockefeller Institute, as well as running his small Neurology group.]

While the Indian Summer lasted I spent Sundays walking in the wooded country on the west side of the Hudson river, at first in the Palisade Park and then further north:

Sunday October 17th. Today was a beautifully clear sunny autumn day. I took a ferry across the Hudson and then a train for about 40 miles up the west bank of the river. For most of the way we were travelling through wooded country which gradually got hilly as we went north. I got out at Stony Point and then walked north over wooded hills for about 5 hours to another station called Bear Mountain. The trees were turning the most extraordinarily brilliant colours. I kept on thinking – there can't be anything brighter than that and then seeing another tree which beat all the preceding ones. Perhaps the most exciting were some delicate maple saplings, turned to a sort of orange crimson and seen against a background of deep shadow. There were lots of birds . . . It was humiliating not to know their names [I had searched in vain for a pocket bird book] and even more annoying (because it was a hot and thirsty day) not to know which of the various edible-looking berries really were edible. Eventually I climbed through thick wood and undergrowth to the top of a hill called Bear Mountain which gave me a good view of the surrounding country. About two miles to the west was the Hudson, which was still about a mile wide, (it's really more like a fjord) and in the other directions you could see wooded hilly

country stretching away for miles. It was getting late so I walked down
to the station on the river . . . a good sunset with the river and hills
very romantic in the red light. The only reminder of New York was a
road along which a stream of cars was travelling at about fifty miles an
hour . . . Back about 7.00 with a good view of New York in
moonlight from the ferry across the Hudson.

On the previous Sunday I went for a shorter walk and had time to
go to the Metropolitan Gallery and discovered the superb Brueghel
Harvesters . . . 'a beautiful mellow picture of autumn with harvesters working in corn in the foreground and a sleepy blue green
landscape behind' – a particularly nice picture to look at if you are
feeling faintly homesick when subjected to what we now would call
the culture shock of New York. On this short visit my initial
(erroneous) reaction was that 'The Italians were a little disappointing – only one rather poor Botticelli and no Piero della Francescas.'

> *Oct 30.* Again to the Metropolitan and discovered a lot more good
> pictures which I had missed the first time . . . A most beautiful Bellini,
> *Madonna and Child*; also several El Greco's including one superb portrait
> of a Cardinal, and an impressive landscape of Toledo with an angry
> tumultuous sky behind. This seemed rather a prophetic picture. [Spain
> was in the middle of the Civil War and there had been fighting close to
> Toledo. The Spanish Civil War was very much in my mind because,
> apart from the newspapers, there was a perpetual quarrel about it in the
> lab between Lorente de Nó, a keen supporter of Franco, and Grundfest
> who was equally enthusiastic about the Government side.]

A week later, I 'discovered' the Frick collection, as can be seen
from a letter written in early November 1937, which I will quote in
full, as it gives a fair idea of the serious-minded way in which I
passed my free time in those days (about all that was left of my
Quaker heritage):

> Dearest Mother:
>   I have had a pleasant week but there isn't a great deal of news.
> Yesterday, I went to see *Cosi fan tutte* done by the Salzburg Opera Guild
> at the 44th Street theatre. I don't know anything about the company or
> about its connections with Salzburg except that all the singers had
> foreign names. The whole thing was very well done with a good cast
> and orchestra. Earlier in the week I went to hear Barbirolli conducting
> the New York Philharmonic. First of all, Gieseking played the
> Rachmaninov first concerto. I liked parts of this but it was mostly too

noisy and sensational. After this there was a Mozart symphony – No. 34 in C Major. I didn't know this at all; I enjoyed it although not as much as some of the better known ones. They also played a Gluck Ballet suite which was quite nice and a rather fidgety Debussy. So the Mozart symphony which was what I went for was fairly dilute by the end.

On Sunday I spent two or three hours wandering about Central Park, sitting in the sun and watching squirrels, and then went to the Frick collection of paintings. This is a most attractive museum; the picture gallery is a one-storey building on the East side of Central Park, and when I went the sun was slanting in through the windows. It is quite a small gallery with only two or three rooms, but there are pictures by a large number of great artists, eg Rembrandt, Vermeer, El Greco, Piero della Francesca, Titian, Holbein, Duccio, Filippo Lippi, Ingres, Monet, etc. It was a nice friendly sort of gallery and I didn't feel so lost as in the giant Metropolitan.

I went to dinner with Gasser on Friday night. Dr and Mrs Lorente de Nó were there and also a very nice nerve physiologist called Helen Graham (married – about 45) who is working in the lab for 3 weeks. Every one but myself had lived or worked in St Louis at one time or another and there was a lot of St Louis gossip. This was interesting although often difficult to follow.

I enjoyed your letter. Yes my spelling does seem to be going to bits, of course I meant Kreisler. [I had spelt it Chrysler!]

I was in fact working pretty hard, partly for the reason given in the following letter (New York 30 October 1939).

All goes well here. There isn't much news in the last week as I have spent most of the time working. I felt that I wanted to do a good deal of work just now because of the encouraging and at the same time rather critical attitude which Gasser has taken up. It is quite natural that he should be critical; he has worked on nerve for twenty years and has always maintained that 'there are no transitional stages in the initiation of a nervous impulse' so that you can hardly expect him to welcome me with open arms when I come along and say 'oh yes there are only you have to work with single nerve fibres before you can see them.'

Work here is much more peaceful than in Cambridge; this is partly due to the excellent apparatus and partly to the fact that most of the 'bottle-washing' is done by assistants . . .

My working day consists in getting up about 8.00 and having a cafeteria breakfast over which I read the *New York Times*. At about 9.00 I walk to the lab. [1½ miles up the East River: very pretty, but regarded as a highly eccentric thing to do] and, with intervals off for saying good morning to other people there, work till 1.00, when we all go to eat lunch together. This is an extremely pleasant meal, and the only really cheap food that I get (two courses cost 25 cents = 1 shilling). We usually sit and talk over this for a little time and sometimes go into the garden after lunch. Then work again till 6 or 7. Sometimes there is a staff meeting which means that tea is provided about 4.00 and some scientist, usually a member of the institute, talks about his work for an hour. At about 6.30 or 7.00 go with Toennies (the German apparatus physiologist) to eat at the Cornell Medical School cafeteria. In the evenings I either read, write letters or work. Later on when there are more concerts, I expect I shall go out more in the evenings. At the moment I am quite happy to be rather solitary and work fairly hard. It isn't really a very solitary life because I see such a lot of people at the Institute who I like very much. At the present rate I don't think I shall acquire an American accent, because any tendency to do so is counteracted by the strong German, Spanish and French accents of many of the lab people.

This letter leaves out several things which stick very clearly in my memory. For example, lunch at the Rockefeller was quite an experience. In those days the Rockefeller Institute was a very distinguished laboratory – as indeed its successor is today. At lunch time the great men led their flocks to separate tables and one would see little processsions headed by Landsteiner, Carrel, Avery, P. A. Levene, Michaelis, van Slyke and so on. The Institute was in fact a pretty formal place, which had become imbued to some extent with the Geheimrat principle, and I missed the free and easy casualness of the Cambridge laboratories. But it was a valuable experience to work in a big, well-organized laboratory and helped to turn me from an amateur into a professional scientist. Apart from Gasser's own group, the people who influenced me most were Osterhout (large plant cells), Michaelis (membranes) and MacInnes and Shed-lovsky (electrochemistry), not to mention Peyton Rous and his family on the personal side.

But the contact which had the greatest immediate effect on my scientific life was with Cole and Curtis at Columbia. I was still anxious to know whether the conductivity of the nerve membrane increased during activity and had obtained some evidence that it did, by showing that a shock applied at the crest of the impulse produced less than half the normal change in potential (Hodgkin

95

1938). I had also obtained a positive effect in preliminary experiments with alternating currents, but the results were untidy because the out-of-balance signal in the bridge was mixed up with a diphasic action potential. I abandoned these attempts after I had visited Columbia and seen the beautiful experiments which Cole and Curtis had already done on *Nitella* and were planning on squid axons (having studied the passive transverse impedance in the previous year). Cole asked me to visit him at Woods Hole in June and at some point during the spring of 1938 we agreed that I should bring up equipment for measuring the membrane resistance of squid axons by a modification of the resistance–length method used by Rushton (1934) (see Cole and Hodgkin 1939). But all that was for the future and I must return to my early impressions of New York.

In thinking over the difference between New York then and now, it seems to me that an important difference is that in 1937 one was much more conscious of the element of water and of the fact that Manhattan was an island and a great port, as well as being one of the most important cities in the world. This may have been partly because my room looked out over the East River and because every day I walked three miles along its shore. But there was more to it than that. When they were in port, you could see the great liners Bremen, Europa, Normandie and Queen Mary towering up at the west end of New York's southern streets and hurried to finish a letter to Europe when you knew that one of them was about to sail. Now they have all gone and their place is taken by airliners, which though infinitely convenient, are to my mind, much less romantic than the great sea-ships.

Of course there have been many other changes, not all of them for the worse. In 1937 it was unusual to see a really prosperous or well-educated black man or woman, whereas now you see and meet them all the time, with little or no self-consciousness on either side – less so than in Britain I would say. By the end of the Vietnam War it seemed to Marni (a New Yorker whom I married in 1944) and me that New Yorkers, and in particular young New Yorkers, had lost a great deal of the tremendous self-confidence that they used to have. Some of that has begun to return, but not completely for New York is beginning to age, as all great cities do as they endure the ups and downs of history.

My experiments did not go quite as smoothly as the impression given by the letters to my mother. In Cambridge I had been isolating single nerve fibres from the common shore crab *Carcinus maenas*. No one at the Rockefeller had heard of *Carcinus* and I assumed that the genus did not occur in America. I tried several kinds of edible crab, but in none were the nerve fibres as robust or easy to dissect as in *Carcinus*.

After several frustrating weeks I visited the Natural History Museum where I learned that *Carcinus maenas* was common on the Eastern Seaboard and that I could obtain a supply from Woods Hole. This failed in mid-winter, but by then I had arranged to have a consignment sent from Plymouth, England, on the Queen Mary.

A few weeks after I arrived, Dr Toennies pointed out that it was essential to use a cathode follower if one wished to make accurate recordings of rapid changes from a high-resistance preparation like a single crab axon. His electronics shop provided the necessary input stage and a new amplifier and it soon became clear that most of my original records were quite badly distorted and that I had to repeat nearly all my experiments. However, although tedious, that didn't take too long and by February I was ready to start writing what turned out to be quite a long paper on the subthreshold potentials in a crustacean nerve fibre.

There were several pleasant and interesting distractions. In those days, the part of the Rockefeller Institute's work that dealt with farm animals, for either veterinary or medical reasons, was housed in a separate Institute at Princeton, and here were to be found some of its most eminent members, for example the brilliant and interesting J. H. Northrop who first crystallized proteins (with Sumner) or W. M. Stanley who did the same for Tobacco Mosaic Virus, less brilliant and interesting as a person, but certainly worthy of the Nobel Prize that he subsequently received. It was sometimes said of Northrop that he worked in Princeton because he couldn't stand New York. But I can testify to Northrop's interest in farm animals because I listened to a lecture that he gave during the annual visit made by the parent institution to their colleagues in Princeton. This lecture was illustrated by a pig which had been inoculated with a virus believed to cause a conspicuous skin disease. The pig was enclosed in a pen of some kind, and for most of the lecture stood demurely with one unblemished side presented to the audience. At the end of his talk, Northrop explained that only one side had been injected, turned the pig around and presented the audience with HELLO NEW YORK neatly spelt out in pustules.

The Princeton visit was combined with attending a scientific conference in Philadelphia whose purpose I have entirely forgotten. On Philadephia, I reported that it was a large industrial town, rather like parts of north London – not at all beautiful but 'after the cold splendour of New York I felt quite glad to see a little homely ugliness'. Also that 'I met a lot of people at the Congress and had a pleasant evening with various physiologists and their wives. It seemed much easier to get to know people in Philadelphia than New York. I suppose that this is partly because it is a university town and everyone lives closer together, and there is altogether a more informal spirit than in New York.'

As if to reassure my mother that I was not entirely friendless in New York the letter continues (10 November 1937):

> Michael Straight, who I knew in Trinity, rang up last week and asked me to stay in their house in Long Island. He took me out in his car on Saturday evening and I spent Sunday there. It was rather a funny party as Michael was host and an actor and wife were the other guests. However, they were both very nice and a change from physiologists. The house was most attractive, full of books and furniture and surrounded by a large garden containing many birds and squirrels. On Sunday morning I was persuaded to play golf with Michael and Actor. I enjoyed this as it was a beautiful day and they were neither at all good. We played on a private course (appalling luxury) belonging to a friend of theirs.
>
> Yesterday I went to lunch with Mr and Mrs Schieffelin . . . both very nice, but at first just a little formidable. They have a beautiful apartment on Park Avenue . . . [As this family will appear from time to time I had better explain that my connection with them was through Barbara Schieffelin, the wife of my cousin Charles Bosanquet, who combined running the family farm in Northumberland with being Vice-Chancellor of Newcastle. Barbara's parents were obviously very well off, to judge from the size of their Park Avenue apartment and ownership of Tranquillity Farm, a summer retreat in Maine.]
>
> I feel quite glad that I am not going to live in New York all my life because I believe that I should want far more money to be content here than in a place like Cambridge. Staying with Michael in Long Island and going to the Schieffelins' house in Park Avenue made me feel this quite strongly.

At about this time I got to know another Rockefeller Fellow from England, Robert Thomson, who had done biochemical research with R. A. Peters in Oxford, and was now working in the general area directed by Avery and MacLeod. We got on well and found we had many Oxford friends in common, including a biochemist, Peter Meiklejohn, who was shortly to visit the Institute. Robert was also living at the Beekman Tower and we soon decided that we would try sharing a double apartment, an arrangement that turned out to have several advantages, one being that it gave us a bath as well as a shower. It was also very pleasant to gossip over a drink at the end of a long day. We got on so well that we agreed to go on a joint holiday to Mexico in April, a visit that I eventually made by myself but which family reasons sadly made impossible for Robert. For me, one pleasant consequence of this friendship was that it introduced me to interesting people such as René Dubos and Mirsky who were working in fields a long way from my own.

At about this time my correspondence became entangled with what I shall call the 'Peterhouse' affair. Briefly, Peterhouse, the oldest college in Cambridge, offered me a permanent Fellowship and teaching position and I had to make up my mind within a few weeks whether to accept the offer. I wrote for advice to various people, but principally to my friend and former tutor, Andrew Gow, whose reply, it seemed to me, advised me to accept the offer. I wrote to my mother saying, 'I expect I shall take the job, but I just hate the thought of giving up my three years in Trinity and settling down in Peterhouse – possibly for life.' But Gow's letter seemed to 'confirm what I had already suspected – that there is no likelihood of a vacancy in Trinity.' In spite of this I decided to refuse Peterhouse and chance getting a teaching post somewhere at the end of my three years at Trinity. And then as I wrote on 21 December 1937:

> All has ended happily in the Peterhouse affair as I had another letter from Gow which completely changed the position. It seems almost incredible but he must have omitted a 'not' in one critical sentence . . . In his second letter he advised me quite strongly to refuse their offer on the grounds that I should do best to keep on with pure research for a bit, and that there was a fairly good chance of my getting a permanent job at Trinity later on.

And this, apart from six years of war, was more or less what happened.

When I got back to Cambridge in August, Gow said that I had been foolish not to understand what he meant in the first place. But

I felt this a bit much, coming from someone who spent his life preaching the benefits of the clarity and elegance that a classical education confers on the writing of English.

I planned to stay in New York over Christmas, and then take a short holiday early in January, when I would pick up Phyllis in Montreal and then go north to the Laurentian hills for a little mild skiing. 'I spent Christmas morning sleeping and reading Edmund Wilson, and in the afternoon went to a large family gathering at the Schieffelins. I enjoyed this but it was very hard work being introduced to masses of Schieffelin relations and trying to sort out people into their right families.' The members of the family whom I saw most of were Barbara's brother, Bayard, and his wife, Virginia, with whom I went to many concerts, and niece, Anne Louise, who invited me to her coming-out dance in February. This was an extremely formal affair with white tie, kid gloves, dowagers at the head of a receiving line and all the rest of it. I can't say I enjoyed it very much, in spite of Anne Louise's pleasant company, but as I wrote to my mother, 'It would probably have been worse in England.'

Early in December (1937) 'our department had to demonstrate to the rest of the Institute and I had to show my single fibres. I had good luck and got out the best nerve fibre I have ever had.' This had one unexpected consequence. During the demonstration I had a longish talk with Peyton Rous, the distinguished pathologist, whom I knew about for his famous pre-war work on tumour-producing viruses. About ten days later I went for my usual Sunday walk in Central Park and then to tea with Dr and Mrs Rous and family. 'Dr Rous is head of one of the departments in the Institute and is a nice, friendly sort of person who has been in England a good deal. I met several interesting people there but the worst of meeting people casually here is that one usually does not see them again.' Not in this case, however. On 4 January 1938, I wrote that I had been to dinner with Dr and Mrs Rous and that 'I met the younger members of the family leaving the house as I arrived, which was a pity as they would have made the dinner a less formal affair. However, the eldest daughter, who is at Swarthmore College, Philadelphia and an entertaining sort of person was there.' And, I might have added, an extremely attractive one too. But even today, that is not the sort of comment that sons are likely to make to their mothers.

We had good snow for the skiing trip in Canada, but it was not a happy time for me as my romance with Phyllis was coming to an end, briefly, because my feelings for her were in no way reciprocated, though she enjoyed an unemotional type of friendship. About a month after the skiing trip I wrote to Charles Fletcher: 'I think that was was probably my last holiday with Phyllis. She is quite sure now that she is not in love with me and doesn't want to marry me, so we have agreed not to see much of one another in future. This was really my doing, because I felt it was just too miserable to go back to a casual unemotional kind of friendship after having had a more intimate relation with her.'

For a time I was very unhappy, but I had several friends who tried to prevent me moping and got me interested again in some of the many distractions that New York has to offer in the winter. My letters speak of going to *Don Giovanni* at the Met. with Robert Thomson – good seats in the gallery for $2.50 – and later *Traviata* with Toennies. During this period I took Marni Rous to see the Lunts in *The Seagull*. I don't seem to have mentioned this in a letter home but Marni and I remember the occasion clearly for we went to the same play, with a different cast, six years later at about the time we decided to get married. But I didn't see as much of her that winter as I should have liked as she was mostly away at Swarthmore.

At the end of January, my Cambridge friends, the Rawdon-Smiths, arrived and I spent quite a lot of time with them, either in New York or in Cambridge, Mass. where Rawdon had a Rock-efeller fellowship at Harvard, I think in the Psychology department. Rawdon's work sometimes took him to the Bell Labs, leaving his wife, Pat, at a loose end in New York, where I enjoyed showing her the sights and she did a very good job in cheering me up for my loss of Phyllis.

All this sounds as though I wasn't doing much work, but I was in fact getting quite a lot done, as appears from a letter to Charles Fletcher, in which I start by complaining about the acute difficulty I was having in writing up two scientific papers.

(*22 March 1938*) . . . I used to write essays and exam questions quite slickly, but now I don't seem able to turn out more than one word an hour. I have this trouble now as I am trying to write two long papers about subthreshold action potentials and doing it terribly badly – I have no idea when I shall get done. Work has gone quite well apart from writing up. I have been doing experiments on the refactory period left by a subthreshold action potential, and it looks as though part of the accommodation process may be due to cumulative refactoriness resulting from the subthreshold activity. Next week I am going to Baltimore to the American Physiological Congress and am reading a short paper there. The congress is a vast one; there will be 5 meetings going on simultaneously and 3 or 400 papers in all. It will be fun to meet the St Louis crowd. [In the end there was only one paper; the ideas about accommodation and refractoriness were tidied up in the Hodgkin–Huxley formulation, 1952].

By the end of March I had recovered my spirits and on 25 March I wrote:

We have had a week of the most heavenly spring weather – clear skies all day and a hot sun and cool wind. New York is really a beautiful and exhilarating place under these conditions and I am getting quite fond of it – particularly of the bit of the East River which I see from my window and the part outside the Rockefeller Institute. I have never seen any paintings of New York and it would probably need a special technique to paint it effectively, but I am sure that it provides a lot of possibilities.

Last Saturday Robert and I spent the afternoon on or near City Island which is the first countrified bit of the Long Island Sound. We had 40 minutes purgatory in the subway because the train (which later becomes 'elevated') runs through desolate slummy country, and it was full of rather depressing people most of whom were chewing hard. However, it was very nice to get out into the sun and country air again and to sit by the sea.

Tonight Isa [Louisa Richardson whom I met in Cambridge] had supper with me and we went to *The Shoemaker's Holiday* at the Mercury theatre . . . very amusing and extremely well done . . . It only lasted an hour and a half, which I think a much better time for a play than 2½ hours. [There seems to have been much interesting theatre around at that time; other letters mention the Orson Welles production of *Julius Caesar* in modern dress, a Socialist play called *Steeltown* and Gertrude Lawrence in *Susan and God*; also the famous Garment Workers Union revue, *Pins and Needles*, but I never could get into that.] Tomorrow, Robert is going to Bermuda and leaving me in sole posssession of our double set. However, the Rawdon-Smiths are coming to N.Y. and then we all go to Baltimore for the congress. I am reading a short paper (max: 10 minutes) and am going prepared to fight as Gasser told me to expect some opposition from the St Louis crowd, who are an argumentative lot.

I have now almost finished my paper but still have a lot to do in the way of preparing figures etc. As soon as it is done I shall go away – first to St Louis and then further west or south.

Chapter 9

Spring 1938: St Louis; Mexico

I HAVE no clear memory of the 'hectic two days' of the Baltimore congress, nor of a trip to Washington with the Rawdon–Smiths and Michael Straight, who very kindly showed us round the city, which was looking beautiful just then with cherry and magnolia blossom. The next letter home, dated 14 April, speaks of spending an hour looking at some beautiful Cézanne landscapes and still-lifes in an exhibition at the Durand Ruel galleries; it also says that I start for St Louis in an hour, that the journey takes 20 hours, that I shall stay there for a few days as Prof. Erlanger has asked me to read a paper to their journal club, and then on to Mexico City, on an express train taking some 60 hours. I added that if I felt very extravagant (which I didn't) I might fly all the way back to New York as there was a daily plane service taking 20 hours from Mexico City to New York. To reassure my mother, I added that neither the British Consulate nor the Rockefeller Foundation seemed to think that there was any objection to my going to Mexico. The only sign of trouble that I came across on the long journey was that after Laredo the attendant on the train was very particular that blinds be kept drawn at night, as otherwise bandits were liable to let off bullets at the lighted windows, without, as he explained, meaning any particular harm.

The passage of 52 years has blurred my recollection of my visit to St Louis. But I am pretty sure that I stayed with the Erlangers, and remember a delightful walk with them through wooded country in the bright spring sunshine. Later, I wrote to Charles Fletcher that 'I spent 4 days in St Louis, arguing politely with Erlanger and Blair. They don't believe that "subthreshold action potentials" exist in frog nerve and think that my crab fibres are peculiar in some way. We did some experiments together and it certainly looks as though the subthreshold action potentials were smaller than in crab nerve. But I don't see how the question can be settled until someone can dissect a single medullated fibre.' This was done after the war by Huxley and Stämpfli (1951a) and Frankenhaeuser (1957), who saw transitional stages very like those that I had observed in crab axons. I am pretty sure that Blair and I did see at least one transitional event at which I said, 'What's that?' to which Blair replied, 'I don't know, but it won't happen again,' which it didn't. If I had been a bit quicker I should have insisted that we take many photographs and stimulate at a higher rate.

The next letter that I have was from Taxco, Mexico, dated 24 April but was written more or less in diary form, starting on 20 April, at about the point where the railway leaves the Mississippi.

> For the whole of that day, 20 April, we travelled across the flat Texas plain. At any other time of the year, I suspect that the country would be rather dull, but when I went through it was looking very green and bright with masses of blue and red lupins, poppies and an increasing number of flowering cactuses as we got further south.
>
> I saw a terrapin sitting rather despondently in a swamp and a number of woodpeckers – one very gaudy kind with a crimson head and black and white wings.
>
> I finished Prescott's *Conquest of Mexico* which I had started in St Louis. I found this very exciting and it gives a vivid impression of this country [Mexico]: so much so that I found it difficult to believe that Prescott never came here, but spent his time writing in his New England home. The Cortes expedition is the most extraordinary piece of history. It seems absolutely incredible that Cortes and a few hundred Spaniards could have marched across a country like this and then conquered a powerful empire like the Aztec one. Prescott is very restrained, but he says enough to make one realize something of the appalling cruelty both on the Indian and on the Spanish side. It is faintly comforting to think that even if modern frightfulness is more extensive, it is at least not quite so ghastly as the 16th century kind.
>
> Well, to get back to my journey. We got in to Laredo at about 9.00 p.m. and I got a glimpse of the Rio Grande del Norte as we crossed the International Bridge into Mexico. We had 3 or 4 hours to wait there, which I spent wandering about the town. It felt hot and damp and I began to think Mexico might be very hot. So far this has not been justified as the weather has been absolutely ideal – they say it always is on the Mexican plateau.

April 21. I got up late to find that we were climbing up through desert mountains in the north of Mexico. There were impressive mountain ranges in the distance, and in the foreground sandy desert and tall cactuses, some of them in flower. There were a few puffy white clouds in the sky, which looked as though nothing in the world, not even an Aztec sacrifice, would induce them to yield any rain. We gradually climbed up to the 6,000 foot Mexican plateau. The climb was interesting because the train wound up a series of hairpin bends and we could see another goods train toiling up behind us in the distance. In the evening we reached St Luis Potosi which has two fine Spanish churches but we got there too late to see more than the silhouette of a dome against the night sky . . .

I had decided to stay in the centre of Mexico City in a hotel called the Canada. This was a Mexican-run place but the people spoke a little English and I got an extremely comfortable room and bath for 6 pesos (1 peso is equal to about one shilling) with a view of the cathedral. Owing to the favourable dollar–peso exchange everything is very cheap here so that you can live like a king on very little. Hence my present luxurious hotel in Taxco. However, I think this kind of luxury is definitely worth while because it is incredibly nice to be able to eat all your meals on a balcony looking out across the valley and to be able to wander about the terraces and garden belonging to the hotel. As it is I get a comfortable room and 3 meals a day for approximately ten shillings which is as cheap as living in New York . . .

I hadn't realized Mexico City was such a big place – did you know it had a population of over a million? [Now 15 million.] It is 7,000 feet above sea level and consquently has a most invigorating and pleasant climate. In the afternoon I took a bus to a village called Xochimilcho which is a show place on account of the 'floating gardens'. The gardens aren't really floating at all, although they were at the time of the conquest when the Indians used to make floating islands on a big lake. Today, the lake has dried up and the floating gardens consist of a bit of land which is irrigated by broad canals. These are attractive as they have tall trees (like poplars but with willow leaves) and flowers growing on either side and you look across them to high mountains which must go up to 11,000 feet or more.

April 23. I first did a little shopping which is very difficult, because you have to haggle for everything and my patience and attempts to talk Spanish are completely exhausted by the time I have knocked one peso off the price.

After this I went to look at the Orozco frescoes in the National Preparatory school. Orozco is one of the 'New School' of artists who flourished after the revolution of 1910. Their painting is usually propagandist and tendentious, but it is very effective and entirely different from any of the other schools of modern painting. The frescoes are in a cloistered courtyard and are spaced over 3 floors. On the ground floor there is a sort of capitalist inferno and then you rise through a purgatory of strikes and civil war to the communist paradise of peace, work and the simple life. I also looked at some of the many Diego Riviera's in Mexico City. (Diego is the most famous 'new school' painter but I don't think he is as good as Orozco.)

At 12.00 noon I caught a bus and went through 100 miles of exciting mountain country to Taxco. The road goes up to a 10,000 foot pass about 20 miles from Mexico City and then winds down 5,000 feet to Cuernavaca. If it was not for the statistical fact that buses travel along that road 4 or 5 times a day with never an accident one would say that the drivers were reckless to take hairpin bends at the terrific speed which they do. Cuernavaca is a gay and pleasant town with parakeets making a deafening noise in the trees, red-tiled houses and another splendid Baroque cathedral. These Mexican churches are very impressive, although I suppose they don't really compete with the very best in Europe. Somehow, red sandstone and ornate baroque seem to fit in very perfectly with this landscape.

After Cuernavaca we drove through 20 miles of rather dull plain and then climbed up a beautiful valley to Taxco. Near the river or in the villages there were the most brilliantly coloured flowering trees – some with lemon and some with scarlet blossoms. The scarlet one was particularly impressive because it was as thickly covered with flowers as an ordinary tree is with leaves.

April 24 . . . Taxco is rather like an Italian hill town, with narrow streets and white houses crowded onto a steep hillside. It must be fantastically beautiful in October . . . but even in this dry season, it is brilliant with bougainvilleas and flowering trees whose names I don't know. The people look very decorative in broad Mexican hats and blue trousers, and they are very good looking, especially the children, which is fortunate because if it wasn't for their beauty one could never forgive them for the way that they pester the tourists. I don't know whether the inhabitants are really as happy and healthy as they look, but you feel that people ought to be happy and healthy in this country where there is always a warm sun and a cool wind smelling of pine trees. The town is very full of birds which vary in size from vultures to humming birds . . . There are two kinds of vultures; I saw about 20 in the sky at once soaring round in vast effortless circles . . .

107

April 25. I decided to go to Acapulco as I felt that having got so close, I must see and bathe in the Pacific. My bus was due to start at 11.09, but it didn't arrive till noon . . . The journey was exciting as the road went nearly all the way through high mountains, but it was also more exhausting as we had to go for about 100 miles along the most ghastly track. It was full of holes and very dusty so that at the end of the day everyone and everything was covered with a thick layer of white dust, which turned the sleek black hair of the young Mexicans into the grey hairs of old men. The last 50 miles of road were better and we went through exciting mountain country – cliffs, gorges and tropical vegetation. It was dark when we reached Acapulco so that I could only see the dim outline of the bay and did not realize until the morning how beautiful it was going to be. I asked at the bus terminus for somewhere to stay, and quite by chance picked on a taxi driver who said that he would take me to Caleta which is a bay about 2 miles from Acapulco. I am staying in a small Mexican hotel, on the edge of this bay and about 2 miles from Acapulco. It is also incredibly cheap. I get a room and 3 large meals for about 5/- a day. As soon as I had settled in I had a bathe. The sea was phosphorescent so that I swam in a great swirl of light and fish were making bright flashes all over the bay where they jumped out of the water. The sea here is so soft and warm that it feels like a new element and I was quite surprised to find that it tasted salt.

April 26. I got up early because I wanted to see what the country was like. The sea was only about 40 yards away and the first thing that I saw was a flock of 30–40 pelicans, sitting on the water or diving after fish. One is so used to seeing pelicans sitting languidly in zoos that I was quite surprised to find what agile and delightful creatures they are when wild. They fly about in a rather owl-like way just above the sea, and then suddenly dart to one side and plunge into the water. At any rate this seems to be their technique when there are a lot of fish; sometimes they behave more like gannets. They open their beaks as they plunge in and come to the surface with an immense mouthful of water ('his beak can hold more than his belly can'); then they slowly filter off the water by opening their beaks a little and are left with a few small fish which they gulp down . . . The sea is such a brilliant colour that the white breasts of the pelicans look deep blue when they fly above it. I spent most of that morning swimming and got quite intoxicated with lying on my back and gazing at blue mountains and yellow sand. Pelicans dive in quite close by as you float in the water, and once a big turtle poked his head out a few yards away. Near a rocky shore you can see brilliantly coloured tropical fish darting about among coral beds. The extraordinary thing about this place is that it isn't hot. Acapulco is well

108

inside the tropics so that I expected it would be terribly hot and that I
should only be able to spend a day there and then race back to the
mountains. Instead of which I feel as if I could spend the rest of my life
here. The sun is very hot in the middle of the day, but there is always a
wind from the sea so that it is cool at night or in the shade during the
day. And even during the day the sun can't really be so very strong
because lots of people never wear hats. I met several people in Caleta; a
Mexican family who spoke English; a USA naturalized Russian from
New York with whom I talked cathode ray oscillographs (he was an
electrical engineer); an ex-RAF officer living in Mexico City; two
youngish Americans who seem to be more or less permanently 'on the
beach'. At any rate they have been here for 2 or 3 months and say that
they want to spend the rest of their lives here. One is an ex-business
man – unemployed and cursing Roosevelt; the other an ex-film actor,
who has what I was told is a typical Hollywood manner. He mostly
talks platitudes – or nonsense, but everything is said with great
emphasis and accompanied by the most dramatic gestures and facial
expressions. He is quite nice, but I should think Hollywood must be a
ghastly place if he really is a typical product.

April 27. I took a rowing boat across to the 'la Roqueta' about 1½ miles
away, climbed about 200 feet to the summit, and got a marvellous view
of the Acapulco bay and the mountains around. In the evening some of
us took a car to a rocky bay about 2 miles away and walked back along
the top of the cliffs – again beautiful views of the coast and big waves
breaking. Felt very lazy all day and began to be rather sore from
sunburn. Bathed a lot.

April 28. I spent most of the morning canoeing with one of the
Americans (the ex-business man). We paddled round the Caleta bay,
and out to the Roqueta Isle. In the evening some of us took a car to the
'Pie de la Cuesta', about 12 miles away. This was a very exciting drive
as the road ran by the sea and was cut out of a precipitous mountain
side. The village of El Pie de la Cuesta is built on a sand bar at the
beginning of a series of lagoons which run for 30 or 40 miles along the
coast . . .
    There were all sorts of duck on the lagoon and big white herons and
egrets flapping slowly across the sky or sitting in the tops of trees. The
waves were making a slow thundering noise on the bar which you
could hear a long way off. The sea looked calm but there was actually a
big swell and the waves were breaking with enormous force. After
looking at the sea for some time, we decided that it was no place to
bathe, and this opinion was confirmed when we noticed the triangular

fins of sharks breaking the surface in the troughs between the waves.
The village was a primitive sort of place; the houses were made from a
skeleton of poles lashed together and then thatched with palm leaves.
The inhabitants were Spanish Indian with a higher proportion of Indian
than you see in most places. They get their living by fishing in the
lagoon and keeping livestock. We went to call on an American called
Ben Todd who lives and works there . . . He said that he was taking
some Americans up the lagoon on the next day and asked if I would like
to come too. I said yes, and decided to stay there that night as he has a
room to put up visitors. Actually I stayed two nights and I think that
this was easily the best part of my holiday. I liked Ben Todd very
much. He comes from Florida and when he speaks which isn't very
often, it is with a slow Southern drawl. He makes a living by
transporting freight on the lagoon, occasional export of copra,
livestock raising, etc. The Mexican government took away his main
commission for transporting freight so that he is now trying to develop
a 'tourist industry' to replace the transport one. He sends money to the
States to educate a son there, but now lives with a very nice Spanish
Indian called Rosita. They haven't any children – I should guess
intentionally – but Rosita has adopted about 5; they all looked very
happy and like so many of the children here, incredibly brown and
beautiful. I quite fell in love with one called Rogelia who was 10 or 11
years old and had a rather wistful, very Indian face. I didn't ask Ben
Todd any questions so that what I have written is based on guesswork
and hearsay but I should think it is probably correct . . .

April 29. Got up in time to see the sun rise behind the mountains and a
beautiful pink sky. Took a boat across the lagoon and explored the
coconut grove on the other side as I wanted to see wild parakeets. (I
have now decided all the ones I saw in Taxco and Cuernavaca were
probably escaped tame ones.) Saw a few parakeets but no parrots (I
have since seen one on the way back from Acapulco). A lot of other
birds – a kingfisher like the English one but the size of a jay, humming
birds, a jet black woodpecker with a crimson crest, bitterns, white
herons and a lot of other birds with absolutely no English equivalents. I
caught a young owl and had the parents which were rather like short
eared owls flapping frantically round me . . . At about 10.30, four
American women arrived having been sent by the American Express to
be taken up the lagoon by Ben T. They rather shattered my illusions
about living in a really primitive place because they were dressy and
incredibly touristy. However I was very glad to see them as it meant
that we should make the trip up the lagoon, and they turned out to be
pleasant company. We went for about 12 miles up the lagoon with

wonderful views all the way. In places the water was covered with acres of a very beautiful pink flower which the Americans called a water hyacinth. I can't begin to describe the birds. Ben Todd took us to a plantation on the edge of the lagoon where they were growing chocolate, coffee, bananas and coconuts. We drank the juice from green coconuts and sampled fresh coconut which is a great improvement on the ordinary kind. Returned to El Pie de la Cuesta at about 4.00 and the Americans returned to Acapulco leaving me to enjoy one more evening there. I took a rowing boat out again and tried to see iguanas and alligators but with no success.

May 1. Bus to Mexico City. This was the only bad day of my holiday – 12 hours in a bus on rough roads is too much and I felt sick and upset for part of the time. So I spent the next day doing nothing in Mexico City.

May 3. Took a bus to the Teotihuacan Pyramids. This was an impressive place and you get a fine view of the valley from the pyramid of the sun which is about 200 feet high. The most interesting thing there was the temple of Quetzalcoatl which is a system of walls, platforms and pyramids extending over several acres. One Pyramid has some fascinating pre-Aztec sculpture – some bas relief and some grotesque dragon heads sticking out from the side of the Pyramid.

May 4 . . . I got up early and took a bus for Amace-Meca which is about 40 miles away and the nearest village to Popocatapetl and Ixtlaccihuatl. On clear days you can see the volcanoes from Mexico City but it was hazy all the time that I was there so that I didn't see them until we got down to Amacemeca and discovered that what I thought was a bit of cloud was really the top of a snow peak. The country round Amacemeca was quite different from anything that I had seen before, and again incredibly beautiful. In the foreground barley fields, aloes, poppies and lupins, in the background the volcanoes rising from 8,000 to nearly 18,000 feet. I climbed to within a few hundred feet of the snow on Ixtlaccihuatl (I should guess about 13,000 feet) and quenched a desperate thirst at a glacier-cooled stream. There were marvellous views of Popo all the way; in the morning its snow cap was a dazzling white and in the evening pink from the sunset. Came back to Mexico City late that evening, very tired, sunburnt and blistered.

May 5. Packed in a hurry in order to catch a specially rapid train for St Louis starting at 12.00. Got to the station and found that it was no longer running. So I had 8 hours to wait for the ordinary train.

111

Everything was closed in Mexico City (May 5th celebrates some revolution) and so had nothing to do and nowhere to go. I took a taxi to the Hotel Reforma (the smartest hotel in Mexico City) as I wanted to look at some Riviera frescoes there. Found that the management had destroyed the frescoes, which is not surprising, because Riviera's communist sentiments don't harmonize with those of an American luxury hotel. Eventually I gatecrashed the lounge and wrote letters there for 2 or 3 hours.

. . . I have forgotten to mention . . . a slight earthquake in Mexico City. Everything suddenly started to sway and for a minute I thought I was going to faint – then I saw other people looking surprised and I realized what was happening. Then the lights went out for a second, then the swaying stopped and everyone went on eating and talking as though nothing had happened.

I have a lot still to see in Mexico, but I am not sorry to be going back to New York now. I haven't exactly been lonely but after a fortnight one gets very tired of one's own company.

I'll post this in New York.

These diary entries from my letters cover most of the events and impressions of my Mexican holiday, but there are two additions I can make now.

I am afraid that I sometimes used to play a mean, somewhat schoolboyish trick on the pelicans. If, as sometimes happened, one dived in close to me, I would swim underwater and emerge a foot or two from the bird which then had to empty its beak in a hurry before it could take off.

On 29 April before the party of American ladies departed they insisted on taking a photograph of me holding an iguana which Ben Todd kept as a sort of pet. I was fairly untidy and probably unshaven at the time and I couldn't help thinking that I would probably appear in countless lantern slide pictures as an Englishman gone to seed in this wild place.

Chapter 10

Summer 1938: New York and Woods Hole

Research

When I was in St Louis I had argued with Joe Erlanger about the local circuit theory and he issued a kind of challenge. He would, he said, take the theory seriously if I could show that altering the electrical resistance of the fluid outside a nerve fibre made any difference to the velocity at which it conducts impulses. I thought about this on the long journey back from Mexico City and realized that there was a simple way of doing this test with my set-up, which I carried out on the day after I got back to the Rockefeller Institute.

A single nerve from the shore crab was mounted in such a way that the stretch of nerve fibre between stimulating and recording electrodes could be immersed either in a large volume of sea water or in oil (Figure 10.1). When the fibre was raised into oil it remained surrounded by a thin film of sea water so that the composition of the external fluid was unchanged throughout the experiment. However, when in oil the cross-sectional area and hence the conductance of the external conducting path was greatly reduced and so a lower conduction velocity was expected. This is borne out by the experimental records in the lower part of the figure which show that raising the fibre into oil increased the time to conduct a fixed distance by about 30%. This was one of the few occasions on which everything went according to plan and this time no hidden snags emerged. I showed Harry Grundfest the records next day and remember that he shook me by the hand, like a character in a novel by C. P. Snow.

When I got to Woods Hole in June, Cole and Curtis very kindly let me use their amplifier and I was able to repeat the experiments

113

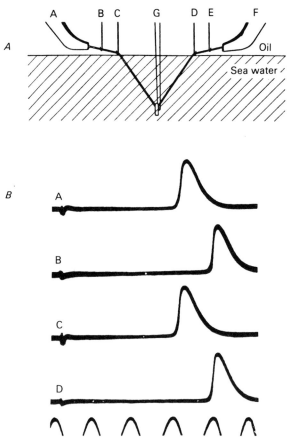

Figure 10.1. *A*, arrangement for comparing the conduction velocity of a single nerve fibre in sea water and oil. As shown, the conduction stretch is in a large volume of sea water; on raising the electrode assembly, the volume of conducting fluid is reduced to the thin film of sea water which clings to the fibre in oil.

B, effect of external resistance on conduction velocity. A and C, action potential with sea water covering 95% of intermediate conduction distance; B and D, fibre completely immersed in oil. Conduction distance 13 mm. Time in msec. (From Hodgkin 1939.)

on squid axons as well as making some other tests of the local circuit theory. Experiments of the type shown in Figure 10.1 gave increases in conduction velocity of about 100 per cent when the volume of conducting fluid outside the nerve fibre was increased from a small to a large value. In this case it was unnecessary to use oil and the external resistance could be changed by the simple operation of raising the fibre out of sea water into moist air. The change in velocity was larger in the squid axon because the

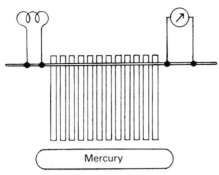

Figure 10.2. Diagram of apparatus for short–circuiting fibre with metallic conductors. (From Hodgkin 1939.)

resistance of the layer of saline which clings to the fibre in oil or air was higher in proportion to that of the axon in a large fibre than in a small one.

The squid giant nerve fibre was also used to determine whether the conduction velocity could be increased by metallic conductors. This was done by placing the stretch of fibre between the stimulating and recording leads on a grid of platinum strips. (Figure 10.2). These were sealed into a moist chamber and arranged so that they could be connected together by means of a mercury switch. When the strips were connected together there was the expected increase in conduction velocity. Such an experiment affords strong evidence because the only agent known which could travel through a metallic short-circuit in the time available is an electric current.

On one occasion a giant axon afforded an opportunity for repeating an experiment carried out previously by Osterhout and Hill (1930) on the large excitable cells of the plant *Nitella*. A fibre which had been placed on the grid of metal strips developed a block between two of the strips. This was effective only when the strips were disconnected; if they were joined by the mercury contact the action potential was able to traverse the injured region and so reach the recording electrodes.

This was a very exciting time to be at Woods Hole. I remember arriving in Cole and Curtis's room there and seeing the increase in membrane conductance displayed in a striking way on the cathode ray tube (Figure 10.3).

After learning to clean squid axons and repeating the velocity experiments, I returned briefly to New York to collect some new

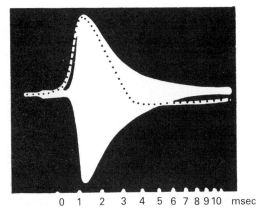

Figure 10.3. Action potential (dotted curve) and increase in conductance (white band) in squid axon at about 6 °C. (From Cole and Curtis 1939.)

equipment and then settled down at Woods Hole with Cole to measure resistance–length curves, which we hoped would allow us to calculate the membrane resistance in the resting state. At that time it was known that the protoplasm of most cells was a reasonably good conductor of electricity and that they were bounded by a membrane with a relatively high resistance and a capacity of about 1 microfarad per cm^2, as might be expected from a bimolecular lipid leaflet, but there was virtually no information about the electrical conductivity of the membrane which was likely to be determined by its ionic permeability. Knowledge of this quantity in squid axons was badly needed because the measurements of transverse impedance made earlier that summer by Cole and Curtis had shown that while the membrane capacity was unchanged by electrical activity, the membrane conductance increased dramatically from a low value in the resting state to about 40 milliSiemens per cm^2 at the peak of activity. However, these measurements with transverse electrodes gave no quantitative information about the resting conductance. At low frequencies nearly all the current flowed outside the cell and at high frequencies, when it entered the cell through the membrane capacity, the membrane resistance was again effectively short-circuited.

We thought we could get around these difficulties by making direct current measurements of the relation between resistance and length and analysing the resulting curve by the method used by Rushton (1934). Our conclusion when the results had been analysed

116

was that the resistance–length curve was well fitted by the equations of linear cable theory, that the conductivity of the protoplasm was about 70 per cent of sea water and that the conductance of the resting membrane was about 1 mS/cm^2, indicating from the results of Cole and Curtis that the membrane conductivity increased some forty times at the peak of the nerve impulse.

Towards the end of this work Cole noticed something like an inductance which showed up in the longitudinal impedance at low frequencies; this was a puzzling observation which did not receive a satisfactory explanation till about ten years later. The inductance is mainly due to the delayed increase in potassium conductance which can make membrane current lag behind voltage provided the internal potential is positive to the potassium equilibrium potential (Cole 1941; Hodgkin and Huxley 1952d). Curtis and I did a few long-shot experiments trying to push electrodes up the cut end of a giant axon. I think we both came away with the idea that it might not be too difficult to record action potentials with an internal electrode; at all events, we both carried out the experiment with different partners in the following year (Curtis and Cole 1940, Hodgkin and Huxley 1939).

On 15 July 1938, towards the end of my stay in Woods Hole I wrote . . . 'Still very little news as I have done practically nothing besides work. This has gone fairly well considering that we bit off a rather tough problem and had only 2 or 3 weeks in which to finish it. Dr Cole has given me a lot of help and if the results are any good we shall write up the stuff jointly. I feel that I have learnt a lot from working with him. This is the first time that I have collaborated with any one, and I never realized till now how much nicer it is than working alone.'

Relaxation: New York

Although I worked hard that summer I had several pleasant expeditions and sociabilities. On 24 May I wrote from New York:

On Sunday I went with Gordon Gould (a biochemist at the Institute) and some friends of his to Oakland in New Jersey and spent the day canoeing down the Ramapo. We drove out there, put 3 canoes on a lorry, went about fifteen miles upstream, and spent the rest of the day and a good deal of the night canoeing back to Oakland. It was most exciting as the river ran through hilly country and consisted of a series of rapids which were just deep enough to 'shoot' . . . a warm spring day and the country very green and beautiful.

Altogether there were six in the party: three boys and three girls because Americans are always very careful to keep the numbers of the sexes even. One canoe went down quite quickly, but the rest of us were so lazy and spent so long lying in the sun that we didn't get back until it was dark. Well, going down shallow rapids is difficult enough in the daytime but at night it is quite impossible to avoid hitting submerged rocks, and ultimately the inevitable happened and our canoe upset. Fortunately we hadn't far to go but we both lost shoes and stockings and so arrived in Oakland soaking and barefooted. However, no one seemed to mind and the rest of the day was so perfect that this did not seem to matter very much. It really was a most beautiful place; both the river and the country reminded me a little of the Tweed. [But the Tweed even at its warmest would have been a great deal colder than the Ramapo.]

I wish I could remember more about that day on the Ramapo, but the only detail that sticks in my mind is the calling of the whippoorwills as they flitted down the river in front of our canoe.

Early in June I spent the weekend with the Dakins, whom I had visited during the winter, at Scarborough about forty miles north of New York. They had a most beautiful garden where I saw humming-birds, goldfinches, orioles, nuthatches, catbirds and, in the evening, fireflies. 'We ate all our meals on a terrace looking out over the Hudson, so that altogether I had a very restful and countrified weekend.' I knew that Dr Dakin was, or had been, an English biochemist (once famous for Dakin's solution) but I was puzzled by his speech and clothes which seemed to be those of an old-fashioned English country squire: knickerbockers, gaiters and a shotgun to shoot vermin, none of which seemed to go with a well-equipped laboratory adjacent to the house. Later that summer my future mother-in-law, Marion Rous, told me Dr Dakin's history, which I find a little sad. Soon after the first world war a wealthy member of the Herter family, who was himself a distinguished biologist, took on Dakin as a well-paid, permanent assistant. Everyone expected that Dakin might one day marry one of the Herter daughters, but when Dr Christian Herter died, Dakin married Herter's attractive widow with whom he was living very happily when I went to see them. With the Herter money, it was not difficult to build a good laboratory for Dakin, who carried on with his own research for a while. But it is very difficult to do research in complete isolation and, gradually, Zyme (as Dakin was known) devoted increasing amounts of his energy to creating a perfect English garden looking out over the Hudson, which was perhaps why I felt so much at home there.

Relaxation: Woods Hole

I came to Woods Hole first at the end of May 1938 and liked it from the beginning. On 14 June I wrote:

> Woods Hole is on Cape Cod and consists of a very large laboratory and a small fishing village. It is a nice place. The village is on a little bay and you look across a smooth sea to islands, promontories and distant sand dunes. It reminds me vaguely of Blakeney or Scolt Head – mainly because of the continual screaming of terns in the harbour. I came because some scientists here (Dr Cole and collaborators) have been getting most exciting results on the giant nerve fibres of the squid. As you know, I spend my time working with single nerve fibres from crabs, which are only about 1/1000 of an inch thick. Well, the squid has one fibre which is about 50 times larger than mine and Cole has been using this and getting results which make every one else's look silly. Their results are almost too exciting because it is a little disturbing to see the answers to experiments that you have planned to do coming out so beautifully in someone else's hands. No, I don't really mean this at all, what I do dislike is the fact that at present English laboratories can't catch squids so that I don't see any prospect of being able to do this myself. [In fact Young, Pumphrey and Schmitt got the squid supply going at Plymouth that summer (1938) and Huxley and I worked on squid there the following year, 1939.]

> *28 June 1938; New York*. . . . I spent ten days at Woods Hole and did a good deal of work there as Dr Cole had finished his experiments and let me use his apparatus for some of my own. I got some interesting results and became fascinated with the giant nerve fibres. I shall go back to Woods Hole as soon as I can. Although I did little else besides work I seemed to absorb much sea atmosphere and felt very well there. In the early morning you can see big kingfishers or little blue herons sitting on posts by the sea or flying above the harbour. The sea was full of little fish which were continually chased by terns from above and big mackerel or squid below.

7 July 1938
Woods Hole

. . . sorry this letter is late again but some new experiments have taken
all my time . . . I came back here a week ago, very glad to get away
from New York as it was hot and sticky. Here, we have had a week of
brilliant sunshine and cool winds. I have a nice room in the dormitory,
looking out onto the 'Eel Pond' which is a sort of sea lake connected to
the harbour by a narrow channel. Independence Day was celebrated
with great energy and throughout the weekend Woods Hole resounded
to the explosion of fire crackers. The harbour was full of yachts with
white sails puffed out by the wind; also a big 3 masted square-rigged
sloop looking very decorative.

On Independence Day I went with a party to Barnstaple dunes on the
north side of Cape Cod: again a beautiful place rather like Ross Links
[Northumberland] with a long sweep of sand and high dunes. We
bathed and walked in the afternoon and in the evening cooked supper
and sat around a fire.

I am getting very fond of this place and would like to work here
every summer. Most of the service, waiting at table in the mess, etc. is
done by students on vacation who come here to earn a little money and
have a holiday so that there is a very pleasant informal atmosphere.

One of the principal amusements is to watch two seals, kept in a sort
of dock in the harbour. They have grown quite tame and play in the
most entertaining way – tail-chasing, fighting etc.

15 July 1938, Woods Hole. Supper and a pleasant evening with the
Curtises. (Curtis works with Cole and I see a good deal of him.) A few
days ago I called on Dr and Mrs Rous, who used to live in Cambridge
and had a long gossip with them about various Cambridge people.
Apart from that, no sociabilities to report, but I meet a good variety of
people in the mess. The seals continue to amuse and are a good excuse
for loitering after lunch.

Late July 1938: New York, saying goodbye, future plans

Before leaving New York I went to say goodbye to Dr R. A.
Lambert at the Rockefeller Foundation who had looked after me
with great kindness during the preceding year. Someone, probably
Herbert Gasser, had suggested that the Foundation might provide
me with an equipment grant and Toennies had helped to prepare a
list of some of the things I might need. I mentioned this to Lambert
and was electrified to learn later that I might expect a sum of £300, a

very large sum for a young scientist in those days. This was a tremendous help, particularly as Trinity had said they would take me on their teaching staff so that it was worth while building up some first-class, fairly permanent equipment in the Physiological Laboratory.

26 July 1938,
New York

Dearest Mother:

I am writing this to catch the Europa which may reach England a few days before my boat, the Ile de France, which arrives in Plymouth on 3 August. But I expect you will get a wire from me before this letter . . . I may stay a day in Plymouth and shall then go up to Cambridge and settle my things there. At present I am dithering about the International Congress in Zurich. Probably I ought to go but I have done so much travelling and also so much work lately that I don't feel very enthusiastic about the congress. One possible reason for not going is that Marni Rous, of whom I saw a certain amount (but not enough) this year, is spending the summer with relatives in Ireland and asked me to stay with them. I should like to do this and thought either of going before Islay (7–15 August) or else of coming to Islay first and then going there. Have you any views as to which would be best?

All goes well here and at present I feel full of energy, which is surprising as New York is like a greenhouse with no flowers in it . . .

. . . I was very glad to hear that Robin [my second brother] was all right, but sorry about the frostbite and that they did not get to the top. [Robin and an Indian Army officer were caught in an avalanche near the top of Masherbrum, a 25,000 foot peak in the Karakoram which a small expedition was attempting to climb. We did not realize for some time how serious the frostbite was; Robin lost all his toes and most of his fingers but continued to climb mountains until he was almost seventy].

The next letter was started on board the Ile de France and finished in Cambridge probably about 4 August. The voyage, in perfect weather, was evidently a most enjoyable one, with the passengers younger and much more interesting than any I had travelled with before.

I was lucky in finding nice table companions – a Yale boy studying architecture and going to Europe to look at cathedrals; 3 American girls; one on holiday, one to teach and the third to study in France; and finally, as a sort of chaperon, an American missionary from Nanking, the Reverend MacGee. The latter is extremely nice and very tactfully

121

conceals his cloth. He is interesting about China, and it sounds as though the capture of Nanking was as bad or even worse than the blackest accounts in the newspapers. He believes the Japanese will lose ultimately, which is comforting, even if his judgement is influenced by sympathy with the Chinese. Other interesting people are a loyalist Spaniard who is a consul in Philadelphia, a Polish scientist with sad blue eyes and a crooked face who works for the Radio Corporation of America, has travelled a great deal and talks about almost anything.

Later, from Cambridge: you will have had a wire about my plans. I decided to go to Ireland first as that fitted best with Marni's plans and also with yours. I shall be staying near Clifden in Connemara where Marni's cousins, the Coffeys, have taken a house for August.

I had a lovely time in Connemara and was very sad to say goodbye to Marni, particularly as I had no idea when we would meet again. I had fallen deeply in love with her and had asked her to marry me. She said no quite firmly and it wasn't till we met again in 1944 that she changed her mind.

Chapter 11

Cambridge and Plymouth, 1938–9

The Munich Crisis, September 1938

When I got back to Cambridge in September, I found myself in the middle of what became known as the Munich crisis. This was the time when, in an unsuccessful attempt to prevent Hitler swallowing the Sudetenland where the main defences of Czechoslovakia were located, Chamberlain made three visits to Hitler, at Berchtesgarden on 15 September, at Bad Godesberg on 20–23 September and Munich on 26 September from which he returned bringing 'peace with honour' on 30 September: a formula which events were soon to turn into war with dishonour as Hitler followed up immediate occupation of the Sudetenland with seizure of the whole of Czechoslovakia in March 1939. At the same time he began to utter threats against Poland to which Britain and France responded by offering military guarantees to Poland which without Russia we had no hope of implementing. This was followed by the Nazi–Soviet pact on 22 August 1939, the invasion of Poland on 1 September and an Anglo–French declaration of war on 3 September after which this country was extremely lucky to survive more or less intact through nearly six years of a most unpleasant war.

This briefly was to be the background against which I had planned a year of research and sociabilities.

My reaction to Munich and subsequent events, a very common one, was a combination of shame at our appeasing attitude coupled with relief that we were not at war. On 4 October, immediately after Munich, I wrote to my mother:

> . . . Life here has been very disorganized. All last week was taken up
> with Air Raid Precautions of one kind and another. I was asked to assist

with the first-aid arrangements in Trinity and had a good deal to do on that account. Now, all this is past and all that is left are the gas masks and two small holes which were dug as trenches in the Backs. The crisis itself was so unpleasant and the present situation in Europe seems so hopeless that I don't want to write about politics.

And in a postcript to a letter in March 1939:

Every one here is very depressed about the political situation and it really does seem almost hopeless sometimes. All the same I am glad we are not at war.

Building equipment, October 1938

Before leaving New York I decided to use my Rockefeller grant of £300 to set up the kind of equipment I had been using in the Rockefeller Institute, with racks, electronic timing, direct coupled amplifiers and so on. I had help with cathode followers and other circuit elements from Toennies and, on his advice, ordered various components that I knew would be difficult to get in Britain. In Cambridge I joined forces with A. F. Rawdon-Smith, K. J. W. Craik and R. S. Sturdy in Psychology, and between us we built three or four sets of equipment, some of which were still in use twenty-five years later. I did some of the wiring but the three psychologists did nearly all the work. Rawdon-Smith designed the d.c. amplifier, and I had help from Otto Schmitt over multivibrators, and from Bryan Matthews over the camera and many other details. I also remember consulting Britton Chance who was then working in the Physiological Laboratory in the Roughton–Millikan suite.

In the late 1930s we were becoming 'professionals' and the objective in designing electronic equipment was not to make some neat miniaturized unit, but to build up as massive and imposing an array of equipment as possible, perhaps with the idea of cowing your scientific opponents or dissuading your rivals from following in your footsteps. These massive units were a nuisance if you wished to move to a Marine Station, but they had the great advantage of being difficult to borrow when you were on holiday or writing up results.

My new rooms; sociabilities

It took three or four months to get all the equipment built and to be ready for experiments again. This suited me for several reasons. I had worked very hard for the previous six years and, as there was obviously going to be a European war, I wanted to leave time for some non-scientific and social activities, which I had neglected more than I really liked. Before I left New York I had heard that I had been allotted what I described as just about the nicest rooms in Trinity. Previously I had lived in undergraduate rooms in Great Court. My new rooms were on the second floor in the north-west corner of Nevile's Court, with the main room looking out on the court and the Wren Library. But what was almost better was that the inner room, where I worked, had an enchanting and little-known view of the river, with St John's College in the background and the Comedy Room wall in the foreground.

Andrew Gow encouraged me to take trouble over furniture, and I spent some time hunting in London as well as in Cambridge antique shops for desks, chairs and the like, buying things that have lasted all my life and have escalated in value far beyond inflation. I only wish that I had been extravagant and had bought Matisse or Picasso drawings which were then obtainable at astonishingly small cost. However, I was fortunate in that when my rooms were nearly finished Victor Rothschild lent me a small Gauguin oil painting of the head of a Tahiti girl. 'Andrew Gow says that it is without merit . . . But I think it is attractive and exciting if a little nerve-racking to have an original.'

My immediate neighbour was Harry Hollond, the ex-Dean, who was on sabbatical in the autumn and let me use his bathroom and guest room while he was away. This meant that I was able to entertain a constant stream of London friends who wanted to spend a weekend in Cambridge or, in the case of medical students, had to return for an exam. (Male friends: other arrangements had to be made for my mother or female friends.)

In the rooms immediately below me lived F. W. Aston, a Nobel prize winner, the inventor of the mass-spectrograph and the discoverer of many isotopes. In contrast to Rutherford, who had died in 1937, Aston was a quiet bachelor with extremely regular habits but, like Rutherford, pleasant and friendly to his junior colleagues. He hated noise and, alarmed at the prospect of a young

Fellow above him, went to the trouble of sound-proofing his ceiling before my return. I thought this unnecessary, but on one occasion I am afraid he was justified, because I remember him appearing in a dressing gown to complain gently about noise when Michael Grant and I entertained some of the Sadler's Wells Ballet at the end of their fortnight in Cambridge.

Hunting for scientific equipment or furniture frequently took me to London and I combined these visits with seeing friends or attending scientific meetings, the most memorable of which being a lecture by the great American physicist Irving Langmuir in early December 1938:

> Langmuir's lecture at the Royal Society was excellent. He is one of the few scientists who has been able to think out immediate practical applications of their discoveries. And his experiments are always so beautifully simple that you wonder why on earth no one thought of doing the same thing before. The Royal Society is always amusing. The rather pompous rooms, the royal mace and the charters going back to Charles II all give it a very dignified air. Langmuir is an American and his address was an amusing contrast to the RS politeness and Bragg's dignified and rather wordy presidential address. After five minutes of flowing sentences and sentiments about the traditions of English and American science it was nice to hear Langmuir begin 'Ladies and Gentlemen, if you put a piece of camphor on water . . .'

I managed to keep up with some of the people I had met the previous year on my visit to America. In mid-November I wrote:

> . . . A French–Polish scientist called Rajchman whom I met on the Ile de France came here on Friday. I got some one to take us over the new Cavendish building which really is the most impressive place . . . vast apparatus 30 or 40 feet high towering above you . . . incredibly Wellsian, with little lifts to take mechanics and physicists to different parts of the machine. I am rather glad I don't do that sort of work because there can't be much scope for originality until you get to a sort of managing-director position.

And a little later:

> . . . Last Thursday to a 'Thanksgiving Day' lunch given by all the Cambridge people who have worked in America to the Americans now in Cambridge . . . enjoyed being in a faintly American atmosphere again . . . ate the proper things – turkey, cranberry sauce, and pumpkin pie . . .

In March (1939) I renewed an acquaintance that I was particularly keen to maintain:

> Dr Rous (Marni's father) was in Cambridge yesterday and came to lunch with me which was nice. He is in England officially to give some lectures in Birmingham and attend the British Empire Cancer Campaign meetings. But I should think the real reason is that he has many friends here and likes to come whenever there is an excuse. He told me that Marni had just got a Henry Fellowship and was coming to Cambridge next year. It is very distinguished of her as there are only ten of these fellowships and women don't often get them. [Marni was in fact the first woman to get a Henry Fellowship.]

That year I frequently visited Victor and Barbara Rothschild who were then living in Merton Hall, an old and very beautiful house adjacent to the St John's Backs. They entertained a lot, especially at weekends, and enjoyed seeing how Cambridge mice (like me) fared with Bloomsbury or Westminster lions. I remember one evening when the two weekend visitors were Aldous Huxley and Isaiah Berlin, from whom I expected a rich feast of intellectual conversation, but instead listened to a long, detailed discussion of the scientific intricacies of the minor novels of Jules Verne.

Early in November I reported another potentially fascinating dinner with the laconic comment '. . . met the Duff Coopers, nice, not at all grand.' This seems a singularly inadequate description of either the witty and highly fashionable beauty, formerly Lady Diana Manners, or of Duff, who had just denounced appeasement in a resounding speech, made when he resigned on account of Munich from the position of First Lord of the Admiralty, a post which he had dignified by the wise and necessary precaution of mobilizing the fleet. At about the same time Victor Rothschild introduced me to two other supporters of Churchill. The first was Bob Boothby, a strikingly good-looking, youngish Member of Parliament of great personal charm, particularly to women.

The other supporter of Churchill and friend of Victor's was Venetia Montagu née Stanley, a lady, then about fifty, with one of the most remarkable histories in British political life. As the beautiful and highly intelligent Venetia Stanley, she became a great friend of the Prime Minister, Herbert Asquith, in the years between 1910 and mid-1915. They met about once a week, but the extraordinary aspect of the friendship was that Asquith wrote to her practically every day, and sometimes more often: for example 141 letters during the first three months of 1915. Quite long letters too. On 30 March of that year there were four letters of a combined length of over 3,000 words. Asquith destroyed her letters but it is thought that she wrote to him almost as frequently. It is not known whether they were lovers, but Asquith seems to have been seriously upset by her marriage to Edwin Montagu, and his political career deteriorated from that point onwards. Roy Jenkins, in his biography of Asquith, concludes that deeply fond of Asquith as Venetia was, she had begun to find the friendship with him a crushing emotional burden, from which a *marriage de raison* to Montagu was an escape, even though it meant adopting the Jewish faith, a major step for a member of the British aristocracy to take.

Edwin Montagu died in 1924, but Venetia retained considerable political influence and entertained Churchill and some of his supporters at her house at Breckles in Norfolk, which became a sort of counter-Cliveden centre for those who opposed appeasement. She was an excellent mimic and had a seemingly unending series of stories about 'The Prof' as Churchill's friend Professor Lindemann was widely known. None of these anecdotes has stuck in my memory, as I think they would have done had I foreseen the fierce arguments that would frequently arise from Lindemann's future position as Chief Scientific Advisor to Churchill.

Teaching

Although I was not working with my usual intensity, I was certainly not idle. For I had the Woods Hole work to analyse and write up, as well as apparatus building and about twelve hours a week teaching to keep me busy. For the most part I enjoyed this. In October 1938 I wrote that I had a lively first-year practical class, and that I had very bright people to supervise; Andrew Huxley, John Gray and Richard Keynes are some that I remember.

The only part of teaching that I disliked was anaesthetising and then decerebrating cats for the advanced class. I had no conscientious objection to this very necessary part of a medical student's training, which inflicted no pain, but disliked killing cats and took no pleasure in the very limited surgical skill that I acquired.

Rushton and Matthews gave the nerve lectures so that slot was filled, so I attempted four advanced lectures on cell membranes. Of course I made the usual mistakes: preparing far too much material and paying too much attention to the expression on people's faces. In late April 1939 I wrote, 'My first lecture went off all right, though lecturing needs a lot of practice . . . at first you are worried by the expressions on people's faces . . . There is one man who looks as though he were trying very hard, but just could not understand one thing. Actually I think it's his normal expression, but it's a bit disconcerting at first.'

Love and friendship

During the year before the war I saw an increasing amount of Tess Mayor, whose looks and acting at the ADC and Marlowe Society I had often admired when she was an undergraduate at Newnham. She was now taking a secretarial course and living in Gower Street with Pat Rawdon-Smith (who had separated from Rawdon), so it was easy for us to meet. Our first date seems to have been in mid-November when I wrote to my mother:

> Yesterday I went to the Arts Theatre with some one called Tess Mayor who used to be up here and knows Aunt Helen and the Bosanquets. The play was *The Frontier* by Auden and Isherwood. The theme was the modern political–social situation – necessarily a very depressing one. The best thing were the rhymed choruses which were sometimes sung and sometimes chanted, The music was modern but quite easy to understand and I thought very effective. It was written by Ben Britten who used to be at Holt.

After that we continued to meet regularly. Towards the end of November, I wrote: 'Train to London this evening in order to have supper with Tess Mayor, and then go on to a party given by some people called Stephen whom we both know.' And in February we went to see *Macbeth* in eighteenth century costume done by the Marlowe Society in Cambridge.

Early in 1939 we planned to take a joint holiday at Easter, thinking in the first place of Italy. However, bellicose noises from Mussolini made us change our plans, and in the end we drove slowly through France to join up with friends at Cap Martin on the Côte d'Azur, on the way visiting some of the enchanting towns that lie between Dijon and Avignon, as well as exploring Provence.

Having both experienced somewhat unhappy love affairs, we decided firmly at the outset that we were neither of us in love and that we were definitely not engaged. Perhaps for that reason the holiday in France, our last before the war, was a great success and deeply enjoyed by both of us.

We continued to see one another in Cambridge and London and sometimes further afield. At the beginning of June I went to see Tess act in the *Merry Wives of Windsor*, done by a group who used to produce a play every year at Ashford Chase in Hampshire, organised by Geoffrey Crump, an English master at Bedales. 'Both he and his wife are very nice and great friends of Tess. The play was good and especially nice to see, because the cast all enjoyed it so much. Donald Beves, a King's don of suitable bulk, made a magnificent Falstaff and Tess was good as a rather young Mistress Ford. The *Merry Wives* is tedious to read but exciting and funny to see acted.'

At about that time Tess became Jonathan Cape's private secretary, a position she enjoyed. She remained in publishing with Jonathan Cape until 1940 when she joined the Security Service in London, doing work in Military Intelligence for which she received the MBE in 1945. In the end we both made happy marriages to other people, and have remained close friends throughout our lives.

Mainly research, 1939: Cambridge and Plymouth

By the end of January 1939 my new equipment was sufficiently near completion for me to start some experiments again. I thought it best to begin with a straightforward problem, partly to run-in the new equipment and partly because I wanted to leave time for other activities. So I decided to use my new d.c. amplifier to check how close the action potential came to the resting potential. At that time the standard form of the membrane theory was that in the resting state, the nerve fibre, which contains a high concentration of potassium ions, is surrounded by a membrane that is selectively permeable to that ion; this gives rise to a resting potential with the inside negative to the outside. When current flow from a neighbouring region makes the inside less negative the membrane breaks down and the resting potential falls to a low value, thus generating the action potential or negative variation as it was once called. On this basis I expected to find that the action potential would approach the resting potential in size, but be slightly smaller, because Cole and Curtis had shown that although the electrical resistance of the membrane fell dramatically during activity it did not fall to zero.

However, as I wrote to my mother on 6 February the result was quite different: 'My experiments have taken an interesting but rather disconcerting turn. I started off to work out a problem and thought beforehand that one particular result was a foregone conclusion. Now I have found the exact opposite. Of course I may discover a mistake or complication of some kind, but if not, this result means a fairly drastic reorganization of current ideas about nerve.'

Andrew Huxley, who was doing the Part II Physiology course, joined me in some of the experiments. We measured external electrical changes in *Carcinus* axons and took the resting potential as the steady potential difference between an intact region and one depolarized by injury or isotonic potassium chloride. We found that the action potential was much larger than the resting potential, for example 73 millivolts for the former as against 37 mV for the latter. Although I wasn't aware of it till much later, Schaefer (1936) had previously reported a similar discrepancy in the sartorius and gastrocnemius muscles of the frog. I got the same result with lobster axons, a preparation which Rushton and I studied later that year in order to calculate passive electrical constants, using the cable-equations which he and others had developed (Hodgkin and Rushton 1946). These results required much analysis and were put on one side until 1945 when the war was nearly over and my part in it was finished.

The results with external electrodes did not give the absolute value of the action potential and resting potential, because of the short-circuiting effect of the external fluid. But there was no reason why this should affect one potential more than another and the difference seemed much too large to be explained by some minor difference in the way the two potentials were recorded. The clear implication was that the membrane potential did not fall to zero but reversed by a substantial amount during the passage of the impulse.

I decided to continue the experiments at Plymouth where I hoped that squid would be available. I bought a trailer which I attached to my car and with some difficulty managed to drag the bulk of my equipment from Cambridge to Plymouth in July 1939.

When I got there I called at Three Corners, Ivybridge, about nine miles away, where I found David, Maurice and Mrs A. V. Hill at home. They very kindly asked me to stay for a few days, which I was delighted to do, as I was not looking forward to going into Plymouth lodgings.

I soon got my apparatus together and working, but . . .

18 July 1939
Plymouth

. . . First and foremost there have been no squid, or only one, which is tantalising as I did one experiment which suggested a lot of interesting possibilities. Last Thursday I went out on the trawler which was fun for an hour or two. Then it got rough and I got very sick. The rolling motion of a trawler is guaranteed to get any one down. However, it was almost worth it as I should not have been able to use the only squid nerve we got unless I had gone out then. [The initial dissection should be done as soon as possible after the squid has died or been killed. A battered squid dies quickly. We had much to learn about the best season and place to catch squid.]

Robin [my brother] turned up and stayed in my digs, which has cheered me up about them. Victorian portraits become more bearable when you have mocked at them together. Other interesting people here are Pantin, J. Z. Young and Lancelot Hogben (*Mathematics for the Million*). I was very surprised by Hogben's appearance as I had expected someone old and distinguished whereas he looks like an undergraduate. I also met Victor's eldest sister, Miriam Rothschild, who is a zoologist (parasitologist) and dined with her and other scientific friends at Yelverton.

Katz, a refugee who works on nerve, has been down here for a few days, and I have seen a good deal of him. He is going to Australia in a fortnight to work with Eccles in Sydney. He is a very good person to talk science with.

In the end the almost complete lack of squid got me down and I went off to Islay for a week and on my return found a large supply of squid waiting for me. On 7 August I wrote:

Work has gone moderately well though I still feel the problem I am working on is a very difficult and almost an unproductive one. But it is so fundamental that it would be silly to leave it until I have tried all possibilities.

Andrew Huxley arrived on Sunday . . . We started another experiment as a side line and it looks quite promising so I think he may go on with it.

132

The experiment Andrew started was to measure the viscosity of axoplasm by seeing how fast a mercury droplet falls down the axis cylinder. He set up this experiment very quickly, using a horizontal microscope and an axon hanging vertically from a cannula. Within a day or two he came up with the unexpected answer that axoplasm (the protoplasm inside a nerve fibre) is normally solid and that the mercury droplet does not fall at all, unless it is in axoplasm which has become liquified as a result of damage or proximity to a cut region. However, this negative experiment was to have an interesting sequel. Huxley said he thought it would be fairly easy to stick a capillary down the axon and record potential differences across the surface membrane. This worked at once, but we found the experiment often failed because the capillary scraped against the surface membrane; Huxley rectified this by introducing two mirrors which allowed us to steer the capillary down the middle of the axon. Figure 11.1 illustrates the technique. The result, illustrated in Figure 11.2, was that the action potential of nearly 100 millivolts was about twice the resting potential of about 50 mV. An alternative statement of this result is that the internal potential of the axon swings from a negative value of -50 mV to a positive one of $+50$ mV. Later work showed that in most excitable cells, and probably in intact squid axons, the resting level is -70 to -90 mV and the active one $+40$ to $+50$ mV.

Andrew Huxley and I were tremendously excited by the potentialities of the technique and started other tests like the effect of potassium ions on resting potential, an experiment later done very elegantly by Curtis and Cole (1942). However, within three weeks of our first successful impalement, Hitler marched into Poland and I had to leave the technique for eight years until it was possible to return to Plymouth in 1947.

Ten days before the declaration of war I wrote from Plymouth on 23 August 1939:

> Andrew and I have been working hard trying to forget that there may be a war next week. On the whole it is the best thing to do – though one feels a bit in an ivory tower – doing abstract scientific experiments at the present time.
> We ran into a good many difficulties, but Andrew is a wizard with scientific apparatus and got over them in an incredibly short space of time.

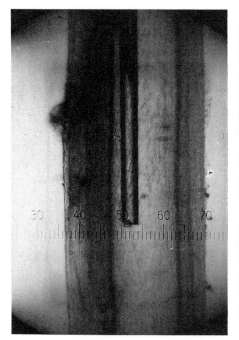
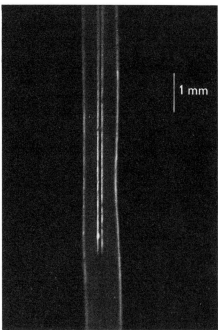

Figure 11.1. *A*, left, photomicrograph of a recording electrode inside a giant axon, which shows as a clear space with small nerve fibres on either side; one division = 33 μm. (From Hodgkin and Huxley 1939.)

B, right, cleaned giant axon of *Loligo forbesi* with glass tube 0.1 mm in diameter inside it; dark ground illumination. (From Hodgkin and Keynes 1956.)

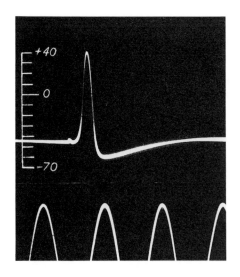

Figure 11.2. Action potential and resting potential recorded between inside and outside of axon with capillary filled with sea water. Time marker 500 Hz. The vertical scale indicates the potential of the internal electrode in millivolts, the sea water outside being taken as zero potential. (From Hodgkin and Huxley 1939; see also Hodgkin and Huxley 1945, Curtis and Cole 1940.)

134

And finally, on 31 August 1939 three days before the beginning of war, when back in Cambridge:

> So far I have spent the day seeing people and talking about the crisis. Today it looks rather like a war, but there is still a good deal of hope. No one here knows what they are going to do in the war and I am in the same position. Some time ago I filled in a form which puts me on the Royal Society Register of Scientists, so I suppose I wait until I am wanted by them.
>   . . . If things blow over Andrew and I will probably go back to Plymouth . . .

They did not blow over. But I was glad to have had so much enjoyment, both scientific and otherwise, out of the preceding year.

Huxley and I wrote a cautious note to *Nature* about our results and for the first few months of the war I tried to work on a full paper. But this didn't get very far as it had to be done in the evenings after a long day at the Royal Aircraft Establishment, Farnborough, where I was working on Aviation Medicine with Bryan Matthews. After I had switched to radar, there were other things to do in the evening, and by June 1940 the war had gone so disastrously and the need for centimetric radar was so pressing that I was forced to bury my interest in neurophysiology for nearly five years.

II

FLIGHT TRIALS
AND TRIBULATIONS

Chapter 12

1939–40: Cambridge, Farnborough and St Athan

IN AUGUST 1939 Andrew Huxley and I were working at the Laboratory of the Marine Biological Association in Plymouth. As I indicated in the last chapter, our experiments on squid nerve had gone almost unbelievably well, for we had developed a technique which enabled us to introduce internal electrodes into a giant nerve fibre without damaging the surface membrane, where the electrical events underlying the nerve impulse are generated. Later, we learnt that a similar experiment had been carried out at about the same time by Cole and Curtis in America. The result was enormously exciting, both because it proved that the nerve impulse really did arise at the surface membrane and not, as some people thought, in long protein molecules orientated axially in the protoplasm, and because it opened up all sorts of exciting new possibilities.

Thus we showed at once that at the crest of the nerve impulse the potential inside the nerve did not fall to zero, as assumed in the classical membrane theory, but swung from a negative value of about −60 millivolts to a positive one of about +40mV. I had already obtained some evidence that the same thing happened in crab and lobster nerve fibres but the result on the giant nerve fibre seemed to put it beyond doubt.

We were also very pleased to obtain absolute values of electrical potential differences and had started some rather obvious experiments like studying the effect of ions on the membrane potential. It was therefore a bitter disappointment when Hitler marched into Poland and we had to abandon work at Plymouth (the laboratory vessel was at once commandeered for mine-sweeping). We left the equipment at Plymouth in the optimistic hope that the war might soon be over and that we could continue our experiments. In the

139

event the war lasted six years and the Plymouth Laboratory was badly bombed so that it was eight years before it was possible to return there and exploit the new technique.

Aviation medicine at Farnborough

Although I had been brought up as a Quaker, contact with the Nazis in Germany in 1932 had removed all my pacifist beliefs and I was anxious to do some kind of military service as soon as possible. I also felt that only by undertaking some physically or intellectually demanding work could I forget the frustration engendered by having to abandon the marvellously exciting nerve experiments, after five or six years very hard work, at the moment when they were most rewarding. I was therefore pleased when Bryan Matthews offered me a temporary unpaid post at the Royal Aircraft Establishment, Farnborough (RAE), where he had been working for several months on Aviation Medicine. There were two major problems to worry about. At that time aircraft were not pressurized and at high altitudes aircrew were kept going by breathing a mixture of air and oxygen. The oxygen was stored at about 100 atmospheres in small cylinders. Many cylinders had to be carried on a long flight, and their weight became a serious problem. What made matters worse was that an oxygen cylinder takes off like a rocket when hit by a bullet. After a few weeks we learned that the Germans had solved the problem of supplying oxygen to aircrew by means of a very beautifully designed lung-controlled device, of which we obtained a sample early in the war. Matthews rightly felt that it would be extremely difficult to get such a device mass-produced quickly in Britain. We also thought that the Royal Air Force, with its brave but individualistic traditions would not take kindly to the strict discipline required in wearing a tightly fitting mask. One of the difficulties at that time was to persuade aircrew to take oxygen at all. They felt that if mountaineers could get to 27,000 feet on Everest without oxygen, they should not have to bother with it at 20,000 feet. These considerations led us to design and build an oxygen economizer that blew oxygen into the pilot's mask when he inspired but not during the rest of the respiratory cycle. This got into service about a year later and was widely used by the RAF. A modified form was used by Hillary and Tensing in their first ascent of Everest in 1953.

140

The other urgent question that concerned us was whether bubbles of nitrogen came out of the blood and produced bends at high altitudes. Matthews and I proved this the hard way by sitting in a decompression chamber evacuated to the equivalent of 40,000 feet (about one fifth of an atmosphere) breathing pure oxygen to keep us going and waiting till something happened. High-altitude bends seem to come on more slowly than the decompression bends that divers experience, but you get them all right in the end and very unpleasant they are too.

Airborne radar at St Athan

I met Patrick Blackett at Farnborough and partly through him and partly through A. V. Hill, who was then Biological Secretary of the Royal Society, I got a job as a junior scientific officer working on RDF (as Radar was then known) with the Air Ministry. The secret of Radar, on which work had been done at Bawdsey Research Station for some seven or eight years before the war, had been well kept and I knew nothing about what I should be doing. I was told to report to No. 32 Maintenance Unit St Athan on Monday 26 February (1940), which I duly did.

The group that I joined, which was headed by Eddie Bowen, consisted of about thirty people working on airborne radar, mainly ship detection (Air to Surface Vessel, ASV) or aircraft interception (AI) or related projects such as long-range ASV, homing beacons or blind approach systems.

When war came in September 1939 the Air Ministry woke up to the fact that all its radar work was very vulnerable to air attack since it was based on Bawdsey Research Station near Felixstowe on the Suffolk coast. Shortly after war was declared many of the scientists and engineers working on ground radar as well as those developing basic techniques were moved from Bawdsey to Dundee, whereas most of those trying to design airborne radar were sent first to Scone near Perth and then to St Athan in South Wales. It was convenient to be at an aerodrome because the best position for mounting aerials on the wings or nose of aeroplanes had to be determined experimentally on each type of aircraft. Although St Athan was good for this type of work it made a poor laboratory. We worked either in a large hangar where the temperature in February 1940 was 0°C at more than six feet from an oil stove or in

141

adjacent offices which were not much warmer. Polar diagrams had to be determined in the open where it was even colder. There was a makeshift workshop with one or two mechanics but nowhere to get anything made with any degree of precision. The *Proceedings of the Institute of Radio Engineers* was available, but there was no library and you had to get books or other journals through Dundee. Fortunately I had bought my own books on Radio Engineering, including George Campbell's collected papers, so I was able to learn aerial and wave-guide theory or revise Maxwell's equations without too much difficulty. Morale at St Athan was not high. This was partly because two scientists, Beatty and Ingleby, had just been killed in an experimental flight, a common occurrence in the days when there were no navigational aids and approaches were made by descending through cloud and hoping for the best. But there were other reasons. St Athan was a large, soulless place with none of the camaraderie of an operational station. Scientists in civilian clothes felt they were regarded by service personnel as shirkers or layabouts and Barry, where we lived, was a dreary town. A further point which I did not realize at once, was that Bowen's group had reason to feel itself neglected by the main group in Dundee and knew that it was liable to be broken up and moved either to Swanage or to Christchurch. I spent the first night with Bernard and Joyce Lovell, to both of whom I became very attached, and then moved into a rather gloomy guest house on the Parade at Barry. This was also inhabited by John Pringle, with whom I had been on that ill-fated undergraduate expedition to the Atlas Mountains, and Robert Hanbury Brown, a brilliant and amusing man already suffering from the effects of several years of intense work at Bawdsey. I saw a great deal of Hanbury during the first years of the war and learnt more about radar and electrical engineering from him than from any one else.

At an early stage Eddie Bowen, who I found an inspiring person, told me that I should give Lovell any help that he needed and explained broadly what he wanted me to do. In a letter to my mother dated 28 February 1940 I wrote, 'I don't yet know exactly what my final job will be. At present I am helping Lovell with a problem which looks as if it's going to be insoluble so perhaps helping is not quite the right word.' In order to explain what Bowen wanted, I must try to summarize the state of play so far as ground and airborne radar in 1940 are concerned.

The chain of long-range radar stations on which Britain depended so heavily in 1940 for its aircraft warning system was reasonably complete and working satisfactorily. It was supplemented by shorter-wavelength coastal stations which gave more accurate information about the bearing and height of enemy aeroplanes, and later by other stations which specialized in directing night-fighters onto the tails of enemy bombers. There were many other naval and army applications of long or medium-range radar, but the instruments developed for ground or ship use were mostly too heavy and cumbersome for installation in aeroplanes. However, there were several promising developments. In the first place all military aeroplanes were fitted, or were supposed to be fitted, with an IFF set, which enabled ground stations to distinguish between friendly and hostile aircraft. (IFF stands for Identification Friend or Foe.) This set had been designed by Vivian Bowden's group with help from F. C. Williams, the brilliant engineer who had come to Bawdsey from Manchester shortly before the war. For obvious reasons those looking after IFF had moved with the ground radar teams to Dundee rather than joining Bowen's small group at St Athan.

The first airborne radar to achieve operational success was the 1½ metre ASV sets which enabled ships to be detected from aeroplanes. This transmitted 1–2 microsecond pulses at a repetition frequency of 1000 Hertz or so, and achieved a range of many miles on large ships. Direction finding was achieved by switching between receiving aerials placed on either side of the nose of the aeroplane. The rotary switch which controlled this operation also determined whether the signal deflected a vertical time-base to the left or right (Figure 12.1). Thus if the left-hand signal was bigger the observer told the pilot to turn to the left in order to home on the target. Coastal Command complained a bit about the serviceability of 1½ metre ASV but on the whole were reasonably well satisfied with it. The complaints about serviceability can easily be understood by anyone who remembers that if one asked for the communications office on a Coastal Command station in 1939–40 the man you were likely to see was often not an engineer but someone whose primary duty was to maintain carrier pigeons in a pigeon loft. Soon after I arrived at St Athan, John Pringle showed me an ASV set operating from a Hudson, an American aircraft recently obtained on Lend-Lease. Everything worked well and I was impressed by what I saw.

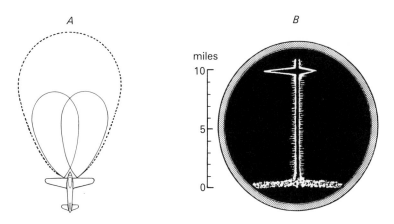

Figure 12.1. ASV Mark II with forward-looking radar.

A. In A the dotted line shows the transmitted beam and the broken lines give the reception patterns for the two alternate receiving aerials which were placed on either side of the fuselage. Direction-finding was achieved by switching between the two aerials with a rotary switch, which also determined whether the cathode ray tube spot was deflected to left or right (B).

B. Appearance of ship about 20° to port at range of 10 miles. The massive deflection at the beginning of the trace is the direct return from the sea. A greater range could be obtained by using sideways-looking aerials with narrow radiation patterns.

Another major piece of equipment developed by Bowen's group was a 1½ metre Air Interception set, AI Mark III, which had been fitted to one or two squadrons of Blenheim aircraft equipped with fixed forward-firing machine guns. The general principle was much the same as that used in ASV, but there were four receiving aerials with broad, overlapping receiving lobes pointing up, down, left or right. This directionality was achieved by positioning one pair of receiving aerials above and below the wing and the other on either side of the nose. An observer looked at two cathode ray tubes and tried to equalize both pairs of echoes (Figure 12.2) by shouting 'up/down' or 'port/starboard' to the pilot.

The RAF had started to fit Blenheim aircraft with AI Mark III but found the radar difficult to use and were beginning to exert pressure for a new set. The complaints from Fighter Command were broadly as follows:

1. Massive ground echoes made it impossible to detect aeroplanes at a range greater than the height above ground. This had come up in an acute form because it had proved impossible to intercept the

144

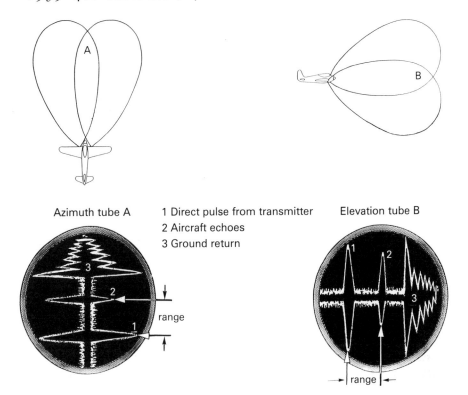

Azimuth tube A
Elevation tube B

1 Direct pulse from transmitter
2 Aircraft echoes
3 Ground return

range

range

Figure 12.2. AI Mark III or IV.

Top. The overlapping reception patterns of the four receiving aerials are shown in azimuth (A) and elevation (B). The transmitted radiation was a broad beam pointing forwards.

Bottom. Cathode ray tube display in azimuth (A) and elevation (B) for a target to the left and above at a range of 5,000 feet. The massive deflection at the end of the trace is the ground return: altitude taken as 10,000 feet.

low-flying minelaying aircraft which had been used extensively by Germany during the winter nights of 1939–40.

Use of a narrow beam might get around the difficulty, but this would require a very short wavelength – perhaps 5 or 10 centimetres (Figure 12.3).

2. The direction-finding arrangement was somewhat unreliable and could not easily be tested on the ground. (The reliability of Mark IV AI was much better than that of Mark III.)

3. The display was difficult for a skilled observer and impossible for an unskilled one. This objection lost much of its force after the famous team of Jackson and Cunningham used AI Mark IV to shoot down several enemy bombers at night.

145

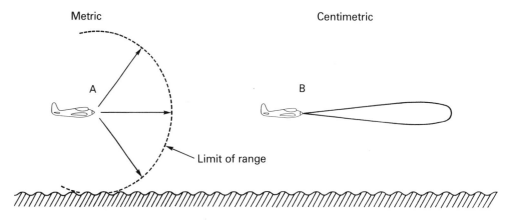

Figure 12.3. Comparison of radiation patterns on metric and centimetric wavelengths.

The left-hand diagram shows the position with 1½ metre AI where echoes from the sea or land make it impossible to detect aircraft at ranges greater than the height above ground. The right hand shows the position with 9 cm AI where a narrow beam is scanned at high speed in front of the night-fighter. A parabolic reflector of about 70 cm is needed to make a sufficiently narrow beam – e.g. 12° at half amplitude; this is housed in a perspex nose in the front of the Beaufighter or Mosquito aircraft – see Figures 17.2. and 17.3, pp. 177–8. (After AHB Monograph *Signals,* vol. 5, p. 146.)

4. The minimum range was too long and the maximum range too short. The RAF wanted something less than 500 feet for minimum range and greater than five miles for maximum range. The minimum range was reduced to an acceptable value by a strong team at EMI who introduced a simple but highly effective modulator that gave a well-defined 1 microsecond pulse. This was used in AI Mark IV.

5. The RAF wanted equipment which could be installed in a Spitfire and would be used by the pilot rather than by an observer. They favoured a 'spot-indicator' type of display on a single cathode ray tube.

6. Aircraft manufacturers did not like cluttering up the wings of high-performance aeroplanes with aerials of any kind.

7. It would be easy to jam a wavelength of 1½ metres.

After outlining some of these difficulties Bowen indicated that he felt the ultimate solution was to use a very short wavelength such as

146

5 or 10 cm. However, before pursuing this line something must be said about the general improvements in the technique of inter-ception at night, which went with the installation of AI Mark IV into Beaufighters.

In the first place the new radar was a more reliable and efficient set than the previous versions of AI. The Beaufighter, with powerful armament and higher speed, was also a far better night-fighter than the Blenheim, really a light bomber which had little claim to be called a fighter at all. Equally important was the installation towards the end of 1940, of special ground radar sets which were used to direct the night-fighter to within a mile or two of the raiding bomber. This equipment, known as GCI, operated on 1½ metre and had a map-like display (PPI) on which the position of fighter and target could be seen. With practice, the controller at a GCI station could direct a night-fighter on to the tail of an enemy bomber. When the fighter was suitably placed the observer would be given the order 'Flash Weapon' which would tell him to switch on the AI and complete the interception.

Such GCI stations were very late in coming. In the autumn of 1939 Hanbury Brown found that although the coastal chain was able to position fighters in daylight it could not reliably get night-fighters within AI range of a bomber. He and Bowen therefore urged the construction of GCI stations, and early in 1940 Fighter Command asked that six stations should be built and equipped by the end of the year. The Telecommunication Research Establishment (TRE) responded quickly and the work was carried through by a team led by Denis Taylor who introduced the necessary modifications into CHL sets. The first of these six sets was completed at Durrington in November 1940 and the last on 6 January 1941. This crash programme was backed up by a MAP order for 120 sets[1]. Another helpful advance was the greatly improved Radio Telephone between ground and air, as well as much better 'intercom' which enabled the observer in the night-fighter to give clear directions to the pilot. This was familiar ground to me because I had helped to design the new oxygen mask, containing a microphone, when I was working at Farnborough in 1939.

With all these changes the performance of Beaufighters equipped with AI Mark IV improved rapidly and the number of enemy aircraft shot down each month rose from two in December 1940, to

52 in April 1941 and 102 in May[2], after which the German Air Force ceased bombing Britain for some time. Apart from the 'Baedeker raids' in the spring of 1942 on Exeter, Bath, Norwich and York, German night attacks on Britain were not of great importance between May 1941 and the end of 1943. It is natural (and not altogether unjust) for an ex-radar technician to associate the cessation of night-bombing in May 1941 with the success of Mark IV AI in Beaufighters on which people like Bowen and Hanbury Brown had worked so hard. However, one should not forget that a much more powerful reason was the enormous military effort incurred by Germany with its invasion of the Soviet Union, which started on 22 June 1941.

As I have already indicated, the RAF was keen that the display for AI should be of the spot-indicator variety. In such a display, which was intended for the pilot of a single-seater fighter, the position of the target is given by the displacement of the spot from the centre of the cathode ray tube and its proximity by the size of two artificial wings whose amplitude varies inversely with the range of the target.

Both Mark V and Mark VI gave a spot-indicator display. In Mark V the target echo was selected manually by an observer; in Mark VI selection was done automatically by an ingenious circuit invented by F. C. Williams, and could thus be used in a single-seater fighter such as a Spitfire.

I do not know exactly when Marks V and VI were designed but remember that R. Hanbury Brown and R. W. Taylor were working on Mark V at St Athan and Swanage early in 1940 and that a powerful team from EMI, including both A. D. Blumlein and E. L. C. White, were developing AI Mark VI at Christchurch in June 1940. They were using F. C. Williams's automatic 'strobe' which must have been designed some months previously either at Dundee or at Swanage.

As in AI Mark V, the echo from the target was switched between four aerials biased up, left, down and right, and the aircraft echo from these positions was fed synchronously into integrating circuits connected to two pairs of cathode ray tube plates. F. C. Williams's circuit for selecting the right echo consisted of an adjustable gate which moved slowly down the time base from a minimum range of 500 feet (1 μsec) to the range equal to the altitude when a massive ground return occurred. If there were no signal before the ground

Figure 12.4. The River Severn taken by H_2S Mark III, a 3 cm rotating Radar with plan position indicator, from 1,000 feet. (Rowe 1948.)

echo, the hunting gate returned to the beginning and tried again. If a signal were found, the circuit locked the gate onto the target, and channelled the output of the receiver into the integrating circuits controlling the position of the cathode ray tube spot. Such tricks would be commonplace nowadays, but in 1940 it seemed marvellous to be doing them in an aeroplane.

In the spring of 1940 many of my colleagues felt that Mark VI would be the answer to AI and thought it crazy to be attempting a development on 5 or 10 cm, a wavelength on which only a few watts of power had previously been obtained. I shared this view to some extent, but pushed on grimly with 10 cm AI, partly because I was told to do so, and partly because after a while I became hooked into the problem. Soon after I arrived at St Athan, Bowen mentioned the idea of scanning a narrow beam of short-wave electromagnetic radiation for detecting ships and the coastline, and pointed out that it would be relatively easy to give a map-like display with a radial time base (Figure 12.4). This of course proved to be the most enduring application of centimetre radar and such equipment is now fitted to any boat larger than a few hundred tons. In 1940, I, or anyone else, would have been delighted to attempt to develop 10 cm ASV but coastal command were satisfied with ASV Mark II and all the operational heat was on the much more difficult

149

problem of centimetre AI. How this situation was rectified will be described in a later chapter.

In the end neither Mark V nor Mark VI AI were used on a large scale in spite of all the ingenuity and hard work that went into their design. Neither could be used at low altitudes and it may be that the Air Staff preferred to go straight from Mark IV to the centimetre sets (like Mark VII and VIII, etc.) which did not suffer from this limitation (see Table 1 for further details).

After explaining some of the background, Bowen told me that wavelengths of 5 or 10 cm were just round the corner, and that they could be used to send out a narrow beam which would have to be scanned in a wide cone in front of a night-fighter. He suggested that I should read up about ways of generating narrow beams – horns, paraboloids and so on, and, if possible, of swinging the resulting beam electrically. He said that I should test out any ideas with a large-scale model on a longer wavelength such as 25 or 50 cm. My friend John Kendrew, who had recently arrived at St Athan, had built a Barkhausen oscillator which gave a weak output at 25 cm, but this oscillator seemed a bit temperamental and I decided to use 50 cm, which was then a reasonably well-understood laboratory wavelength. Lovell's assistant Chapman, from Blackett's laboratory, helped me to build an oscillator at this wavelength which we used for a month or two.

After talking to Lovell and Hanbury Brown, and reading the literature, I concluded that one way of swinging a beam was to transmit and receive radiation from an array of open-ended pipes with the phase of the radiation varying steadily across the array. If all the pipes are in phase the beam is normal to the array; if there is a progressive phase shift it comes out at an angle. The beautiful diagram given by Campbell which I had found in my copy of his collected papers gave the detailed answer to this problem. There are, of course, great practical difficulties in applying such an idea and it was not used or tried out for Air Interception in World War II. The Americans did, however, use this principle in a highly successful shell-splash indicator that they developed for the US Navy.

Following Bowen's advice, I tested out the beam-swinging method at 50 cm using three large wave guides. The physicists at Dundee, whom I was soon to meet at Swanage, felt that my tests were a great waste of time since the result could be calculated with

Table 1. *AI used by RAF in 1939–45*

Mark	Wavelength	Aircraft	Aerial system	Display	Period of use	First kill (date)	Total kills (E)
I	1.5 m	Blenheim	1T, 4R, H	2CRT	1939	—	—
II	1.5 m		1T, 4R, H	2CRT	1939	—	—
III	1.5 m	Blenheim	1T, 4R, H	2CRT	1939–40	18 Jun 40	1
IV	1.5 m	Beaufighter Mosquito	1T, 4R, V	2CRT	1940–45	19 Nov 40	400
V	1.5 m	Beaufighter Mosquito	1T, 4R, V	Manual spot indicator	1941–45	25 Jun 41	50
VI	1.5 m	Defiant Hurricane	1T, 4R, V	Auto: SI	1941–45	17 Mar 42	<10
VII	9 cm	Beaufighter	Spiral scan	RTB	1941–45	5 Apr 42	100
VIII	9 cm	Beaufighter Mosquito	Spiral scan	RTB	1942–45	3 Feb 43	350
IX	9 cm	Beaufighter Mosquito	Auto-lock	Spot indicator	—	—	—
X	9 cm	Beaufighter Mosquito	Helical scan	Azimuth range Azimuth elevation	1943–57	20 Feb 44	200

Notes:

1T, 4R, H (or V) means 1 transmitting and 4 receiving aerials with Horizontal (or Vertical) polarization.

2CRT here means 2 cathode ray tubes with left–right deflexion in azimuth tube and up–down deflexion in elevation tube (Fig. 12.2).

RTB, radial time base with arc-circle type of display. (Fig. 17.4). SI, spot indicator.

Azimuth elevation, map-like display. Azimuth range, crude television picture; requires manual strobe set on azimuth range.

Dates from AHB Monograph Vol. 5. *E against* Total kills *means that these are crude estimates; in some cases little more than guesses, based on information in AHB Vol. 5.*

AI XI–XIII were 3 cm equipment designed for and used by Fleet Air Arm; AI XI and XV used American 3 cm equipment, XV was installed in Mosquito aircraft and used successfully by the RAF at the end of the war. Several higher Marks of AI exist; they are probably classified and do not come into this story.

certainty on the back of an envelope. However, the tests made us organize turntables and other equipment for measuring polar diagrams and it taught me something about electromagnetic radiation. It should also be said that the reason Bowen wanted everything tested was that the 90° side lobes from a reflector or array giving a narrow beam are nearly always larger than those calculated by simple wave theory. This is important in Air Interception because the echo from an aeroplane at the same range as your own altitude is liable to be swamped by the enormous ground return.

Chapter 13

The move to Swanage; break-up of Bowen's group

CONDITIONS AT St Athan were far from ideal but those at the Dundee establishment seem to have been even worse – to judge from the description given by our superintendent, A. P. Rowe, in *One Story of Radar*. On arriving at the building allocated to them in Dundee, the Bawdsey team were told that after all there was no accommodation and that their equipment was being dumped in the open. After a struggle, two rooms were made available and in a few days the Principal of the Dundee Training College gave part of that building to the radar group. Rowe must have realized quite soon that Dundee was a disaster because he and others arranged that the whole of the Air Ministry research effort, both Dundee and St Athan, should move to a stretch of land near Worth Matravers on the Isle of Purbeck in Dorset. Here huts had to be built, roads made and a working chain station erected with several 240 foot wooden towers as well as some 350 foot steel ones. In retrospect it does not seem clever to mark an experimental station with such very conspicuous objects. But I imagine the argument was that there was still some research to do on the long-wavelength chain stations and the powers that be thought this should be done in close proximity to the newer short-wave work. At that time no-one imagined that the Germans would soon be only sixty miles away across the Channel. A more immediate disadvantage was that Christchurch aerodrome where our airborne equipment had to be tested was separated from us by twenty miles and a ferry, or a longer distance through Bournemouth.

The station itself was situated between the picturesque village of Worth Matravers and the sea, in some of the most beautiful countryside in the world. A short expedition would take you to fields filled with bee orchids and a longer one to Kimmeridge where the cliffs are packed with ammonites as well as other much rarer fossils.

On one occasion, when I got as far as Lulworth, I watched one of the early air raids in which the port of Weymouth was attacked – with some success apparently, to judge from the columns of black smoke rising from ships and oil tanks. Later on there were often air battles overhead and I remember one in particular that I watched from the beach at Windspit where we used to swim at the end of a long day. The date was in mid-August and our fighter squadrons had orders not to intercept beyond the coast – so as to save pilots whose planes were damaged. So we were not surprised when a large fleet of German bombers flew towards the coast unopposed at an altitude of perhaps 20,000 feet. As we had hoped, a squadron of Spitfires approached from the north and we expected to see a great air battle. But to our consternation, the Spitfires appeared to do nothing as they passed through the bombers before making a great wheel out to sea in order to catch up with their enemies. And then, nearly 20 seconds after the initial contact, we heard a long burst of firing which had taken all that time to travel from the fighters to ourselves. We realized that an engagement had taken place after all, and soon, that it had not been without effect when we saw an aeroplane descending with a long trail of smoke behind it. Later on, the German bombers were escorted by fighters and then, if atmospheric conditions were right, the whole sky might become covered with an interlacing pattern of vapour trails of incredible complexity. All this seemed to emphasize the apparent futility of messing about with gadgets that few people thought would be ready in time for Hitler's war. At the suggestion of Pringle, who was himself a glider pilot, a few of us applied to learn to fly, on the somewhat specious grounds that this would help us to design practical instruments for aircraft. This request got nowhere as one might expect.

We arrived at Worth Matravers in early May but did not have to report at once and so had a chance to explore the countryside or hunt for digs in the pleasant unassuming town of Swanage. This did much to reconcile me to some rather unpleasant administrative decisions associated with the move.

Although I had not seen a great deal of Bowen at St Athan I had liked what I did see and thought him a good leader of a scientific team. I admired his optimistic but incisive approach and his ability to cut through red tape. Like many of my colleagues at St Athan, I was upset to learn that Bowen was to be removed from leadership of the airborne team and that John Pringle was to take his place. I had no quarrel with Pringle who had not intrigued to achieve this promotion, but I was and remain indignant at the way that Bowen was treated. Whatever the provocation, his demotion seemed a poor reward for the hard work and ingenuity expended in getting airborne radar going, as well as for his pioneering research on the radar chain itself.

Matters became worse after we settled in to Worth Matravers and began to get going on centimetre radar. Bowen was isolated from the many discussions about the future of airborne radar and seemed to have nothing to do except kick his heels in a small office. After a month or two he was sent to America where he helped to build up the magnificent radar group in the Radiation Laboratory at the Massachusetts Institute of Technology. To begin with, Bowen joined Cockcroft and others as part of the Tizard Mission whose job it was to disclose military secrets to the USA in return for help with technical developments and production. Bowen and Cockcroft sailed for America at the end of August 1940 taking with them a cavity magnetron and other important gadgets in the famous black box.

On 13 November Cockcroft wrote, 'We have done our set job – to get the US moving and now Fowler, Bowen and Walker can stay to mind it for a while.' He also reported that American firms had 'pushed ahead at a surprising speed and delivery of a large amount of gear for the first five experimental 10 centimetre airborne sets is expected by 23 November.'[1]

In peace-time a team of British scientists might be aggrieved when they learnt that their American colleagues were pushing ahead fast on a line that they felt was a British invention. But in the autumn of 1940 the position of this country was so desperate that none of us felt anything but pleasure when we learnt of the rapid progress that the Americans were making with centimetre radar. In the end the Americans pushed well beyond us in many branches of radar – for example with the magnificent gun-laying radar SCR 584. The rapid development of American radar may not have been good for the post-war position of electrical industries in Britain but it certainly helped the Allied war effort and, for that, much is owed to the contribution of Bowen and other members of the Tizard

155

Mission. In that sense the Bowen story had a happy ending though it does not do this country much credit. If asked why none of us tried to do anything about it, my defence would be that the prospect of imminent invasion by Hitler removed any feelings people may have had about their own status or that of their friends.

Chapter 14

An early visit to the Christchurch aerodrome

FOR SOME time in May and June 1940 I was not sure whom I was working for or what I was supposed to be doing, except in the general sense that I should design a scanning system for a 10 centimetre Air Interception set. Someone, perhaps Pringle or Hanbury Brown, suggested that it would be helpful to use a 1½ metre set to compare the size of the echo from an aeroplane with the direct return from the ground. I don't remember much about these tests, except that I used a modified 1½ metre AI set in a Blenheim or Anson aircraft. I turned in a report but the results were of little interest to anyone except myself. The main conclusion was the well-known point that at 1½ metres the direct return from the ground was several orders of magnitude greater than the echo from an aeroplane at the same distance as the height above ground. This result was supposed to strengthen the hands of those who believed that all efforts should be concentrated on development of 1½ metre AI rather than going for centimetric radiation. They felt that even though the direct ground return would be much smaller at 10 cm it would still swamp the aircraft echo. On the other hand, enthusiasts for centimetric radar like Lewis, Dee, Skinner and Lovell considered that I had been wasting my time because results obtained at 1½ metre were not relevant to those likely to be obtained in the 3–10 cm range. This would now be my view, but the visit to Christchurch aerodrome, where I was later to spend a great deal of time, may have had some incidental advantages.

In the first place I found (as Hanbury Brown had done in 1938) that a town like Bournemouth[1] was a much better reflector of radiation than the ordinary countryside. Similar observations at 10 cm by Hensby and others were to form the basis of the town detector H_2S about which there will be more to say in a later chapter.

Another incidental benefit was the confirmation of the great importance of reducing the 90° side lobe from the radiator employed. I evidently devoted a lot of time to this for my technical notebook for that period is full of calculations, notes on Mott's work on wire grid arrays (a possible alternative to paraboloids), as well as conversations with the famous team of Ryle and Wiblin who had just arrived at Worth Matravers. Someone must have told me that side lobes could be smoothed out altogether by using a Gaussian distribution of intensity across the front of the radiator, but as far as I know this elegant idea was never put into practice in World War II and I certainly could not see how to apply it.

A useful aspect of the trip to Christchurch was that I learnt a good deal about testing radar equipment in a military aeroplane. This was perhaps more exacting than it would be today. In the first place the frequency of the power supply was determined by the engine frequency and varied between 1,200 and 1,500 Hertz. The high frequency was helpful in reducing the weight of transformers; its main disadvantage was that faults which appeared while flying often could not be reproduced on the ground. The worst example of this that I remember occurred about a year later when we were testing a prototype of AI Mark VII or VIII in a Beaufighter. We were using a radial time base synchronized to a 28 inch paraboloid spinning at 1,000 revolutions per minute in the nose of the aeroplane. On the ground the display was clear and free from interference, but in the air it was horribly ragged, perhaps, we thought, because something in our electromechanical linkage was vibrating mechanically. At the time we were under great pressure from the top brass in Fighter Command and given more or less any priority that we needed. But no change that we made cured the horrible mess that appeared when we were airborne. However, there was one experiment that we were not allowed to do because it was dangerous to run a stationary aeroplane at full power on the ground if the engines were air-cooled, as they were in a Beaufighter. Eventually we[2] talked the pilot into carrying out such a test and immediately reproduced the ragged display on the ground. It then did not take long to discover that, unknown to us, our power leads had been routed quite close to the multi-core cable which synchronized the observer's cathode ray tube with the scanner in the nose of the aeroplane. Both cables had been thoroughly screened but this had not eliminated magnetic interference, which we removed by increasing the separation of the two cables.

Another hazard when testing unpressurized aeroplanes was altitude. Until 1945 virtually all aeroplanes[3] were unpressurized as was our radar equipment. On the other hand, in order to get a range of five miles or so on an aeroplane we needed to apply a microsecond pulse of 15,000 volts to the magnetron. Such a voltage difference in a small box is bad enough at sea level but it flashes-over much more easily in the thin air of high altitudes; 15,000 V at 25,000 feet being equivalent to 50,000 V at sea level. To make matters worse the observer in a Beaufighter sits forwards of the tail with his back to the equipment, and (as happened on one occasion) the first thing he knows about a flash-over is that the aeroplane is filling with smoke and the pilot asks him to use a fire-extinguisher as soon as possible.

Another unpleasant feature of flight trials was the noise and vibration of military aircraft and in rough weather the bumpiness which could nauseate even quite experienced observers. When testing centimetric AI there was also a slight danger that the unusual nose on the aeroplane might lead friendly pilots to mistake our experimental plane for an enemy. Something like this must have happened to a close colleague, Downing, and his pilot who were both killed in an accident in December 1942 in which their Beaufighter was shot down by two Spitfires who had been ordered to intercept an unidentified plane. A description of this tragic accident, which also very nearly resulted in the death of the physicist Derek Jackson, is given in Chapter 20.

About eight months before this accident I was involved in a similar encounter which I found more frightening than anything else in the war. Few people will now remember the history of the Boulton Paul fighter aeroplane known as the Defiant. This was a two-seater single-engine fighter in which an air gunner sat behind the pilot in a turret equipped with four 0.303 inch machine guns. Squadrons of Defiants were first used at Dunkirk and initially had a resounding success. However, as soon as the Germans found out what they were up against, they modified their tactics and shot the Defiants out of the sky, with the result that the latter were withdrawn from Dunkirk and eventually used as night-fighters. During the following summer one of our Blenheims, fitted with a most unsightly perspex nose, was intercepted by a Defiant which

159

circled round with guns trained on us until we were forced to land at a neighbouring airfield.

In spite of some discomfort and a mild amount of danger, both much less than those encountered by my friends who worked in London during the Blitz, I enjoyed flight-testing radar and felt that it suited my training as a biologist better than the business of trying to be an engineer. One also seemed to meet a greater range of people on experimental airfields than when stuck for month after month at the Telecommunication Research Establishment (TRE),[4] as our parent station was eventually called. For a time I shared digs with John Kendrew who was responsible for liaison between the RAF and civilian research workers, a difficult job which he did outstandingly well[5]. When Italy lined up with the Axis, John was joined by his mother who had got out of Florence only just in time to escape internment. She turned out to be a somewhat colourful person who wrote distinguished articles on Italian painting under the pseudonym Sandberg Vavalla. She was clearly irritated by the way in which the war had mucked up her life's work and destroyed the pleasant British community in Florence. It is pleasant to record that she got back to Florence in 1946 and continued her distinguished work on Veronese and Florentine painting until her death in 1961. I think it may have been on one of his periodic visits to his mother that John Kendrew, with Max Perutz, thought up the European Molecular Biology Organization.

At the hut in which I worked, Blumlein and White of EMI were testing AI Mark VI, and although I never got to know them well, it was fun to meet the owners of these somewhat legendary names. Blumlein, who was clearly an inspiring person to work with, tended to keep up a noisy running commentary on any work that he was doing. On one occasion when he was calibrating a dial I became intrigued with the extreme accuracy of his measurements which apparently resulted in a series of numbers such as 0.8977, 0.9011, etc. I crept up expecting to see a very fancy dial indeed and was paid out when I found that it consisted of a circular bit of plastic with nothing more than zero and unity marked on it.

Chapter 15

Centimetre work at Swanage in 1940

WHEN I got back to Worth Matravers at the end of June I found that Hut 40, where I occupied part of a bench, had filled up with interesting people. I have already mentioned Ryle and Wiblin, as well as Lovell, and we were joined by the distinguished physicist Philip Dee and his younger colleague William Burcham, both of whom had arrived in mid-May. Dee had got a Klystron made at Cambridge to Mark Oliphant's design and he and Herbert Skinner were trying to extract as much power from it as possible. We had several visits from Oliphant and he told us about the work done in his laboratory at Birmingham. In particular we heard about Randall and Boot's cavity magnetron, of which we received an engineered version from Megaw of the General Electric Company on 18 July. Atkinson and Burcham got this going in a few weeks while Skinner, Ward and others designed a 10 centimetre receiver with a crystal as detector and a reflection Klystron as local oscillator. The Klystron came from the team built up by J. H. E. Griffiths in the Clarendon Laboratory, Oxford, of which Professor Lindemann was the redoubtable head. The Klystron was known to us as a Sutton tube after its inventor, Dr R. W. Sutton.

It now seems extraordinary that all this high-powered work went on in a small isolated hut, reached by a muddy track, with no gas, and only makeshift water and electricity supplies. Nevertheless a great deal of work got done and an echo from a Battle aeroplane was recorded at a range of two miles before the end of August using a cavity magnetron.

Someone, probably Dee, had suggested that I get a prototype scanner going, first on the ground and then in an air-to-air test to take place by Christmas. This was a tall order, as we then had no way of transmitting and receiving on the same aerial. Nor did we know the answer to much simpler questions such as how to make a satisfactory rotating joint in a concentric cable or wave-guide carrying electromagnetic waves at centimetre wavelength.

With some difficulty I persuaded the workshop to build a crude scanner in which the dipole emitting or receiving power was vibrated vertically at about 5 per second through the focus, while the whole paraboloid was oscillated over 120° in the horizontal plane with a period of 2 seconds or so. This gave the scanning pattern shown in Figure 17.1A (p.175) and the intention was to combine this scan with a map-like display (azimuth range) and a very crude television picture (azimuth elevation). In this mechanism the aerial was vibrated by means of a powerful high-speed wind-screen wiper that I had obtained from Farnborough.

The problem of transmitting and receiving on the same aerial at centimetre wavelength was not solved satisfactorily until May 1941 when Ward, Skinner and Starr produced an elegant switching device with a Klystron-type resonator filled with gas at a low pressure that flashed over when the radar pulse was transmitted but not when it was received (p.182). Until then we were forced to use an arrangement in which three-quarters of the power was wasted in a dummy load. I thought we could do better by using circularly polarized radiation in a kind of bridge circuit. However, in spite of strenuous efforts this arrangement never worked well enough to protect the crystal and had to be abandoned. A year or two later I learnt that Martin Ryle had a similar idea independently and did get it to work on a lower power, wave-guide system, where it was impossible to use a Klystron-type spark gap.

In mid-July someone asked me if I could see a visitor, A. W. Whitaker, from the firm of Nash & Thompson. Apparently I was the only person available as everyone else was off in a grand conference with the superintendent, A. P. Rowe, and his deputy, W. B. Lewis. All I knew about Nash & Thompson was that they manufactured a high-powered sports car, the Fraser–Nash. I had no office in which to entertain Whitaker, so we sat on the chalky hillside looking out over the sea and well placed to watch any air battles that might take place. We had plenty to talk about as Whitaker had been at Cambridge in the early 1920s and had read Physics Part II before taking up a career in engineering. He at once saw the point of centimetre radar and understood the mechanical problems that I was encountering in designing a scanner. He had good advice about how to obtain light paraboloidal mirrors as well as about the best way of getting a perspex nose made for our first trial aeroplane. I explained that it would be better to oscillate the paraboloid in two planes, fast vertically and slow horizontally, rather than to rely on moving the aerial vertically for the fast motion since this defocused the beam for movements greater than about ±10°. Whitaker was keen to have a shot at such a scanner and we left it that Nash & Thomson would produce one to our specification if TRE would provide the necessary design contract, which they did very promptly.

This discussion with Whitaker led to Nash & Thompson becoming involved in some twenty contracts and the production of tens of thousands of scanners of various kinds. The contact proved invaluable, not only because Whitaker and Barnes, his number two, were first-class engineers, but also because they knew their way around the aircraft industry. Without the Nash & Thompson contact I am pretty sure that if I or anyone else in Dee's team had gone straight to the Bristol Aircraft Company and had asked them to fit a scanner and perspex nose into a Beaufighter, we should have been met with much opposition and procrastination. This was even more true with De Havilland's whose success with the Mosquito, a high speed bomber or night-fighter, depended on ignoring the multiplicity of Air Ministry requirements and building their own idea of a beautiful, streamlined aeroplane in a free-lance venture. As it was, I had a rough time at first with De Havilland's chief designers, Bishop and Wilkins, but our conversations became much easier after April 1942 when the RAF had its first operational success with 9 cm AI in a Beaufighter.

After the discussions with Whitaker I had regular meetings with Nash & Thompson, usually at their factory in Tolworth, Surbiton, but also at Christchurch or at Farnborough, Boscombe Down or GEC's research laboratories at Wembley. Before pursuing this line I must refer briefly to various other contacts, some useful some not, that I made away from Swanage and Christchurch.

Early in the summer of 1940 some one high up at TRE, perhaps Rowe or Lewis, had arranged that RAE should equip a Blenheim aircraft with a perspex nose large enough to house a 30 inch mirror. As I had worked at Farnborough and was vaguely in charge of scanning I was also made responsible for seeing that the aeroplane was ready for a flight trial round about Christmas. My inquiries at

Farnborough showed that the job of making a perspex nose had been delegated to a firm called Indestructo Glass located on the western outskirts of London. I paid several visits to Indestructo and although not impressed by the appearance of the Manager, Mr Moon, who had a cauliflower ear and looked like an ex-pugilist, I was at first prepared to accept his assurance that Indestructo would have no difficulty in making us an aeroplane nose from one piece of perspex, as we wanted. I must have paid several visits to Indestructo but remember only the last somewhat bizarre visit which took place after one of the great London air raids. Indestructo Glass was one of those transparent factories that were fashionable in the 1930s and my last picture of Mr Moon is of him standing ankle deep in shattered glass and wringing his hands with regret that he would not be able to produce what we wanted. My friend Mark Pryor who worked at RAE told me later that Mr Moon was unreliable and I was right to be suspicious of him. In the end we didn't get a satisfactory perspex nose until the following spring and the first flight trials were carried out with a makeshift job made from several pieces of perspex reinforced by a wooden ring. This was to cause a lot of trouble. In defence of RAE and ourselves it should be said that the nose of an aeroplane has to withstand considerable aerodynamic forces and that no one had previously attempted to mould perspex sheets into a structure anything like as large as the one we wanted.

Another difficulty was that Nash & Thompson needed to know a good deal about the aeroplane in which they were supposed to install a scanner. This was apt to entail troublesome meetings at Farnborough which, however, had the incidental advantage that on one of the visits Bryan Matthews lent me a lathe which was used more or less continuously by the centimetre group during the war.

The reason the Farnborough visits were time-consuming is illustrated by my notes of the conference held at RAE on 23 October 1940.

> *Object of Conference.* To settle questions relating to installation of scanner, such as: drive available for hydraulic pump: mechanical fitting of scanner.
> *Present* Noble (Main Drawing Office, RAE, acting for Wilkinson)
> Watson, White (Armaments department RAE).
> Long (Engine department RAE)
> Whitaker, Barnes, Gibson (Nash & Thompson) Hodgkin (TRE)
> The conference had been requested by Whitaker and arranged between

myself and Noble who said that he would get hold of appropriate people. I had mentioned Sims and Brewer.

Whitaker stated that the scanner required 7 gallons/min at 300 lb and that the standard pump on the Blenheim was inadequate. He wanted a standard Beacham pump fitted and inquired what drives were available. No one seemed very certain. At this point it transpired that the Blenheim N3522 was not at RAE; it had been moved to Exeter. No one knew on whose authority. A long discussion followed from which it appeared that there probably was a spare drive. This was the drive for the RAE low pressure pump which looks after the Automatic Pilot. But it was impossible to find out whether this ran at engine speed or half-engine speed.

We also tried to find out who would be responsible for the actual fitting of Nash & Thompson's scanner, without success. Noble said that he believed Childs would be responsible, but was not sure of this. He was sorry Brewer and Sims were not there. He had intended to bring them over from Bagshot when we arrived but could not reach them on the telephone.

After 1½ hours we went over to Bagshot, saw Wilkinson who lectured us for some time and ended by ringing up all the people we had already seen at RAE. He called Brewer and asked him to look into the position at RAE.

[ALH Radar Notebook I, pp 211–213]

On 24 October, after a night in London with bombs falling all round, I spent the day at GEC Wembley in a conference discussing scanners with Marris, Espley and Edwards of GEC; Barnes and Gibson of Nash & Thompson, and Skinner, Atkinson and myself from TRE.

On returning to Swanage on 25 October, Rowe, our superintendent, rang up de Burgh apropos Blenheim N3522 and found (i) it had returned to RAE, (ii) I should contact Barton not Wilkinson at RAE. At the height of the Blitz, which this was, long-distance telephoning was most difficult, but after two frustrating days I managed to chivvy up the fitting of the Blenheim and give Nash & Thompson the electrical and mechanical information they needed.

I usually combined meetings at Nash & Thompson with a day or two at the research laboratories of GEC Wembley. This meant staying in London during the Blitz, which I enjoyed because it gave me a chance to keep up a friendship that I particularly valued. After being bombed out of their flat in Gower Street, Tess Mayor and Pat Rawdon-Smith had managed to rent an apartment in a house where the owner had had the foresight to reinforce the basement and turn one room into a reasonably secure air raid shelter. During air-raids, which then occurred practically every night, occupants of the various flats in the house spent the night in the basement where we were much safer than upstairs, though not immune from a direct hit. From my point of view the worst thing about air raids was being caught in one while driving a car. All lights were supposed to be out and one had to get off the street as soon as possible – which was easier said than done.

On one of these London trips, I think in late September 1940, Herbert Skinner took me with him to talk to Air Chief Marshal Dowding, the Head of Fighter Command. After our passes had been cleared we were taken into the main operations room at Stanmore to wait until Dowding was ready to see us. The scene, which will be familiar to anyone who has watched films of the Battle of Britain, was reminiscent of Monte Carlo – or so it seemed to me. Here calm and sad-eyed WAAFs were pushing counters representing friendly or enemy aircraft around a large map, using long rakes like those employed in a casino to transfer stakes from one player to another. And every now and again a senior officer would give an order to a fighter wing, starting the action in a manner reminiscent of a croupier calling, 'Faites vos jeux messieurs et dames'.

After watching these operations for some time we were taken to see Air Chief Marshal Dowding for a meeting lasting fifteen minutes or so. In order to appreciate the flavour of this encounter it is necessary to say something about the personalities involved. Lord Dowding, as he later became, was a man of dominating presence and, although I didn't fully appreciate it at the time, a figure of outstanding military ability. During the rapid German advance through France in May 1940, the RAF which had suffered very heavy losses was under great pressure to send in every available aeroplane. According to Churchill, Dowding considered that with twenty-five squadrons of fighters he could defend Britain against the German Air Force, but with less we would be defeated. He therefore resolutely refused to reduce the strength of Fighter Command below this minimum in spite of moving and pressing appeals to the contrary. Again, at the height of the Battle of Britain, Dowding withdrew seven fighter squadrons from the south, in order to protect cities in the north like Newcastle, which he rightly guessed would soon be under air attack. This daylight phase of the Battle of Britain was nearly over in late September but the night raids were increasing in severity and except on bright moonlight nights, there was little that Fighter Command could do about them. Anti-aircraft guns made a great noise, and were said to boost morale, but in the autumn of 1940 the number of 'rounds per bird' was something like 30,000, i.e. thirty thousand 3.7 inch shells per bomber shot down.

Air Chief Marshall Dowding seemed desperately tired, but nevertheless was kind and even encouraging to the two mad scientists who had invaded his office at a critical moment. It would, in fact, have been difficult to find two more unsuitable envoys. None of my friends have ever accused me of having a military appearance and Herbert Skinner looked even more like the RAF's idea of a Boffin. As I wrote in a letter to my mother, 'He (Dr Skinner) is an extraordinary looking individual – very unkempt and often unshaven, very bright eyes which are actually blue, but so deep set that you think they are black.' Although a wealthy man, his clothes were extremely shabby and usually covered in cigarette ash. Dowding didn't seem to mind and listened to Skinner's account of the magnetron. However, it soon became evident that Dowding was primarily interested in a device which could be used in a single-seater aircraft and would be ready in a few weeks time. Herbert Skinner was always optimistic about dates, but even he could not hold out hope that centimetre AI would be ready as soon as that. Dowding's exact requirements were never met in World War II but the general operational requirement was partly satisfied by the use of 1½ metre AI in Beaufighter aircraft and later by 9 cm AI in Beaufighters or Mosquitoes.

Chapter 16

Another move

TOWARDS THE end of September 1940, all the centimetre work and much of the rest of TRE were moved from Worth Matravers to an empty girls' school, Leeson House, in the village of Langton Matravers, about half-way between Swanage and Worth. Two other schools were requisitioned: Durnford House and Forres, the latter being used as a radar school. It is characteristic of the period that A. P. Rowe later wrote that he could not remember why we moved.[1] But, as he says, the probability was that Headquarters thought that the establishment at Worth Matravers with its 350 foot towers was a very conspicuous target for enemy air attack. Oddly enough, although most radar stations on the south coast were heavily bombed during the Battle of Britain, TRE suffered only slight damage that summer.

At Leeson House the centimetre group was housed partly in a classroom and partly in stables. On finding that the electricity supply was completely inadequate we installed heavier fuses and drew ever-increasing loads until the walls became dangerously hot. This was appreciated by the mice which multiplied to a vast population and caused havoc by eating any notebooks that were left lying around.[2]

In his biographical memoir of Dee, Curran, who arrived during the autumn, describes his horror at the absurdly dilapidated surroundings in which we worked. However, conditions were no worse than those at Worth Matravers or St Athan so the move was not particularly unpopular. It was a nuisance to be separated from the main workshop, which remained at Worth, but we made a small workshop of our own where we installed a watch-maker's lathe lent us by Blackett as well as the larger one belonging to Bryan Matthews.

One very distinct advantage of the new site was that it looked down across the town of Swanage to the sea and in the distance to the Isle of Wight. This area, sometimes called centimetre alley, provided an excellent place for testing 9 or 3 centimetre radar on aeroplanes or ships and submarines. After our experience in the bitter cold of the previous winter, Lovell was rightly determined to be able to measure polar diagrams in reasonable comfort. He achieved this by ordering enough perspex to build a large greenhouse-like structure. By then we had mastered the business of filling in Air Ministry forms in pentuplicate and Lovell, who knew that it was as easy to order several thousand square feet of perspex as a smaller quantity, built himself a fine crystal palace. Unfortunately this particular order was queried by headquarters and Lovell received a reprimand to which he very sensibly paid no attention. I remembered this incident twenty-five years later when Lovell got into a similar row over the building of his largest radio telescope at Jodrell Bank.

During the late autumn of 1940 we had three trailers fitted with 10 cm radar. In the middle trailer, run mainly by Burcham and Atkinson with help from Lovell, every effort had been made to maximize the range on aeroplanes or marine targets, the cliffs of the Isle of Wight at a distance of forty miles providing an excellent standard signal. It was from this trailer that 9 cm radar was 'sold' to the Navy on the occasion to be described in Chapter 23. Some 30 yards to the south another trailer had been fitted out with second-line centimetre equipment and the crude oscillating scanner described earlier. Here, the intention was to mock-up a complete AI set and to mimic in a crude way the display that we might see in an aeroplane. The latter objective was virtually impossible to fulfil since radar echoes obtained by an aeroplane in the air are utterly different from those seen on the ground. However, apparatus must be tested on the ground before taking it into the air so the hard work we put into the trailer was not wasted, particularly since it focussed attention on the formidable problems to be solved.

The third trailer was equipped a little later in order to demonstrate the potential of 10 cm radar for controlling anti-aircraft guns. The story which went around the establishment was that someone at a Sunday policy meeting had said that if TRE were given the job they could shoot down an enemy bomber from Wareham Heath within a fortnight. Certainly the gun-laying application looked easier than centimetre AI – no serious weight or size restrictions, no absolute necessity to transmit and receive on the same aerial – and so on. But of course the project had its own massive difficulties and work in that trailer went on for many months instead of a few weeks as had been hoped for originally. In the end the British centimetre gun-laying set was very much inferior to the American SCR 584 and I have sometimes wondered whether the project didn't suffer from a premature attempt to force the pace at the early stage.

After the move to Leeson House the centimetre group expanded rapidly with people of the quality of S. C. Curran, A. D. Starr, and G. Hensby joining us during the autumn and winter. The Establishment was also beginning to recruit some first-class people at a more junior level; A. E. Downing, J. V. Jelley, E. J. Denton and J. B. Adams are some of the names that come to mind (we were all pretty junior but I mean people who had come straight to TRE after a curtailed wartime university course). A curious feature of the build-up was that Dee's team was largely composed of physicists and we had hardly any mechanical or electrical engineers. However, we could always obtain help from people like F. C. Williams and Hanbury Brown and the imbalance was partly rectified by a close collaboration with GEC of which more will be said in a later chapter.

Gradually some sort of administrative structure emerged. Bowen's group was split into one section containing Pringle and Hanbury Brown, which dealt with 1½ metre airborne projects, such as long range ASV, or various programmes involving radar beacons which became especially important several years later, when the Allies invaded Normandy. The 9–10 cm applications were put in Dee's charge while Skinner's group concentrated on basic problems associated with the development of 9 and 3 cm wavelengths.

Within this framework people naturally specialized on different subjects; for instance, Burcham was the expert on the magnetron and transmission in general, Curran looked after the development of the high-power modulators that we needed – a 1 microsecond pulse of 50 kilowatt power was required to run the strapped magnetron. Skinner, Ward and Starr worked on the crystal receiver and helped to solve the all-important problem of transmitting and receiving on one aerial. Bernard Lovell was initially in charge of work on aerials and paraboloids but did a great deal to promote the use of centimetre radar for ship detection and later for detecting cities (H$_2$S). He was also involved in the design of a lock-follow scheme which later became AI Mark IX. I was supposed to look after scanning and display systems for AI and for the initial air trials, a job which Downing, Jelley and I could not possibly have carried out without much help from the research laboratories of GEC. Some other important activities at TRE were navigation by means of Gee and Oboe (R. J. Dippy, A. H. Reeves, F. E. Jones), Counter-measures (R. Cockburn and M. Ryle), Ground-Controlled Interception (W. B. Lewis, D. Taylor, F. C. Williams), trainers (G. W. Dummer and F. C. Uttley), radar aids to invasion (W. B. Lewis, R. A. Smith, J. W. S. Pringle). In addition there was a large section devoted to techniques, including a mathematics pool where H. Booker, M. V. Wilkes and G. McFarlane did high-powered work on anomalous propagation but were always prepared to help with *ad hoc* problems as they arose. Much centimetre radar was pushed into service long before it would have satisfied any peacetime criterion of reliability. This was only possible because J. A. Ratcliffe, who controlled about one third of TRE's personnel, had built up an excellent post-development service, whose job it was to 'wet-nurse' equipment during its early service career, and often for long periods thereafter.

Although we naturally grumbled a good deal, I now consider the administrative structure of TRE to have been flexible and imaginative, particularly in the way that it allowed young scientists and engineers to take responsibility for major ventures. In 1940 when I was still a junior scientific officer earning £300 per annum I was encouraged to write and sign letters to firms on technical matters, and Dee's authority was required only if some point of major policy were involved. In a large establishment such a system would rapidly have led to chaos, as some people write too few letters and others too many. The system was rationalised by circulating a weekly summary of all incoming and outgoing letters (except top-secret ones in subjects involving counter-measures or military plans). If you had failed to answer a letter or had officiously moved into someone else's territory, you were politely chased, or chastened, as the case might be. I seem to remember that Rowe's central office had the good sense to assign the chasing to a lady of considerable tact and charm.

A striking feature of life at TRE was the meetings known as Sunday Soviets which Rowe organized in his office every Sunday. These meetings, which started in 1940 and continued throughout the war, were attended by many influential people, including very senior officers from all three services, cabinet ministers, government advisers, senior civil servants and university professors. They provided important feedback from service users and enabled TRE scientists to learn about service needs as well as to sell their own ideas. As a relatively junior scientist I didn't often attend a Soviet but this didn't worry me as my boss, Philip Dee, looked after the interests of centimetre AI in a highly effective manner. But the whole establishment was expected to be there on Sunday so that the top brass could see the hardware working.

After the German invasion of France in May 1940 we started the practice of working a seven-day week so that we didn't mind the Sunday Soviets. They were also popular with our visitors since it gave them a breath of country air and a rest from London bombing and administration. Although undoubtedly of benefit to the war effort, TRE's Sunday meetings were disliked by the parallel Army and Navy institutes and by RAE who thought that TRE used the meetings to oversell its wares and to gain special priority for its projects. Since I rarely attended a Soviet I am in no position to form a balanced opinion on this point but would guess that in general they were valuable. A defect of the Soviets was that industry was rarely represented, except in an indirect way by financiers like Sir Robert Renwick.

Few people can stand a seven-day working week indefinitely and when the immediate danger of invasion had receded we were supposed to take one day off a week. However, since Sundays were banned and other days were psychologically difficult when you were under pressure from the RAF or industry, many of us went on working more or less continuously. After several years, Saturday became a regular day of rest but by then the damage had been done and the old hands were so tired that they had little left to contribute. I don't know whether Rowe was aware of this but if so I think he would have justified his somewhat ruthless administration with the proposition that his objective was to help defeat Germany and not to build a large post-war empire.

Chapter 17

Preparations for the first flight trials

FOR SOME years after the summer of 1940 Dee's team worked closely with the General Electric Company's research laboratories at Wembley and later with its factory at Coventry. The person from GEC with whom I had most contact was George Edwards, a radio engineer who spent several years at Christchurch and the other airfields where radar was tested. Edwards, who was observer in many of the early test flights, was responsible for writing and editing the excellent log-book on which these chapters are largely based. He was a member of the team led by Espley,[1] a brilliant but somewhat moody engineer whose primary interest in peacetime was the development of television. Espley's group included a number of able people; for example, Cherry, Hawes, Clayton and O'Kane. His team had access to an excellent model shop which produced two-thirds of the units used in the first centimetre flights, sometimes to TRE's design and sometimes to their own. This collaboration was quite distinct from the earlier and more fundamental liaison between Randall at Birmingham and Megaw at Wembley over the development of the cavity magnetron. TRE personnel were also involved in this development, as they were in the very successful collaboration over the development of cathode ray tubes for ground and air use.[2]

Relations between Espley's group and the TRE physicists were at first somewhat uneasy. GEC had an Air Ministry contract to develop an Airborne Interception radar at centimetre wavelengths and were planning to do this at either 25 or 10 cm. They had also gone some way in designing a gunnery radar at 25 cm for the Army. Espley felt that the jump from 1½ metres to 25 cm was quite big enough and was inclined to favour 25 rather than 10 cm. On the other hand, Dee and Skinner considered that airborne radar would only really become effective at 10 or 3 cm and that it was essential to capitalize on Randall's magnetron. The controversy was partly an argument between engineers, who were very much aware of the difficulties of putting a new idea into practice, and the physicists who believed that any technical snag could be overcome provided the basis of the invention was theoretically sound. In this respect my training was helpful since both sides felt that a biologist was too ignorant to be committed one way or the other.

Before starting to equip an aircraft we had to settle on a scanning and display system. After much discussion, some at TRE and some in London, we agreed on the following action at a conference in Wembley on 25 October 1940: GEC would develop a helical scanner covering 45° in elevation and 180° in azimuth (Figure 17.1). In this scheme, which was used successfully by the American radar team at MIT, the mirror spins rapidly about a vertical axis and therefore wastes half the time looking back into the aeroplane. We none of us liked this and GEC eventually designed an elegant model with two mirrors back to back. They planned to use this with a device which ensured that only the forward-facing mirror transmitted or received power. It was clear that this whole development would take some time and that it certainly would not allow TRE to keep its promise of flight trials by Christmas. At that meeting Skinner and I saw more immediate hope in the oscillating scanner which Nash & Thompson were designing and we agreed that they would complete one model as well as looking into the design of a spiral scanner, in which they were interested.

About a month later Nash & Thompson reported that they had completed the oscillating scanner and that the design of the small-scale spiral was well advanced. Philip Dee and I then visited Surbiton to see the scanners and RAE Farnborough where the Blenheim aircraft was being given a new nose. At Surbiton, Nash & Thompson showed us the oscillating scanner which looked elegant but for some reason or other could not be tested. We gathered that the slow oscillation was satisfactory, but that the fast vertical oscillation was noisy and caused a good deal of vibration. Afterwards we learnt that this was the understatement of all time and that the reason we hadn't been shown the scanner in action was that it made the most appalling noise and vibration. To begin with we had discussed the possibility of balancing the reflector dynamically but this looked difficult in the restricted space of the aeroplane nose and

Scanning patterns

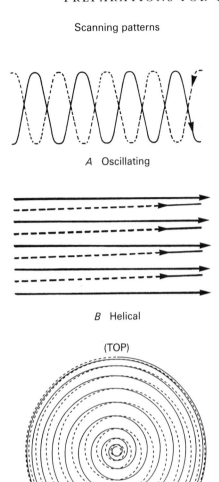

A Oscillating

B Helical

(TOP)

(BOTTOM)

C Spiral

Figure 17.1. Alternative scanning patterns.

A, Oscillatory.

Here there was a slow horizontal oscillation and a rapid vertical one. In the crude model built at TRE the rapid motion was obtained by oscillating the aerial at about 3 per sec with a high-speed aircraft windscreen wiper; the horizontal position was moved at a constant speed and reversed every second, giving the pattern shown in A. In one model built at Nash & Thompson, the mirror moved in both vertical and horizontal dimensions, but the rapid vertical scan gave too much vibration and the system was abandoned in favour of the spiral scan (C). The cover planned for A was ±60° elevation and ±60° azimuth.

B, Helical (GEC model flown August 1941 and SCR 720, later AI Mark X).

Details of Mark X. Rotation rate 360 per min. Cover ±90° azimuth and +50° to −20° elevation (could be reduced giving faster scan).

C, Spiral scan (as in AIS, AI Marks VII and VIII).

Characteristics of scan employed:
1,000 rpm; period of complete cycle 1.2 sec; period of rotation 0.06 sec.

Let ϕ = angle of rotation; θ = angle of tilt of mirror (both in degrees) and t = time (in seconds) then

$$\phi = \frac{360t}{0.06}$$

$$\theta = 1.5 + 15\left(1 + \cos\frac{360t}{1.2}\right)$$

and

$$\theta = 1.5 + 15\left(1 + \cos\frac{\phi}{20}\right)$$

In the figure θ is plotted radially against the angle ϕ in polar coordinates, starting with $\phi = 0$ at vertical and continuing for 1 complete cycle to $\phi = 7{,}200°$; continuous line for 0 to 3,600° (decreasing θ) and broken for 3,600 to 7,200° (increasing θ). Since the aerial was fixed the beam was deflected about 50% more than the axis of the mirror. (with some defocusing). Hence the total cover was ±45°.

175

Whitaker thought that by using a very light mirror he could get away without it. Which he couldn't.

The GEC research laboratories were electrical rather than mechanical so it was not surprising to find that their helical scanner would not be ready for some time – in fact about eight months. This left us to decide between going ahead with the oscillating scanner in spite of its mechanical defects or commissioning a full-scale spiral scanner which Nash & Thompson thought they could complete in two months. I remember that Dee and I sat up half the night in a hotel bedroom arguing the pros and cons of the two courses of action. In the end we decided to give the spiral first priority and Nash & Thompson managed to complete their first model within 6 weeks or so. The 28 inch reflector was spun at 1,000 revolutions per minute with an eccentricity varying between 1.5 and 31.5° in about 1 second; thus giving the scanning pattern shown in Figure 17.1C. Since the aerial was fixed the beam covered a cone of rather more than ±45° with some defocusing at the edge. The scanner was dynamically balanced and there was no serious vibration, but when run up at full speed it had a devastating effect on the observer. As Rowe wrote later, 'It must be confessed that when RAF personnel at Christchurch saw the first AI scanner system installed in an aircraft, doubts were cast on the sanity of the scientists. Before the system reached a speed of rotation greater than the eye could follow, it could be watched rotating in a curiously irregular fashion with the one apparent desire of escaping from the aircraft altogether. When however the system was found to do its job and to give less trouble than many devices of greater apparent respectability the RAF personnel soon learnt to regard it as a normal piece of equipment'.[3]

Figures 17.2 and 17.3 illustrate the installation of the scanner and centimetre equipment in the nose of a Beaufighter.

Although Nash & Thompson's rapid success in designing and building the spiral scanner offered some hope that TRE might keep within a few months of its target of 'flight trials by Christmas', we had to think hard about the display system. For either the helical or oscillating scanner we had decided that the observer would hunt for the target on an azimuth range display, which would in effect be a sector of the plan position indicator now familiar to most sailors. For the final pursuit we would probably use a crude television picture with an azimuth elevation display. Neither the azimuth

Aerial system

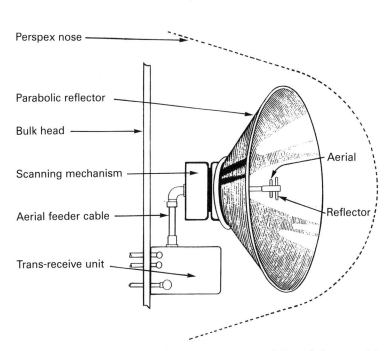

Perspex nose

Parabolic reflector

Bulk head

Scanning mechanism

Aerial feeder cable

Trans-receive unit

Aerial

Reflector

Figure 17.2. Arrangement of aerial system in nose of Beaufighter or Mosquito aircraft in AIS, AI VII, and AI VIII.

In these systems the aerial remained in a fixed position while the tilt of the mirror varied between 1½ and 31½°. The mirror was 70 cm in diameter; it was dynamically balanced and rotated about a horizontal axis at 1,000 rpm; its tilt was controlled by two ball bearings which ran on a tilted circular plate that rotated at 1,050 rpm (not shown in the diagram). (After AHB Monograph *Signals, Vol V,* p. 146.)

range nor the television display were as well suited to the spiral scan as to a helical one, but we soon realized that there was another somewhat unorthodox alternative. By using a special type of rotary transformer known as a magslip it was relatively simple to couple a radial time base to the fast rotary motion of the mirror. With such an arrangement the target aircraft appeared as an arc of a circle whose radius corresponded to the range and whose position gave the direction in which the pilot had to turn (Figure 17.4). When the target was far off-course the arc was short, but it grew in length as the pursuer turned towards the target and became a circle when the fighter was pointing in exactly the right direction. Although unorthodox, most observers liked the display and found it easy to

177

Figure 17.3. Installation of spiral scanner (AI VIII) in Beaufighter. Top, Scanner. Bottom, Perspex nose housing the scanner. (From AHB Monograph *Signals*, Vol. V, p. 148.)

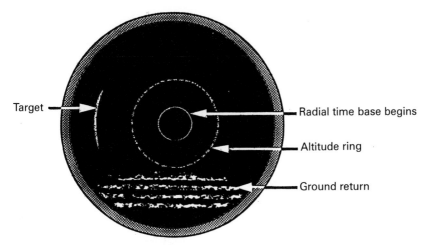

Figure 17.4. Arc circle display for AIS, AI VII and AI VIII.

The figure shows the target aircraft at a range of 10,000 feet and 20° left where it appears as a bright arc. Radiation scattered straight down to the ground and echoed back gives the altitude ring at a range of 4,000 feet, the height above ground.

Where the main beam strikes the ground it gives a straight horizontal line which rises and falls as the tilt of the beam increases or decreases; this acts as an artificial horizon and is a convenient feature of the display.

The reason for the ground approximating to a straight line on a forward looking radial time base is as follows. If θ is the angle of tilt and ϕ is the angle of rotation about a forward pointing axis of a narrow beam which hits the ground at a range r from a height h, then

$$h = r\sin\theta\cos\phi \qquad r = \frac{h}{\sin\theta\cos\phi}$$

which on a radial time base display gives a horizontal line of range $h\mathrm{cosec}\theta$ when the beam is pointing down, which is here defined as $\phi = 0$.

use. A point in its favour which we didn't appreciate until the first successful flight trial was that the main beam return from the ground appeared as a horizontal bar which acted as an artificial horizon and made interception of a target easier. However, before we could get anywhere near flight trials there were several months hard work to be done in getting ready electrical equipment for the aeroplane. I have only a confused recollection of the intervening period but would guess that TRE and GEC Wembley were each responsible for about half the electrical hardware used in the first centimetre flights. At TRE some of the people most closely

179

involved were Burcham (transmitters), Ward (receivers) and Downing, Jelley and myself (display).

The Blenheim we were to use arrived at Christchurch with a makeshift perspex nose at the end of December 1940 and had to be fitted out with hydraulic and electrical connections. The electrical side was not straightforward as any one who has been involved with wiring up an aeroplane will know. To give one example: in the early model of centimetre AI we ran the magnetron with a 10 kilovolt pulse lasting 1 microsecond which had to be brought from the back to the front of the aeroplane through plugs and sockets. Remembering the altitude factor this meant designing and making new compact plugs, preferably with moulded insulators to take a 50 kV pulse. This would take a few weeks at best.

The senior physicists at TRE were keen that we should bring in some other firm besides GEC and I remember going to Manchester in May 1941 to talk to Ferrantis and Metropolitan Vickers about the development of AI. The immediate effect of this visit and of another made during the autumn of 1941 was to make GEC Wembley give more priority to the centimetre AI project. At about the same time S. C. Curran started an extremely successful collaboration with Metropolitan Vickers which led to the introduction of the spark gap modulator for supplying power to the magnetron. Unfortunately this development came too late for AI Marks VII and VIII and could not be introduced without a major revision of the whole system.

Chapter 18

Initial flight trials of 9 centimetre AI

THE FIRST successful flight trial was made on 10 March 1941 from the small airfield at Christchurch. Barrington was pilot and Edwards (GEC) and I were the observers. We never made visual contact with our proper target but picked up another unknown aeroplane at a range of 7,000 feet when flying at 5,000 feet. The display was more or less as shown in Figure 17.4 (p.179). As expected, the aeroplane appeared as an arc but we were surprised and pleased to find that the ground echo from the main beam showed up as an artificial horizon which rose and fell as the pitch of the spiral scan increased and decreased. Later, when making interceptions, it proved helpful to know from this 'radar horizon' how much the pilot had banked the aircraft in response to the radar operator's instructions. We hadn't anticipated this useful property of the display, but as soon as we thought about it we realized that the result could have been predicted by an elementary exercise in solid geometry. In addition to the main ground return there was an altitude ring produced by radiation scattered directly downwards. In the early trials this was unpleasantly large and one of our first tasks was to reduce the altitude return to manageable proportions.

During the next six months Edwards, Downing and I made 70 trial flights with various forms of the basic prototype. Most of the flights lasted an hour or two, some were over sea, some over land. The altitude varied between 500 and 15,000 feet. We didn't go much higher because this early equipment tended to flash-over in the thin air of high altitude. One hundred hours' flying time in six months doesn't sound much but in those days we had no navigational equipment and on a cloudy day our flight home often involved a cautious descent through cloud with the pilot hoping he would see ground before flying into a hill, so we didn't fly in really bad

181

weather. It should also be said that the point of the tests was to try out possible improvements, and any modification is liable to introduce new faults which may not appear until the equipment is airborne.

A small group under Burcham played an essential part at Christchurch by running a ground-based set on the airfield against which the airborne set could be tested, a necessary precaution when dealing with new and temperamental equipment. Burcham also ensured that any technical improvements made at Swanage were quickly introduced into the airborne set.

Improvements to the equipment

Introduction of common TR. In the first trials we obtained ranges of only two or three miles when viewing a small aeroplane from behind. The physicists had calculated that we should get a range of five to ten miles and were anxious to achieve that distance in our trials.

The most serious defect of the equipment was that at least three-quarters of the power was wasted in the 'bridge circuit' used to transmit and receive on the same aerial. This arrangement also had the drawback that unless the balance between aerial and dummy load was perfect, the transmitted pulse was liable to burn out the crystal in the receiver, a fault that could not be rectified in flight.

For obvious reasons, several laboratories in Britain and America were interested in devices that allowed transmission and reception to take place through a common aerial. Eventually a number of successful 'circuits' were invented, all depending on using a cavity resonator filled with gas at low pressure that flashed over when the radar pulse was transmitted but not when the weak echo was received. At TRE, Skinner, Ward and Starr had been working on the problem for some time and in June 1941 they presented the flight-testing group with a 'TR' box which incorporated this feature as well as a better receiver and which still fitted into the very small space left in the nose of the aeroplane. (Figure 18.1).

Downing and Edwards tested this on 2 July 1941 and at once obtained a greatly improved performance with ranges of four to five miles on a medium-sized aeroplane.

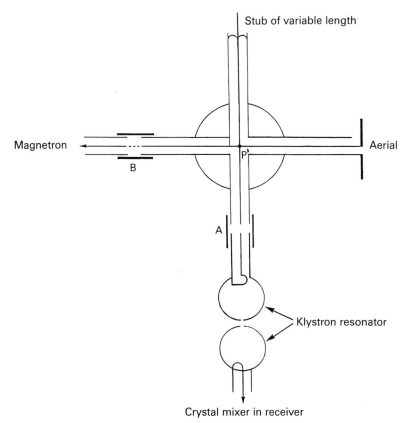

Figure 18.1. Common transmit–receive unit using soft klystron.

Diagram based on sketch made by Downing in his notebook and dated Sunday 29 June 1941 showing essential features of common Transmit–Receive Unit using a soft Klystron. This was designed by Ward, Skinner and Starr, and was later built into a compact unit, the T^2R, together with the magnetron and crystal receiver.

Downing gives the following (correct) description of the way that the device works. When the magnetron is pulsed microwave power causes the soft-klystron (i.e. one containing gas) to flash-over; this effectively puts a short-circuit at the end of line A. Using the slider, the total length of line A is then made an odd number of quarter wavelengths so that from the point P no power goes down A, the crystal is protected and all the power goes to the aerial.

When the magnetron is cold its admittance is largely reactive and, by adjusting the slider in B, that branch can be given a high impedance so that no power from P passes into it from the aerial. On the other hand, since the Klystron spark is quenched as soon as the magnetron pulse ceases, the two loops on either side of the klystron resonator are tightly coupled and all the power passes from the aerial to the crystal mixer in the receiver.

The adjustments were: (1) adjust A for maximum power output from aerial; (2) trim impedance with variable stub to increase power; (3) adjust B for maximum signal/noise from receiver.

183

The altitude ring and other possible defects. When using radar in an aeroplane it is difficult to avoid getting an echo from the radiation scattered by the ground directly underneath the aeroplane. In the early tests this effect, which gave rise to the 'altitude ring' in Figure 17.4, was so large that it might have concealed the signal from the target aeroplane at the moment when the range of the target was the same as the altitude. A series of changes made during the next four months reduced the artefact to an insignificant level. Some of the changes recorded in Edwards's log were (i) use of vertical rather than horizontal polarization, (ii) careful matching of the aerial, (iii) metal screening below the scanner, and (iv) use of perspex nose of uniform thickness. During the first three months we used a botched-up nose, partly made by 'Indestructo Glass' which had been strengthened with a wooden ring. When this was replaced with a perspex nose with clean joints and no ribs the altitude ring disappeared altogether overland at heights of more than 2000 feet and though visible was negligible over sea. These observations were made with a set on which an aeroplane was clearly visible at a range of five miles. I do not remember any complaints from service personnel about the altitude ring in the production versions of AI Marks VII and VIII.

There were other minor defects which were either put right or which rectified themselves. In the early tests the set came out of tune rather easily, perhaps because the magnetron frequency had drifted. This trouble disappeared when we used the more powerful and stable strapped magnetrons.

Minimum range. In World War II the standard way of attacking a hostile bomber with a night-fighter fitted with radar was to use the AI to get to a range where the bomber could be seen against the night sky. On a dark night this might require an approach to less than 1,000 feet so the RAF set 500 feet as their operational requirement. This meant that the radar set should be capable of receiving an echo about 1 microsecond after emitting a pulse with a peak power of 10 kilowatts or more (about 10,000 volts on the magnetron). To begin with, the minimum range was more than 1,000 feet which was unacceptable. After shortening the pulse and making numerous other small changes, Edwards and Downing achieved a minimum range of 200 to 500 feet some six months after the start of the flight trials. The Beaufighter was very heavily armed with four 20 millimetre cannons and six machine guns in our version, and a good minimum range not only helped the pilot to get a visual but made it likely that he would destroy the enemy bomber before it had a chance to open fire.

Potentialities of airborne centimetre radar

It soon became clear that one of the things we had to do was to explore what might be called the natural history of airborne centimetre radar. For example, we found that if you were flying through a heavy rainstorm the display was obscured by reflection from raindrops. This effect occurred too rarely to matter, but we concluded that it would become more serious at a 3 centimetre wavelength and would rule out even shorter wavelengths for use in air-interception equipment. Another question to be answered was the optimum repetition frequency of the radar. Up to a point, the higher the frequency the better the display, particularly if the night-fighter is pursuing a weaving target. However, if too high a repetition frequency is used, the display will be confused by 'second time base' returns. For instance, with a recurrence frequency of 2,000 Hertz an echo from a cliff at a distance of fifty miles will show up 536 μsec later on the next time base at an apparent range of three to four miles. This effect limits the repetition frequency of an airborne radar set to about 2,000 Hz. A frequency of 2,500 Hz was used in AI Mark VIII and apparently gave little trouble, but lower frequencies were employed in AI Mark X or the ship or town detecting sets (ASV and H_2S).

Other interesting observations were that echoes from flocks of birds were occasionally seen on the ground, for example if a large flock of gulls was disturbed on the aerodrome, but that bird echoes never interfered in the air, that towns gave larger returns than the ordinary countryside and that a rough sea gave larger echoes than a smooth one.

As soon as the centimetre radar set was working fairly reliably, many flight trials were devoted to demonstrations to visiting scientists or RAF personnel. These led to useful criticisms and did not waste time because they could be combined fairly easily with tests of the equipment. Between March and October 1941 those taking part in flight trials were first the members of the scientific team who were anxious to see the equipment in action after all the hard work they had put into it; from TRE these included Dee, Burcham, Ward and Skinner and from GEC, Marris, Espley, Clayton and O'Kane; others from TRE were D. M. Robinson, R. W. Taylor (AI Mark V) and J. A. Ratcliffe, who was then arranging the major effort that his organization put in to making centimetric AI work in practice. Physicists or electrical engineers from outside TRE included Bedford from Cossor, Bowen from the Radiation Laboratory at MIT, Watson Watt from London and Oliphant from Birmingham. The visitors from the RAF included Air Marshal Sir Philip Joubert, Wing Commander Cunningham, Squadron Leader Derek Jackson and Group Captain Hart, and Wing Commanders Gilbert Smith, Hiscock and Horne. In July the last four made thorough trials to check the genuineness of interceptions, using the sort of tests that would be applied later at the Fighter Interception Unit. On the whole, the reaction of service personnel was more favourable than Edwards, Downing or I had expected. Partly as a result of the tests, headquarters in London decided to order a hundred sets in a crash programme to be installed in Beaufighter aircraft the following autumn.

In addition, Edwards and Downing made an important series of tests in April and August 1941 on the ease with which submarines could be detected. The first trials were carried out in Flight 24 on 30 April 1941 on submarine N72, HMS Sea Lion, and the second set in Flights 53–55 on 10–12 August 1941 using HMS Sokol. These established that even if only the conning tower was exposed, a submarine could be located accurately and reliably by airborne centimetre radar. Partly as a result of the April trial, TRE applied formally to headquarters in June 1941 for permission to develop a ship (or submarine) detecting radar using a centimetre wavelength. This and related developments are considered further in Chapter 23.

Chapter 19

Preparations for AI Marks VII and VIII

DURING THE spring and early summer of 1941 members of Dee's team spent much time helping to organize the initial trials of AI Mark VII in Beaufighters. One of the most interesting things I had to do was to persuade the Bristol Aircraft Company to fit a perspex nose containing a spiral scanner onto a Beaufighter. In this I had the support of Barnes from Nash & Thompson as well as high-level pressure from Dee and others at TRE. To begin with, the engineer in charge of the workshop at Bristol Aircraft was reluctant to muck about with his nice aeroplane but once he decided to move, the job was done in a very short time. Subsequent flight trials proved the system airworthy and Beaufighter X-7579 arrived at Christchurch in September 1941. Earlier on, perhaps in May 1941, the Ministry of Aircraft Production ordered one hundred 9 centimetre AI Mark VII sets from GEC Coventry to be installed in Beaufighters, of which thirty-six were eventually fitted. The sets were to be built on a crash programme basis and installation of the first set was to begin at Christchurch in September.

Somewhat remarkably, the first aircraft was ready for flight testing on 2 October 1941 and moved to the Flight Interception Unit (FIU) at Ford in late November 1941. However, before describing these trials, I should say something about the tests of alternative scanning systems that had been made during that summer. To do this without prejudicing the Mark VII programme, we kept our front-line equipment in the original Blenheim N-3522, fitted that aeroplane with a new perspex nose and transferred the old nose to Blenheim V-6000 which we used for testing alternative systems: first a manual scan and hold, using a rotating aerial and then the GEC's helical scanner. After a lot of hard work we decided that it really was desirable to have a high-speed scan and that

considerable practice would be needed to make and hold interceptions successfully with a manual system. We guessed that the
system would be improved by replacing the manual hold with an
automatic lock, as was done in AI Mark IX, but that it would still
be desirable to provide a high-speed scan for the initial stages of the
search.

Edwards and O'Kane made preliminary tests of the helical
scanner but there was not enough time to reach any firm conclusion
about the merit of this system. It soon became clear that the only
practical way of equipping the Beaufighters shortly to arrive at
Christchurch was to stick to the spiral scan and fit the equipment
that went with it. This may account for Bowen's comment about
the apparent lack of interest shown by TRE personnel in the
American prototype AI which arrived in Britain in the summer of
1941:[1]

> Edwards' log refers to 'separate comments' about flying with the
> American AI equipment in a Boeing on 20 August 1941 and much later
> (January 1943) gives an assessment of the American SCR 720 installed
> in a Wellington Bomber. I have no recollection of the 1941 trials
> perhaps because I was briefly in hospital and then on leave in August
> 1941. Pressure to deal with the incoming Beaufighters may also explain
> why I have completely forgotten a flight on 8-Nov-41 that Lovell's
> records show I took in his Blenheim equipped with the first air borne
> lock-follow type of AI.

The effort expended in equipping and testing Blenheim V-6000
with a helical scanning system was by no means wasted because,
after some modification, the equipment was used by O'Kane and
Hensby in the very important tests of the town-finding equipment,
H_2S.

The Beaufighters allocated for AI Mark VII began arriving at
Christchurch aerodrome in September 1941 and for the next year I
had to divide my time between several disparate activities. There
was still some design and testing to be done in connection with AI
Mark VIII, the production version of the crash programme Mark
VII. Mark VII itself had to be flight-tested, first at Christchurch and
then at the Fighter Interception Unit at Ford. I had also acquired the
job of carrying out technical liaison with a number of firms, notably
the Bristol Aircraft Company, De Havilland's, Nash & Thompson,
GEC Wembley, GEC Coventry and E. K. Cole in Malmesbury. It

was necessary to ensure that the size, weight and external appearance fitted in with RAF requirements. For this we had to consult a curious department called RDQ whose job it was to ensure that the proposed installation did not conflict with other operational plans for the aeroplane. The head of RDQ was a regular RAF officer of air vice marshal rank but the moving spirit was an efficient, rather bossy squadron leader who was a highly fashionable decorator in private life.

I took on all this liaison work fairly cheerfully because it gave me an opportunity to travel and to spend an occasional night in London, but at times the attempt to get two firms to make the same piece of equipment was very trying. It turned out that there was considerable rivalry between E. K. Cole and GEC and each firm was determined to engineer AI Mark VIII in its own way, whereas the RAF rightly thought it essential to have identical sets of equipment. The reason why two firms were involved was that the senior people at TRE, Dee, Skinner and Lewis, felt that GEC would always drag its feet because it hankered after its 25 cm project and that the only way to get things moving was to inject some competition into the system.

At first the E. K. Cole engineers showed little enthusiasm for centimetre radar and it soon appeared that in addition to their dislike of GEC they were equally suspicious of interfering scientists. When I arrived at Malmesbury for a three-day visit I was subjected to continuous abuse about the damage done by scientists who had opted for a soft job rather than fight for their country. I think the trouble was partly that in 1939 E. K. Cole had been suddenly evacuated from its home in Southend to the small town of Malmesbury where everyone, including the work-force, was thoroughly unhappy. In contrast, GEC's production factory in Coventry which had been badly bombed was relatively easy to deal with. Of course the only thing to do was to ignore personal abuse but this wasn't easy as complaints from Malmesbury were sometimes directed towards a high-level chasing committee set up by the Prime Minister. Eventually I made friends with the senior engineer at E. K. Cole and felt better when I learnt that one of my persecutors was known to be very difficult and the other was subject to acute attacks of religious mania.

Beaufighter X-7579 which arrived at Christchurch early in September was wired up, ground-tested and ready for its first airborne trial on 2 October 1941. The equipment was the first off the production line of the crash programme at Coventry. From the start we knew that some of the electrical equipment would break down at high altitudes but it was fairly all right up to 10,000 feet which was all the RAF wanted initially. There were numerous teething problems but we got over them and after eighteen flights at Christchurch this and another Beaufighter which had been fitted and tested flew to the Fighter Interception Unit in late November 1941. Edwards (GEC) and Burcham (TRE) accompanied the aircraft and spent ten days doing further operational tests and training RAF personnel.

The FIU trials and various design problems kept us busy until we were ready for Mark VIII, the production version of the early centimetric system.

The first squadrons of Beaufighters were used in Britain against enemy bombers during the Baedeker raids in the spring of 1942, after which some aircraft flew to the Mediterranean in order to take part in the air defence of Malta. We heard later that a Beaufighter fitted with Mark VII had shot down an enemy plane on 5 April 1942. Subsequently the total score for Mark VII rose to about 100.

Tests of the prototype AI Mark VIII

The Ministry of Aircraft Production eventually placed contracts with GEC and E. K. Cole for several thousand AI Mark VIII sets to be fitted in Beaufighter and Mosquito aircraft. The radar set was intended to be an improved version of the Mark VII spiral scanning equipment and would operate on the same wavelength of 9 cm. It was to use the more powerful and stable strapped magnetron initially developed by J. Sayers at Birmingham and subsequently produced by GEC as CV 64 (Figure 19.1).

When operating properly, this magnetron produced a peak power of 50 kilowatts, but it then required about 10 amps at 15 kilovolts. No one liked the idea of piping such voltages around an aeroplane so it had been decided to generate a 3 kV pulse in the modulator and transform up to the higher voltage in a pulse transformer close to the magnetron. These units were housed in the same relatively small box as the crystal mixer and the common TR system using a soft Sutton tube which made it possible to transmit and receive on the same aerial. This last part of the system was

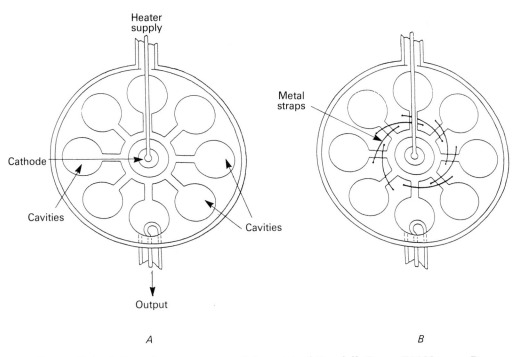

Figure 19.1. *A*, 8-cavity magnetron of the original Randall–Boot, E1198 type. *B*, strapped magnetron of the type introduced by Sayers and manufactured as CV 64.

In the strapped magnetron, which operates more stably and with greater efficiency than the unstrapped type, alternate tongues between cavities are connected by metal strips in the manner shown: i.e. 1 to 3, 2 to 4, etc. This is believed to promote the π-mode of oscillation in which adjacent cavities are 180° out of phase and alternate ones approximately synchronized.

The anode of the magnetron was made out of a copper cylinder about 3.5 cm in diameter and 1.5 cm long; it was cooled by metal fins welded to the anode which are not shown in the diagram. From Callick (1990), p. 69.

basically similar to the resonant T illustrated in Figure 18.1, but was a good deal easier to set up as the whole box contained only two pre-set adjustments. The unit was known as the T^2R because it combined the functions of the transmitter with that of the system used for transmitting and receiving on the same aerial. Ideally the unit should have been pressurized but shortage of time made this impossible. It was made in large numbers for several different applications, principally AI Mark VIII, H_2S/ASV and AGLT. It is important to notice that the unit contained only one soft Sutton tube and not two as in the devices described in Dr Callick's recent

191

book from *Metres to Microwaves* or in the articles published in the 1946 volumes of the *Proceedings of the Institution of Electrical Engineers*. Addition of a second Sutton tube in the T-position might have helped setting up, but it would have increased the bulk of the unit and might have had a devastating effect on minimum range.

Priorities are always difficult but it seems generally agreed that A. H. Cooke of the Clarendon suggested using a soft Sutton tube for common TR work to H. W. B. Skinner, who then developed it with A. G. Ward and A. T. Starr. Arrangements with two spark gaps were used earlier, perhaps initially by Banwell in 1940 (described in Banwell, 1946). Whatever its origin, the soft Sutton tube certainly shook the Germans who, when they shot down a bomber fitted with an H_2S set, could not imagine what function could be served by half a klystron unsupplied with power. Several years ago Professor Reichert of Tübingen, who worked on German AI during the war, told me that this was the one device in the whole outfit that they did not understand and Lovell had the same reaction from Professor Hachenberg of Bonn.

The equipment also had to include facilities for interrogating friendly aeroplanes fitted with IFF[2] Mark III as well as for homing onto a centimetre beacon. When using 'IFF', the equipment sent out pulses on a carrier wave frequency of about 180 megahertz at a repetition rate of 500 Hertz with each pulse synchronized with every fifth pulse of our centimetre transmission. For identifying a friendly aeroplane the answer came back on the IFF frequency (180 MHz) and was fed into our display. In theory if you pressed the identifying button a rising sun display (Figure 19.2) appeared around an echo from a friendly aeroplane. In practice the RAF night-fighters usually depended on visual identification because IFF sets never seemed to be fitted sufficiently widely to be of much use. Still one had to keep trying and the RAF requirement was that our production sets must be able to interrogate IFF Mark III.

The beacon system, on the other hand, was certainly useful as it enabled the night-fighter to home on its base with accuracy from a range of fifty to a hundred miles.

According to Burcham's diary, what was done was worked out at TRE between Clegg and myself and agreed at informal meetings between Dee, Burcham and myself on Sunday–Monday 13–14 July 1941. Airfields using centimetric equipment like AI would be equipped with wide-band centimetric beacons which would

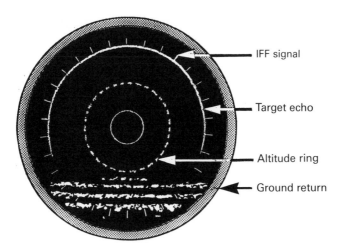

Figure 19.2. Appearance of IFF signal from friendly aircraft on AI VIII.
A friendly aircraft, fitted with IFF III, is 2–3° above, at a range of 10,000 feet, altitude 4,000 feet; the target appears as a crescent of length 180°. On pressing the interrogator button a rising sun display appears at the same range as the target.
The scanner rotation rate is taken as 1,000 rpm, the repetition frequency of the radar as 2,500 Hz and of the IFF pulses as 500 Hz.

receive, amplify and return an enlarged signal of frequency close to that received: close rather than equal, to help the return from the beacon stand out against the echoes from the ground. Clegg's team were responsible for the design of the beacon and I looked after the modifications that had to be made in the aircraft equipment.

What we decided was that on Beacon operation the repetition frequency would be dropped from 2500 Hz to 930 Hz: 930 because a radio, or light, echo takes $\frac{1}{930}$ second to travel 100 miles there and back. At the same time the single 1 microsecond transmitter pulse used in the AI mode would be replaced by a pair of pulses each 1.25 μsec long and separated by 4 μsec. The point of this arrangement was first, that on beacon operation the power drained by the transmitter was roughly the same as on AI, in spite of the large change in recurrence frequency, and second, that the double pulse provided a crude form of coded key which might help to prevent intruders homing on the beacon.

When operating in the beacon mode a multiple switch converted the radial time base into a simple vertical trace, covering 95 miles, with left or right deflection synchronized by a commutator switch on the scanner.

193

After noting these general points, Burcham comments laconically, 'Can probably be done OK but is some job.'

However, after the usual teething troubles the beacon system worked well, and, as George Edwards and others showed later, could be used for a blind approach to the runway.

Another important refinement was that there should be two ranges of radial time base for use with AI: one with a maximum range of 8 miles for the initial search and another, faster, time-base, with the periphery expanded to 2 miles for the final pursuit and interception. The RAF required marker pulses giving the range. On AI these took the form of concentric circles spaced at 2,000 foot intervals on the 2 mile range and 2 mile intervals on the 8 mile range; in both cases the marker pulses were to be less than 1 μsec in duration and limited in amplitude, so that they did not obscure the radar echoes from aeroplanes. For the 95 mile beacon scan, which was exponential, the marker pulses lasted 5 μsec and occurred at 10 mile intervals.

After carrying out many tests with the spiral scanning system in Beaufighters, my colleagues and I decided that we had developed an excellent if somewhat unorthodox method of intercepting night-bombers. I was particularly impressed with a flight that I made on 29 March 1942 with Flight Lieutenant Goddard as pilot and D. O. Hawes of GEC as second observer. Before our target was airborne our radar picked up another aeroplane at a range of three miles, which later turned out to be a friendly Hurricane fighter spiralling up in the clouds from 1,500 feet to an altitude of 9,000 feet. We didn't know this initially and as Christchurch aerodrome had recently been shot up by a stray Messerschmitt, we set off in pursuit with cannon ready to fire. After a long series of climbing turns in which I found myself shouting, 'Left . . . left . . . up . . . left . . .,' we emerged in bright sunshine on the tail of a Hurricane fighter. It seemed to me that this was about as severe an operational test as one could devise and finally removed the doubts that had worried me for so long.

The serviceability of the equipment was less satisfactory and I believe the high-voltage modulator continued to give trouble after the equipment got into service. However, Edwards did manage to carry out successful trials at an altitude of 21,000 feet in September 1942.

By the autumn of 1942, Edwards, Burcham and their colleagues were ready to check the production models of AI Mark VIII before they went for full trials at the Fighter Interception Unit. I had hoped to take part in these myself but was switched to a Bomber Command project in September 1942. This was a disappointment which became more acute when I heard that one of the early trials of Mark VIII, in which G. G. Roberts, was the observer, resulted in the destruction of a raiding enemy bomber.

Burcham took over much of the work that I had been doing and became the 'leader' of the group working under Dee on centimetric AI.

Chapter 20

Alternative scanning systems; Window

THE EVENTS of this chapter cannot easily be described without referring to a character for whom my feelings remain a mixture of affection, admiration and exasperation. Derek Ainsley Jackson FRS[1] was a wealthy man who entered Trinity College, Cambridge, as a scholar in 1923 and obtained first-class honours in Natural Science in 1926. His physics supervisor at Cambridge, H. W. B. Skinner FRS, whose name appears frequently in this book, stimulated Jackson's interest in spectroscopy. When Jackson was asked why he chose to work with Lindemann at Oxford, rather than with Rutherford in Cambridge, Jackson is reported to have answered, 'Oxford bought me just as you might buy a promising yearling.' By that he probably meant that Oxford gave him facilities and allowed him to work on a spectroscopic project of his own, rather than in nuclear physics as Rutherford had suggested. The reference to racing is characteristic as Jackson managed to combine his scientific work with a social life in aristocratic and artistic circles. He rode his own horse three times in the Grand National, first in 1935 and again in 1947 and 1948; on the last occasion he was lying second when his horse refused at the penultimate fence. Part-ownership of the *News of the World* enabled Jackson to finance his scientific research as well as his social and racing activities.

Jackson's matrimonial arrangements were as adventurous and varied as his racing career. In 1931 he married Poppet John, one of the daughters of Augustus John who cannot have been easy as a father-in-law. The marriage to Poppet was dissolved in 1936 and in the same year Jackson married Pamela Mitford, one of the daughters of Lord Redesdale, the 'Uncle Matt' of Nancy Mitford's novels. This marriage which made Jackson the brother-in-law of Sir Oswald Mosley as well as of Unity Mitford, an admirer of Hitler, gave rise to the belief that Jackson was, or had been, a fascist. I agree with Jackson's biographer, H. G. Kuhn, in thinking the accusation incorrect, though Jackson certainly was strongly opposed to the socialist ideologies in which most liberal scientists then believed.

The marriage to Pamela lasted until 1951 when Jackson married Janette Woolley; in 1957 he married Consuelo Regina, the widow of Prince Ratibor; in 1966, Barbara Skelton, and finally Marie Christine, daughter of Baron Reille, a happy marriage which lasted from 1968 until Jackson's death in 1982. After the war Jackson spent much of his life abroad, first at Tullamaine Castle in Eire and later in Paris and Lausanne. My last contact with him was in 1975 when I wrote to thank him for a generous gift to the Royal Society.

Jackson's friendship with Professor Lindemann who was then Churchill's scientific adviser, helped to give him a powerful voice in the military applications of science. But there were more valid reasons for the extremely influential position that he occupied in the years from 1941 to 1945. In 1939 Lindemann was unwilling to release him from work that the Clarendon Laboratory in Oxford was doing for the Admiralty, but Jackson soon got round that and joined the RAF as an air gunner, a well-known way of committing suicide. Early in 1941 he was transferred from Bomber Command to a night-fighter squadron where he was highly successful in using the Mark IV AI to direct the pilot of his Beaufighter on to enemy bombers. This gave him a formidable reputation as a radar tactician.

In the spring of 1942 there was fierce controversy as to whether Bomber Command should be allowed to confuse enemy radar, both ground and airborne, by dropping metallic chaff, or 'Window' as it was called.[2] Lindemann supported Fighter Command's request that Window should not be used until its effects on our own night defences had been investigated and suitable remedies developed. One result of this decision was that Jackson, who frequently visited TRE at Malvern, was put in charge of trials of radar at an aerodrome at Coltishall in Norfolk. The trials were carried out by Fighter Command with TRE personnel present. They confirmed the vulnerability of our night defences to Window and showed how it could best be dropped from bombers.

All this was naturally kept very quiet and I don't remember having seen a report on Window trials of Mark VII or VIII. My guess, long after the event, is that Jackson decided that the spiral scan did not perform well against Window, perhaps because its use required a stern chase of the bomber. There seemed to be two possible alternatives, the first being the AI Mark IX 'lock-on' scheme and the second a helical scanner developed in the USA and known as AI Mark X or SCR 720.

During 1941–2, F. C. Williams and his group at TRE designed and built a highly efficient system for making a radar dish lock on to a target aeroplane. This contained advanced features in the control system, such as feedback from velodyne motors which measured the speed at which the radar mirror was moving. A radar known as AI Mark IX which incorporated these features, as well as other highly desirable improvements such as a spark-gap modulator, was developed by Lovell and Ritson. Early in 1942, A. E. Downing who had worked with Edwards and myself at Christchurch was switched to the development of a prototype 'AI Lock' in Beau-fighters. One of the advantages claimed for the new system was that it would be less vulnerable to jamming from Window. I think now that the high-ups at TRE who hardly ever went in an aeroplane had persuaded themselves, or been persuaded by Jackson, that a lock-on system would be subject to less interference than one depending on scanning and visual display.

Trials of a lock-on system were made in late 1942 at Coltishall with Jackson as navigator and Downing in charge of the centimetre equipment. Jackson made thirteen trials but found that Window threw the equipment off lock and that the target could not then be recovered. Downing then modified the set and arranged to fly with Jackson in a further test. Jackson agreed but found that there was no-one available to discharge Window from the target – also a Beaufighter. He took over this job while Downing operated the radar set in the pursuing night-fighter. The control station was warned, but this did not avert the subsequent tragedy described by Jackson in the following words:

> I heard the control order two Spitfires to scramble: they were to be ready to intercept an unidentified plane. I was somewhat dismayed at hearing this, particularly because the Controller gave no indication of how far away the unidentified aircraft was, and he appeared to be about to ignore my request that the fighters stationed at Coltishall should be kept well away from the two Beaufighters making the Window trial. When we were over the sea I saw one of the Spitfires coming towards us in the most sinister manner, and at once said to Squadron Leader Winward on the intercom that a Spitfire was approaching us in a

manner which I did not like. Winward immediately started a very sharp diving turn; almost simultaneously we were hit by cannon fire from the Spitfire. Although neither the pilot nor I were aware of it, this had severed the intercom leads, so each thought the other had been killed, as there was no reply on the intercom. It seemed to me that the only thing for me was to bale out; but the turn was so tight that the g was so strong that I could not reach my parachute; and I had a decidedly unpleasant wait of about ten seconds before what I knew would be the end; but I was in error; a few hundred feet above the water, the Beaufighter levelled out; I saw the fault in the intercom and remedied it just in time to hear Bill Winward say on the R/T '. . . and he has killed my observer!' Our aeroplane was damaged so we had to return to Coltishall immediately; but as we turned Bill Winward saw an aeroplane burning, on the surface of the sea; our fears that this was the other Beaufighter and that both the pilot and Dr Downing had been killed were well-founded.[3]

Because I was moved to a Bomber command project in September 1942, I didn't hear of Downing's death for some time. Indeed I don't believe that I knew the details until I read Christopher Hartley's account in *Biographical Memoirs*. Downing had just got married. I don't know what happened to his widow except that she sensibly decided to join the WAAF and start a new life there.

As Hartley says, the loss of Downing and the only prototype AI Mark IX might well have endangered the release of Window to Bomber Command, but fortunately Jackson had survived and the prototype of the American SCR 720 was available for trials in a Wellington early in 1943. I never flew with SCR 720 but G. W. Edwards made a long flight test on 27 January 1943 on which he commented that the equipment worked well without trouble but was more difficult to use than the spiral scan of AI Mark VIII. He added that it would probably be 'OK with training'. Soon afterwards, Jackson satisfied himself that the operator of an SCR 720 set could make an interception in the presence of Window. But he still had to convince Fighter Command that 720 was satisfactory in other respects.

Because this proof was critical to the timing of the Window release Jackson used his by now considerable influence to see that the usual procedure was short-circuited so that when the only set at that time in existence was installed in a Mosquito, it was at once sent to an experienced front-line unit (the Fighter Interception Unit).

Jackson was in charge of the trials and flew on many of them himself. At the same time, it was necessary to establish that an 'average' operator could use the set, as it had more controls than British equipment. One of the earlier 'guinea-pigs' commented on the complexity of the equipment and said that after his first flight he felt 'physically exhausted'. Jackson was incensed and his speculations about the private life of the unfortunate navigator were as uncomplimentary as they were accurate. But the trials at FIU, under Jackson's guidance, established that SCR 720 (later known as AI Mark X) was not only Window-resistant but also a first-class radar for night combat. The final obstacle to Fighter Command's agreement to the release of Window was thus removed, when AI mark X went on its first operational patrol in May 1943, with Jackson as navigator, and its availability in quantity from American sources by the autumn of 1943 already assured.

The use of Window was finally authorized by the Air Ministry on 16 July 1943 and it was first dropped by Bomber Command in a raid on Hamburg on 24 July 1943, with results that were devastating to the German defences.[4]

Chapter 21

Operational performance of centimetre AI[1]

AI Mark VII in the United Kingdom

At the end of November 1941 the first two Beaufighters fitted with AI Mark VII flew from Christchurch to the Fighter Interception Unit at Ford for operational trials by the RAF. G. W. Edwards of GEC and W. E. Burcham of TRE accompanied them and spent ten days there doing further operational tests and training RAF personnel in the use of the equipment. The two aircraft remained at FIU for tests of the serviceability and maintenance of AI VII as well as for further operational trials. After practice with target aircraft the trials took the form of operations against German minelaying aircraft over the Thames Estuary. During the first trial on 7 December 1941 the FIU picked up an aircraft which they chased and damaged, after recognizing it as a Ju 88.[2] In connection with this episode Burcham's diary records a message from Dee saying that X-7613, one of the two Beaufighters at FIU, had a shot at a Hun over the estuary on 12 December 1941. The fighter got in to 1,000 feet at 1,000 feet altitude and fired, but there was no definite kill. It seems likely that this is the same episode as that described for 7 December and that one of the dates is incorrect; 12 December is more likely to be correct as an earlier combat would probably have been recorded by Edwards or Burcham.

The Radar observers rapidly became adapted to the radial time base display and obtained maximum ranges of 3½ miles; they reported that the equipment 'gave far less trouble than any other prototype AI set'.[3] In its unmodified form the set was unsuitable for use above 8,000 feet because of internal arcing. Nevertheless FIU recommended early introduction into two night-fighter squadrons for use against enemy low-flying aircraft. As a result of this

201

favourable report it was decided to equip forty Beaufighters as soon as possible from the hundred sets of AI Mark VII that would shortly come from the crash programme production at the GEC factory in Coventry. This was the first occasion on which centimetric equipment went into service with the RAF, and great trouble was taken by TRE's Post Design Services as well as by Fighter Command to make sure that it was a success. Between January and May 1942 personnel from the four squadrons who were to be using Mark VII were trained in centimetric techniques, and TRE service stations were set up at Christchurch to deal with installation and maintenance. RAF radio mechanics joined with the installation of Mark VII so that later on they might deal with Mark VIII. Installation, first at Christchurch and then at Hurn, took longer than expected, but two aircraft were ready by mid-March after which the rate of completion was about three a week.[4]

The first definite operational success was on 5 April 1942 when an FIU pilot took off in a Beaufighter against enemy minelaying aircraft under CHL control and established an AI contact at a range of four miles. A 'visual' was obtained on a German Dornier 217 aircraft which the night-fighter shot down in flames, after closing to about 300 yards.[5]

From its introduction at FIU in December 1941, AI Mark VII was 'outstandingly successful'.[6] Before 15 May 1942, when the four squadrons were still not fully equipped, seven enemy aircraft had been claimed as destroyed, in addition to several damaged and 'probables'. At this time enemy activity was small and consisted mainly of low-flying raiders and minelayers. Previously this type of enemy operation had been almost immune from interception at night.

AI Mark VII in the Mediterranean

The requirement for centimetric AI became acute in June 1942 when the Germans in Sicily started to jam 1½ metre AI in Malta where it had been operating successfully since 1941. It seemed likely that the jamming would soon be extended to the North African ports, which would make it difficult to provide air defence at night for the forthcoming invasion of Sicily and later the mainland of Italy. Although only forty Beaufighters had been fitted with Mark VII it was decided to dispatch five aircraft fitted with Mark VII together with five spare sets of AI. These were to be accompanied by a technical officer, two radio mechanics and one hydraulic fitter (for the scanner), all experienced with Mark VII.[7]

For security reasons the AI equipment was taken out of the Beaufighters, four of which were then flown out via Gibraltar. The radar equipment and ground personnel were then transported by the same route in two Sunderland flying boats arriving at Malta early in August 1942. Aircrew who had been operating in Malta with 1½ metre AI were trained and began night flying with the new Mark VII sets. This equipment, which had been ordered before the design had been stabilized, was, as FIU had noticed, liable to flash-over at any altitude greater than 10,000 feet. In Malta, at that moment, it was needed to deal with high-flying bombers coming in at 20,000 feet, at which altitude the voltage of 10 kilovolts in the modulator pulsing the magnetron would behave like one of 25 kV. However, the Technical Officer, or someone around, must have been a talented engineer because the ground crew of No. 89 Squadron introduced modifications to enable the sets to be used up to 20,000 feet. It is quite difficult to see how this was done! Nevertheless, in the six weeks from 1 September 1942, fifteen AI contacts with enemy aircraft were made at night and resulted in six visual sightings. These led to the destruction of five enemy aircraft and one 'probable', all between 19,000 and 22,000 feet.[8]

Within two months of the beginning of the Malta operations with centimetric AI, three of the original Beaufighters fitted with the equipment were written off by crash landings or taxiing accidents. However, a steady but small stream of replacements trickled in during the rest of 1942 from the few Mark VII-equipped Beaufighters available in Britain.

Early in 1943 the Chief of the Air Staff approved the transfer of twelve Beaufighters fitted with AI Mark VII sets together with experienced pilots, air and ground crew as well as 100 per cent spares. The aircraft reached North Africa in February 1943 and when fitted had an immediate effect, eighteen enemy aircraft being destroyed in April and thirteen in May. For the future it was decided that the Mark VIIIA sets, made by GEC, should go to North Africa whereas Mark VIII, made by E. K. Cole, would remain in Britain.

By the end of the war AI Mark VII had been fitted in about forty Beaufighters and, either in Britain or in the Mediterranean, this led to the destruction of more than a hundred enemy planes, about one for each set of equipment.[9]

The AI Mark VIII series

The Mark VIII programme was a major production effort intended for the whole night-fighter force; the set was of higher power than Mark VII and included Beacon and IFF facilities, which were omitted in the hurried design of Mark VII. Efforts were made to have a uniform, standardized set, but hurry and changes in operational requirements soon resulted in a multiplicity of versions:[10]

1. AI Mark VIIIA was an interim program for 500 sets, later increased to 1,000 sets, to be made by GEC Coventry as soon as they had finished Mark VII.
2. AI Mark VIII was the main production set for which a contract for 1,500 sets was placed with E. K. Cole.
3. AI Mark VIIIB contained 'Lucero' navigational facilities which enabled the night-fighter to use the 1½ metre chain of beacons.

There was also a high-altitude version which allowed lightly pressurized Mosquitoes to operate up to 45,000 feet.[11]

Since there was no great difference between the operational efficiency of the different kinds of AI Mark VIII, I shall not attempt to distinguish between the various subdivisions, but will refer to them all as Mark VIII, specifying the manufacturer if this seems necessary.

After tests had been made at FIU, the GEC's version of AI Mark VIII started going into service at the beginning of 1943. In mid-January, Technical Training Command took over the maintenance of Mark VIII so the equipment used for instruction was moved from 'TRE's airfield' at Defford to the Signals School at the Science Museum, South Kensington, where larger courses began at the end of January. By May 1943 five squadrons had been equipped.

In spite of difficulties with the modulator, where the CV57 valves were being over-run, the set was 'very successful operationally'. The first aircraft to be shot down by the new set was a Dornier 217 on 3 February 1943 and by May some thirty enemy aircraft had been destroyed.[12] . Contact with one of these was made at a range of seven miles.

In assessing the efficiency of different kinds of AI, a useful method is to calculate the average ratio of the number of aircraft destroyed to the number of attempts. This was appreciably higher for Mark VIII than for other marks of AI used up to the spring of 1943.[13] During October 1942 George Edwards had tested a prototype installation of AI Mark VIII in a Mosquito with satisfactory results. Fitting of Mosquitos began in March 1943 and, after installation troubles had been cleared up, aircraft so equipped showed an effectiveness on night operations similar to that of the Beaufighter.

A pre-production model of the E. K. Cole version of Mark VIII went for service trials on 23 December 1943 and was subjected to 120 flying hours of tests. The maximum range averaged only four miles which was disappointing as the new strapped magnetron should have produced ten times as much power as the one used in Mark VII. The general reliability and serviceability of the new equipment was good though troubles with the CV57 modulator valve occurred as in the GEC version. However, FIU considered 'that the efficiency of the night-fighter defences would improve out of all recognition with the general introduction of the new Mark.'[14]

The first success of the E. K. Cole production version was claimed by No. 488 Squadron at Drem in Scotland when one of its night-fighters destroyed a Dornier 217 that was making a low-level attack on the north-east coast.[15]

German air activity over Britain was fairly low throughout 1943, but increased markedly in the period before and after D-Day when all available types of AI were pressed into service. However, there was constant pressure from the Mediterranean theatre for centimetric AI.

AI Mark VIII in Malta and North Africa, 1943–5[16]

In North Africa and later Italy the Allied supply lines depended on the availability of the ports and was vulnerable to night-bombing. In the spring of 1943 squadrons in North Africa and Malta each had a few Beaufighters equipped with Mark VII AI, with the remainder fitted with Mark IV. Jamming of the 1½ metre band and the extremely limited supply of Mark VII made it impossible to continue this arrangement, so it was decided that all future Mark VIII coming from GEC's Coventry factory should go to the Mediterranean and that Fighter Command's home requirements should be met from the E. K. Cole production line. To strengthen the defence in North Africa further, No. 219 Squadron which had been fitted with Mark VIII at the beginning of 1943 was flown out early in June with the ground party sailing on 19 May (1943). Fitting of Mark VIII proceeded more slowly than hoped because almost every set arrived with some part missing. In spite of these difficulties a total of about 140 aircraft were equipped in time for the Sicilian campaign (July 1943) when RAF Beaufighters based on Malta destroyed forty enemy aircraft during July. Rather surprisingly several American squadrons of the Mediterranean Allied coastal Air Force were fitted with AI Mark VIIIA, as they relied on Beaufighters until Northrup P61s fitted with American radar (SCR 720) arrived about a year later.[17]

The GEC's version of Mark VIII in Beaufighters remained the principal AI used by the RAF in the Mediterranean for the rest of the war. Unfortunately, although Signals V continued its excellent analysis of night-fighter activity in Britain and Western Europe until the end of the war, it says little about such activity in the Mediterranean theatre after the invasion of Sicily in July 1943.

From information in recently published volumes of the official history of the war in the Mediterranean, Miss Janet Dudley[18] of the Royal Signals and Radar Establishment has shown that six night-fighter squadrons of Beaufighters, fitted with AI, downed some seventy enemy aircraft during 1944 and the latter part of 1943. They were much involved in convoy protection and to some extent in attacks on barges or small ships. Thus on 8–9 July 1944, No. 255 squadron intruded over the Danube and destroyed eight oil barges and 102 other craft.

Window and the development of AI[19]

The losses of our bombers over Germany early in 1942 brought up the debate, referred to in Chapter 20, as to whether bombers should confuse enemy radar, both ground and airborne, by dropping Window. Churchill's scientific adviser Professor Lindemann or Lord Cherwell, as he later became, supported Fighter Command's request that Window should not be used until its effects on our own defences had been investigated and suitable remedies developed. Derek Jackson had shown that Mark VII AI with its spiral scan and radial time base did not perform well against Window perhaps because its use required a stern chase by the night-fighter.

The problem was too urgent to wait for long-term solutions such as the use of Doppler radio to distinguish between nearly stationary chaff and a target moving at roughly the same speed as the night-fighter (though I do not remember that being suggested at the time), but there were other more or less ready-made remedies. One was to use the elegant lock-follow system, devised by F. C. Williams and developed by A. C. B. Lovell and their colleagues. In this system, which became AI Mark IX, a mirror in the front of the aircraft automatically tracked the target with considerable accuracy, thus enabling the fighter to fire blind if necessary. The hope was that once locked-on, the mirror would not be thrown off by Window and thus allow the fighter to follow the bomber through the cloud of metallic chaff. However these hopes proved illusory, as Downing and Jackson showed in the air trials described in Chapter 20. The chaff did throw the mirror off lock; modifications were tried, but never properly tested because of Downing's death. The loss of Downing and the only prototype, as well as the difficulty in holding lock through a Window cloud, meant that Mark IX was not going to be a quick solution to the Window problem, so hopes were pinned on the American SCR 720, with a helical scan and a different type of display, which Jackson and others found less susceptible to interference from Window. With modifications to make it suitable for Beaufighter and Mosquito, SCR 720 became AI Mark X, which with AI Mark VIII was an important element of air defence at the time of the Normandy Landings.[20] It had its first operational success when a Beaufighter of No. 25 Squadron shot down two enemy aircraft on the night of 20 February 1944.[21]

Window was first used by us during a heavy raid on Hamburg on the night of 24–25 July 1943, with results that were devastating to the defences. The Germans first used Window over Bizerta (Tunisia) on 6 September 1943 and over Britain on 7 October 1943.[22]

After Downing's death and the loss of the prototype of the lock-follow AI, it did not receive high priority and Mark IX was not ready for operational use before the end of the war.

AI Marks VIII and X and the Normandy Landings[23]

Before deciding to use Window the Prime Minister had been assured that a hundred aircraft could be fitted with Mark X before the end of 1943.[24] Shortage of SCR 720 (made by Westinghouse) and modifications needed to adapt the equipment to RAF use caused some delay, but a more serious hold-up came from interference between the spark gap modulator of the AI and the aircraft's radio telephone. This proved most difficult to eliminate, so that instead of a hundred aircraft equipped by the end of December 1943 there were only five at the end of January 1944. However, no great harm was done and after January re-equipment with Mark X proceeded reasonably smoothly.

In the first six months of 1944 the Germans increased night-bombing to an extent that led to its being known as the little Blitz. The attack on 28 January was said to be the heaviest since May 1941 and the Window released had a serious effect[25] on the GCI stations controlling the night-fighters, though this became less serious after the introduction of the new centimetric GCI. During the whole period 144 enemy bombers were destroyed, with 21 probables and 30 damaged.[26] These results were achieved by Mosquitoes and Beaufighters, fitted with either Mark VIII or Mark X.

Statistical analysis showed little difference in the operational efficiency of the two types, about 12 per cent of interceptions being successful in both cases. Nor, surprisingly, did there seem to be much difference between the effects of Window on the two systems.[27]

No. 85 Group of the RAF was made responsible for the day and night defence of the base area and ports. The night-fighter component, which consisted of six squadrons of Mosquitos fitted initially with Mark VIIIB, accounted for 62 enemy bombers claimed as destroyed between D-Day and the end of June.

After the Allied break-out from the bridge-head the enemy air effort decreased, but the Ardennes offensive was accompanied by considerable night activity. By then three squadrons were operating with Mosquitoes fitted with Mark X. A comparison between the two AIs showed that Mark X had maximum and average ranges of 10 and 4.9 miles while the comparable ranges for Mark VIII were 8 and 4.16 miles. 'The superiority of Mark X was reflected in the figures for the conversion of interception attempts to AI contacts, being 68 per cent for Mark X as compared with 52 per cent for Mark VIII.'[28]

During the period from D-Day to VE-Day, 298 enemy aircraft were destroyed at night, with 15 probables and 44 damaged.[29]

AI against flying bombs

For six months after D–Day the Germans relied on pilotless V1 and V2 weapons for attacks against Britain.

From 13 June to 6 September 1944 Ground Radar plotted 8095 V1s flying towards London from the Pas de Calais. A total of 3752 were destroyed, 1904 of them by fighters and 1570 by gunfire.[30] Mosquitoes fitted with AI accounted for a substantial proportion at night, but as the jet flame was clearly visible at night the contribution of AI must have been less than usual.

After losing the Pas de Calais, the Germans relied either on V2s, against which there was then no defence, or on V1s launched over the North Sea from Heinkel 111 aircraft. Between mid–September 1944 and mid–January 1945, the period of air launching, 881 bombs were plotted, of which 317 bombs were destroyed by gunfire and 70 by night-fighters; 26 Heinkel 111 were also destroyed by night-fighters.[31]

AI in Bomber support

From the end of 1943 Mosquitos supported the bomber offensive either by attempting to intercept enemy night-fighters or by patrolling German airfields. Mark IV was used to begin with and achieved considerable success as did the centimetric equipment, Mark X and VIII, which replaced it later.

AI at the end of the war

In the spring of 1945 German air activity at night was so small that some AI squadrons were already being disbanded. Those remaining were often converted to AI Mark X either before or immediately after hostilities had ended. The main reasons for preferring it were its greater range, horizontal cover and discrimination against Window. The direct nature of the display may also have been an advantage, provided the observer was able to deal with two cathode ray tubes.

No airborne radar system can be perfect, particularly not an AI, with its many conflicting requirements, and the scanning mechanism of Mark X was found to suffer from various electrical and mechanical troubles, as well as lacking adequate downward cover.

A new scanner with improved cover and better serviceability was made and sets incorporating it were known as SCR 720D or AI Mark XA, but only a few arrived in Britain before the ending of the Anglo–American Lend-Lease agreement.[32]

Chapter 22

Life at Christchurch, Swanage and Malvern

TOWARDS THE end of 1941 I moved from Swanage to Christchurch where I spent six months staying in the Nelson Hotel, a small pub on the edge of the airfield, where Hensby, Fortescue and I shared a sitting room with one of the RAF officers who piloted our night-fighters.

In the mornings we walked across the grass airfield to a hut with perspex windows where we could check and modify our radar equipment. Most of TRE's flying was done from the nearby aerodrome at Hurn, but we stayed at Christchurch because Beaufighters could just get down there and no satisfactory radar hut had been built at Hurn. To begin with, security arrangements were rather casual as can be seen from a letter written to my mother on 7 February 1942.

> . . . We have now got some very energetic soldiers guarding the aerodrome. They creep about at night and challenge you with the most blood curdling yells of 'Halt, who goes there'. Fortescue who lives with us was marched off with much clanking of bayonets to the guard room because he had not found out the countersign. All this is, we are sure, a good thing but we sometimes pine for the more lackadaisical times of which the following scene is typical: Time: 7.30 p.m. Pitch dark. Soldier, 'Halt, etc.' Self, 'Friend.' Soldier, 'Advance Friend and give the countersign.' Self, 'We aren't told the countersign but I can show you my RAF pass.' Soldier, 'Have you a torch?' Self, 'No.' Soldier (who has never seen me before in his life and anyway it is dark), 'All right Friend, I recognize you, carry on.'

This relatively casual state of affairs came to an end after the British Commando raid on the German Radar station at Bruneval on 27 February 1942. Soon everyone became convinced that the

Germans would retaliate with a raid on one of the radar establishments on the south coast of England. The commanding officer sent for me and asked us to ensure that everything was securely locked up as he didn't want Christchurch to become 'a second Pearl Harbour'.

But there was worse to come. An RAF sergeant gave us lessons in firing a tommy gun from the hip and we took it in turns to sleep in the hut with the gun by our side. I suppose that the real objective of this night vigil was that we should set off the detonators with which the magnetron was fitted, but the tommy gun is the only detail that I remember.

While we naturally obeyed orders, I am afraid that we didn't take the security business very seriously. But it must have been difficult in the home station at Swanage of which A. P. Rowe gives the following account:

> Then came a bombshell, unheralded by rumour. There were, we were told, seventeen train-loads of German parachute troops on the other side of the English Channel preparing to attack TRE. The Prime Minister, we were told, had said that we must move away from the south coast before the next full moon. A whole regiment of infantry arrived to protect us; they blocked the road approaches to our key points, they encircled us with barbed wire, they made preparations to put demolition charges in our more secret equipments and, in the execution of their lawful duties, they made our lives a misery. Nor was this all. Our Home Guard, of considerable strength, was expected to be on duty all night and to work all day. My own time was spent less in dealing with the work of TRE than in discussions on whether we should die to the last scientist or run and, if the latter, where. These events made us co-operate whole-heartedly in the task of finding a place where we could get on with the war in peace.[1]

Rowe goes on to explain that perhaps the real reason for the move from Swanage to Malvern was that even in 1942 military planners were thinking that the south coast of England might be a likely place from which to launch an invasion and that TRE should be moved before it had grown too large. There may be something in this but my conclusion is that fear of reprisals for Bruneval was the dominant motive.[2]

After much discussion, Malvern College[3] was chosen as the future home of TRE in preference to several unsuitable locations such as a site for an underground factory where it would have been impossible to make field tests of radar at any wavelength.[4]

On paper, Malvern College was an excellent choice though few of us liked the hasty move. It was only ten miles away from RAF Defford, a large aerodrome close to Bredon Hill (too close for safety) where nearly all TRE's experimental flying could be done. From the school buildings on the steep side of the Malvern Hills it was easy to distinguish aeroplane echoes from radiation scattered from the ground. Aerials, paraboloids and even gun turrets were soon sprouting from every window or balcony of several buildings.

During the Easter holidays of 1942 it was decided that the pupils of Malvern College should be switched, at zero notice, to Harrow, and that TRE should move its establishment of about a thousand to Malvern on 25 May 1942, leaving wives and children behind. A week after the move, Swanage, which had been left alone until then, was badly bombed.

Although the move to Malvern College brought many benefits, it was as unpopular with the TRE scientists as it was with the inhabitants of the town of Malvern. There was much to be said on both sides. Until that moment, Malvern, which contained many elderly and retired people, was a pleasant backwater that had escaped any bombing. Billeting is never popular and the small, quiet town suddenly found that it had to put up both the airforce and army radar establishments, in the end numbering several thousand. It was natural for Malvern wives with husbands serving overseas to resent the presence of research personnel who could not talk about their work and did not appear to be doing anything dangerous or important.

The situation was aggravated by TRE's practice of working on Sundays which added Godlessness to the general reputation for eccentricity and gave rise to the belief that we were exclusively of the Jewish faith.[5] A. P. Rowe tried to rectify this by arranging a service in the school chapel every Wednesday morning which we were supposed to attend. There was an element of hypocrisy in all this. As I wrote in a letter to Tess in September 1942:

> . . . Nothing much has happened here today except that we have all
> been nauseated by the Superintendent's latest notice about chapel-
> going. He is quite irreligious himself but thinks, or pretends to think,
> that we shouldn't omit anything which might help to win the war and

that there is a chance that there might be something in religion. At least that is the general implication of the notice.

I was lucky in my landladies but billeting raised all sorts of problems for them as well as for us. A graduate member of my small team was asked to enter the house by the tradesmen's entrance and many of those billeted had to be in by 10 p.m.

We had to wait for all meals in an unpleasant, overcrowded cafeteria where there was soon to be an outbreak of food poisoning. To begin with there was enough to eat, but after about a month I wrote that 'everyone is hungry and talking about food.'

I went on a few flight trials of AI Mark VIII at Defford but most of my time was spent in sorting out teething troubles with Mark VIII, trying to ensure that the GEC and E. K. Cole equipment at least looked the same on the outside and that their boxes would fit into both Beaufighter and Mosquito aircraft. This involved frequent visits to Bristol, Malmesbury and Coventry as well as several firms in the London area. I was also beginning to get involved in discussion about the use of radar in defending night-bombers. As a result I became very friendly with one of the Bristol Aircraft engineers to whom the following letter to Tess, refers:

Malvern
4 December 1942
. . . On Saturday I went to Bristol to see a turret designer call Fonsecca. He is an extremely nice and interesting man. He spent fifteen years in S. America, then came back to England and got a job in the Clifton Instrument company, a firm which makes various types of physiological and medical instruments; in 1938 he went into Bristol Aeroplane Co. where he has been most successful as a designer. I went home to lunch with him and discovered that his main interest in life was collecting butterflies and insects. He said that he hoped to go back to S. America after the war to collect for museums which have lost their specimens in the blitz. He showed me some of the collection which was stored in cases piled ceiling high in the kitchen. It seemed strange to open these on a cold and grey day in a thoroughly British suburban house and see the brilliant irridescent blue wings of Morpho butterflies and the enormous Emperor moths which are so large that they seem almost like small owls. Fonsecca of course had grown rather tired of the very big and showy butterflies and most of his affection was devoted to the smaller and more delicate ones – brilliant multicoloured hairstreaks which make you think of elaborately set brooches, and the elaborate

clearwings which remind you of tiny stained glass windows. I of course was thrilled with it all and I think you would have liked Fonsecca and his butterflies too. Usually I am not very keen on collections and should hate to keep them myself but I was quite excited to see these. I don't know whether this was the result of Fonsecca's enthusiasm or the contrast between the thought of the English winter on the one hand and of brilliant butterflies shimmering in tropical sunlight on the other .

The news certainly is very good, especially from Russia. I often wonder whether these Russian offensives are really the product of a master stroke of strategy by Stalin or whether they are not after all produced in the sort of way which Tolstoy describes. Can't you imagine a chapter of *War and Peace* starting like this:

'The official Russian historian uses the battle of Stalingrad to illustrate the brilliant strategy of Stalin and Timoschenko. Realizing the great strength of the enemy Stalin lured the enemy to Stalingrad, held him in three months of bitter fighting and then launched the great counterattack which destroyed so many of Hitler's divisions. This description,' Tolstoy might say, 'is like that of a child who says that it grows cold because the trees lose their leaves. Stalin and Timoschenko may have thought that they planned a brilliant campaign but in reality it was determined by the will and spirit of the Russian soldier.' And Tolstoy would go on to show how events shaped themselves round the seasons, the terrain, the growing anger and offensive spirit of the Russian people and the growing weariness and disillusionment of the Germans. He would point out that the decisions taken by Stalin were merely, as it were, ratifying those involuntarily taken by the millions whom Stalin thought he was controlling.

I never cared much for Malvern which for me retained a strong flavour of artificial gentility. However, like Swanage, it is set in beautiful country where it was possible, and indeed absolutely necessary, to spend an occasional day away from work. I had to make flight tests at RAF Defford and this gave me the opportunity to visit a favourite aunt and uncle, Margaret and Richard Barrow, who escaped from Birmingham at weekends to stay in a seven-teenth century mill on the River Avon at Fladbury. Here the noise of the mill race drowned out the sound of aeroplanes and my friends and I spent the warm summer evenings swimming or learning to paddle a coracle over the weir without upsetting. It was also possible to catch eels in a kind of long wicker lobster-pot known as a 'putchin' which I think dates from the Middle Ages if not earlier.

I spent one Saturday at the mill with my friend Geoffrey Hensby

215

and had arranged to return with him a week later. But this happy plan was prevented by a particularly nasty aeroplane accident which clouds my recollection of the early period of Malvern.

In January 1942 a TRE team led by Philip Dee and Bernard Lovell was given very high priority to develop a 10 centimetre target-finding device known as H_2S for Bomber Command. Lovell set about this with great energy and imagination, and in collaboration with a team led by A. D. Blumlein of EMI had an experimental prototype installed in a Halifax bomber. This was ready for flight trials soon after the move to Malvern. It took off from Defford on 7 June but crashed in the Wye Valley, probably as a result of a fire in one engine, killing all eleven people on board and destroying the only prototype. Those killed included Blumlein, Blythen and Browne from EMI, Hensby from TRE, the pilot, Barrington, who had flown on many of the early centimetre flights, and a crew of six. By then we were fairly used to aeroplane accidents but this one was particularly painful for me because I was a close friend of Hensby's and in addition had the sad task of breaking the news to his parents. This involved driving overnight to Farnborough, a journey on which I narrowly missed a bad traffic accident, and was nearly killed myself.

In spite of this major setback, Lovell's team and EMI made rapid progress and a further prototype was ready for testing in November 1942. The subsequent use of this equipment by Bomber and Coastal Commands is considered further in the next chapter.

Chapter 23

Anti-submarine and town-finding applications

A LTHOUGH I was not directly involved in anti-submarine research, I learnt a good deal about the scientific background to this important development. As my account is likely to be biased towards work done for the RAF, it is worth while saying at the outset that the Navy was the first service to use 9 centimetre radar operationally and that the Admiralty did an excellent job in organizing the development of centimetre valves on a large scale for all three services.

In the early summer of 1939 the Air Ministry put into effect Sir Robert Watson Watt's proposal that several teams of physicists should learn about radar, and contribute to its development by working in the chain of stations that was being set up in the south and east of England. Professor J. D. Cockcroft took his physicists to the main research station at Bawdsey where they studied the detection of low-flying aircraft and also of surfaced submarines, which they found could be detected at a range of several miles. As a result, Admiral Sir James Somerville, who was in charge of Naval Radar and had already stressed the need for a radar watch on submarines approaching Scapa Flow, asked that the radar sets should be put up on Sumburgh Head, Fair Isle and the Orkneys; if possible this work was to be completed by Christmas. Herbert Skinner was in charge of the development at Sumburgh Head[1] and perhaps partly for this reason became enthusiastic about the application of centimetre radar to anti-submarine warfare.

His somewhat uncompromising attitude, which did not endear him to everyone, is illustrated by a poem written by him in 1941, called *Wartime Research*,[2] which ends:

> . . . And so alone
> we, fighting every inch of the way,

217

against those ingrained elephants of inertia,
against the prejudice and the hardened pride
of self-established, self-supporting systems
we fought (through forests thick with self-satisfaction)
to shorter electromagnetic wavelengths.

Skinner's enthusiasm for the anti-submarine application of radar was shared by Dee and Lovell and indeed by everyone working with centimetre waves at TRE. As I have mentioned in an earlier chapter, Burcham, Atkinson and Lovell had assembled a 9 cm radar set in which every effort had been made to maximize the range on aeroplanes or marine targets. The set, which used two 36 inch paraboloids, looked out over Swanage bay with the Isle of Wight forty miles away in the background. Early in November 1940, a party including Naval officers and Admiralty scientists came to see how centimetre radar would perform against a submarine in Swanage bay. At that time radar research for the Navy was done by the Admiralty Signals Establishment (ASE) under the joint direction of a naval officer, Captain Willett, and a civilian scientist, Dr Horton. Both Willett and Horton thought it right to make an extremely conservative estimate of the range at which an echo from the submarine could be seen against noise. Radar echoes fluctuate as the aspect of the target changes, so there was room for disagreement, which is what we had. As the submarine approached Swanage, Lovell insisted that the echo from it was plainly visible when it was quite a long way away, but Horton and Willett refused to make such an entry in their notebooks until the echo was so large that even the most casual observer could detect it. I don't think I actually witnessed this scene, but Lovell's subsequent indignation is clearly engraved on my memory. I also remember meeting Horton and Willett in the Skinners' house near Corfe where they were staying with him. They clearly felt that TRE scientists were a pretty rum lot (I rather think that Herbert Skinner was cutting his own hair when they arrived) but at the same time could not fail to be warmed by Herbert and Erna Skinner's hospitality and charm.

In spite of their cautious and conservative approach, Willett and Horton did not waste any time and the Navy had type 271 sets with magnetrons at sea within six months of the initial trial at Swanage.

In order to understand the anti-submarine application of centimetre waves by Coastal Command, one must remember that

Bowen and Hanbury Brown had developed a successful 1½ metre radar known as ASV Mark II which was installed in several types of Coastal Command aircraft in 1939–40.[3] This set, which enabled patrolling aircraft to find convoys far out in the Atlantic, could also detect submarines, and the first kill with ASV Mark II was recorded on 20 November 1941.[4] Perhaps because of the early success of Mark II there was no particular priority for developing airborne centimetre ASV and for a long time the much more difficult project of AI (Air Interception) was given priority over everything else. However, as the physicists at TRE[5] realized, centimetre ASV had several tremendous advantages in an anti-submarine role. In the first place, the plan-position indicator, which goes naturally with the radially scanning beam of centimetre radiation, is a most impressive aid to navigation as well as an excellent ship-detector. The second advantage is that for technical reasons a short-wavelength radiation is much better at detecting a small floating object than a long-wavelength one. Also, as our Blenheim trials with the magnetron showed in the summer of 1941, centimetre wavelengths give the hunting aircraft sufficient accuracy to bomb blind if that is the order of the day. And even if the submarine hunter has to depend on a visual recognition before releasing his bombs, it will be much better that he should know exactly where to expect the submarine when he emerges from the low clouds that so frequently cover the Western Approaches. Finally, in 1941, 9 cm was a relatively unknown wavelength and there was a good chance that it would take the Germans a year or two to develop the necessary listening or jamming equipment after we had started to use it. A radial scan and narrow beam also made jamming or listening more difficult for the submarine. One or two senior Air Force officers, for example Joubert, Tait and Slessor, realized these advantages, but Coastal Command was felt to be a poor relation and it was hard to get things moving in spite of pressure from Patrick Blackett who was chief scientific adviser to Coastal Command in 1941–42.

After the successful airborne trials of the AI-equipped Blenheim against a submarine in April 1941, TRE applied to headquarters for permission to develop ship and submarine detecting equipment at a wavelength of 10 cm. Hensby, O'Kane and later Kinsey were involved in this development at Christchurch, using first a Blenheim and later a Wellington aircraft. Matters were proceeding at medium priority until tremendous pressure was put on TRE to develop something very like ASV to help Bomber Command navigate and find towns. I remember that Professor Lindemann[6] (Lord Cherwell) visited Christchurch and became airborne after squeezing his large bulk into a small space beside Hensby in a Blenheim equipped with forward-looking centimetre radar. He saw for himself what we all knew, that if you point a mirror at a town like Bournemouth or Southampton, you get a larger echo than if you point it at the ground, and a much larger one than if you point it obliquely at the sea which gives practically nothing. Of course this did not prove that it would be possible to make a real radar map, on which towns and large rivers would show up clearly, but it was encouraging and we were pleased to find that someone as influential as Lord Cherwell should come to see for himself.

To explain why Cherwell was there at all, it is necessary to go back a bit and refer to historical events[7] of which at the time I heard only wisps of rumour.

Late in 1940, Cherwell (then Professor Lindemann) raised doubts about the accuracy of our night-bombing. Subsequent investigation of air photographs by Bomber Command showed that many bombs fell in fields and that two-thirds of the bombers failed to find the target. The navigational aid known as Gee, which had been developed at TRE, would certainly help, but it could easily be jammed and in any case would not reach as far as Berlin. Somewhat similar objections could be raised to the more accurate device, Oboe, then being developed by A. H. Reeves and F. E. Jones at TRE.

In the autumn of 1941, A. P. Rowe, who knew of Lord Cherwell's concern, held a Sunday Soviet on ways of helping Bomber Command find unseen targets at greater ranges than those covered by Gee and Oboe. As a result of this meeting, Dee, Skinner and, I imagine, Lovell, urged that a 9 cm anti-ship set with a radial scan might well be able to provide a map on which one could see towns and big rivers. O'Kane and Hensby's trial flights gave an encouraging answer and Cherwell must have seen enough in his flight to make him give high priority to the town-finding project which subsequently became known as H_2S.[8]

Early in January 1942 Bernard Lovell at TRE was directed to develop the H_2S target-finding equipment for Bomber Command. This was intended for Halifax and Lancaster bombers and the electronic development was to be carried out in conjunction with a strong team led by the well-known radio engineer A. D. Blumlein of EMI who, with HMV Co., were later to be involved in the large scale production of H_2S.

Much of the scientific engineering needed for the centimetre 'ship detector' or target-finder is different from that in the Air Interception equipment. There are also differences between what is required in an aeroplane looking down from the sky and that in a small vessel on the surface of the sea. The matter is somewhat technical but the simple upshot is that in a ship you can use a flat half-cylinder or 'cheese' like those you see today on every small boat, whereas for an aeroplane you need a paraboloid of a rather curious shape in order to ensure that the size of echoes from the ground does not vary greatly with range. Then there were major problems to be solved in synchronizing the scanner with the radial time base in the cathode ray tube, as well as in providing a satisfactory modulator.

With the approval of W. B. Lewis and P. I. Dee, Lovell farmed out these problems to several very able people: J. B. Adams, A. T. Starr and E. J. Denton on the paraboloid; F. C. Williams on the circuitry controlling and synchronizing the radial time base; S. C. Curran on the modulator. He also had powerful support from Blumlein and White at EMI and aeronautical help from the aircraft manufacturers (A. V. Roe and Handley Page) as well as from Nash & Thompson in the construction and installation of a fairly massive scanning device.

After great efforts had been made, a working prototype in a Halifax bomber took off from Defford on 7 June 1942 but crashed and burnt out within 20 minutes of take-off. This is the accident to which I referred in the previous chapter.

Undeterred by the loss of the prototype and several key people, Winston Churchill summoned Lovell and others concerned to the Cabinet Office on 3 July 1942 and demanded that we should have two squadrons equipped with H_2S in operation by October. This was not possible, but TRE, EMI and others in industry managed to have another Halifax equipped and ready for trials by Bomber Command in November 1942, a remarkable achievement considering the tragedy of 7 June.

Fears about using the magnetron over enemy territory delayed the operational use of H_2S until the night of 30–31 January 1943 when it was employed by the Path-Finder Force against Hamburg. Later in 1943 an improved version of H$_2$S was again used by Path-Finders to lead the large scale bomber attacks on Leipzig, Berlin and other cities well outside the range of navigational aids like Gee.

Meanwhile an opportunity occurred to promote the case for using centimetre waves in detecting submarines from aircraft. Unlike the night-bombing of Germany, about which scientists and historians continue to argue, there is general agreement that in 1943 a relatively small number of centimetre radar sets played a decisive part in the Battle of the Atlantic and thus had a major influence on the general course of the war.

In order to appreciate what happened, it is again necessary to go back a few years. After the fall of France in 1940, U-boats began operating from French ports in the Bay of Biscay. By the summer of 1942 shipping losses had reached the alarming figure of 600,000 tons per month.[9] U-boats could be detected by 1½ metre ASV radar but this was not accurate enough for attacks in darkness; U-boats therefore submerged by day and recharged their batteries at night. However, in June 1942 Coastal Command introduced the 'Leigh light'[10] with which U-boats detected by ASV radar could be illuminated and then attacked with depth charges. By August U-boats had given up surfacing at night and usually charged their batteries during the daytime, thus exposing themselves to attack by Coastal Command, in many cases assisted by 1½ metre radar. As a result, shipping losses fell and by the end of 1942 were down to 200,000 tons per month.

The Germans retaliated by fitting the U-boats with a radio receiver which detected the 1½ metre ASV and allowed them to dive before they could be illuminated by the Leigh light. The U-boats then reverted to their practice of submerging by day and recharging batteries at night so by 1943 shipping losses were once more rising at an alarming rate.

Partly as a result of a TRE initiative, a small number of H$_2$S sets were diverted from Bomber Command and modified for operation with the Leigh-light Wellingtons of Coastal Command. These operations, which started over the Bay of Biscay in March 1943, had an immediate effect. The shipping losses which had been increasing at 400,000 tons per month in March, were reduced to less than 50,000 tons in June[11] and remained fairly low for the rest of the war.

According to Blackett,[12]

> . . . the U-boats seemed more frightened of the night attacks than the
> day attacks, for they began again to surface by day and attempted to
> fight off their air attacks by gunfire. In this they had little success and
> they exposed themselves to very heavy and continuous air attack. Later
> they were forced to spend much of both day and night submerged, so
> enormously reducing their freedom and range of action.

The Germans eventually introduced a 10 cm listening receiver,
known as 'Naxos', but by then the number of escort vessels and
searching aircraft had increased to such an extent that the U-boat
campaign never recovered its former intensity.

There is general agreement about the important effect of centimetre ASV in the battle of the Atlantic. Churchill[13] mentioned its
potential importance in a letter to Roosevelt written on 20 November 1942, and other authors (Rowe,[14] Bowen[15]) quote Hitler as
saying in late summer 1943, 'The temporary set-back to our
U-boats is due to one single invention of our enemies'. Subsequently, Admiral Doenitz[16] made clear that radar was what Hitler had in
mind.

The decision to divert a few H$_2$S sets from Bomber to Coastal
Command was a compromise between those who wanted Coastal
Command to have first use of centimetre radar and those who
favoured Bomber Command's claim. The Prime Minister's chief
scientific adviser, Lord Cherwell, and several influential staff-
officers at Bomber Command were so convinced that the war could
be won by 'dehousing' the German working class population with
night-bombers, that they were reluctant to see any diversion of
resources from Bomber Command. Sir Henry Tizard, who no
longer had any great influence in Whitehall (to the regret of many
scientists), and Patrick Blackett who had recently been put in charge
of operational research at the Admiralty, took the opposite view.
Blackett thought that Cherwell's arguments were phoney statistically (they probably were) and believed that the decision to adopt
them was unethical as well as being militarily unsound. Twenty
years later he wrote:

> . . . So far as I know it was the first time that a modern nation had
> deliberately planned a major military campaign against the enemy's
> civilian population rather than against his Armed Forces. During my
> youth in the Navy in World War I, such an operation would have been
> inconceivable.[17]

And on Lindemann, Blackett wrote that he 'even suggested that the building up of strong forces for the projected invasion of France was wrong. Never have I encountered such fanatical belief of the efficacy of bombing . . .'[18] But Cherwell was not alone in his beliefs, for I remember a highly intelligent Wing Commander at Bomber Headquarters expressing much the same view in the spring of 1944.

All the scientists that I knew supported Blackett's point of view and were delighted at the conversion of some H_2S sets to centimetre ASV sets for Coastal Command. Nowadays we should nearly all support Blackett and condemn Cherwell, but one should not be too holy about this. We may not have approved of night-bombing but I can't think of any case where a scientist at TRE refused to help Bomber Command when asked to do so. There is a strange paradox in this whole affair, as Lovell points out in his *Biographical Memoir* of Blackett, where he writes:

> . . . It is ironical that the very policy which Cherwell proposed and pushed through, and to which Blackett and Tizard so fiercely objected, both on statistical grounds and because they wanted the effort to go into the anti-U-boat campaign, quickly led to the development and production of the centimetre H_2S device, which when modified for us as ASV in Coastal Control, had such a decisive effect on the campaign.[19]

Chapter 24

An attempt to defend night-bombers; AGLT

IN THE autumn of 1942, P.I. Dee gave me the general task of seeing how radar could be used to defend bombers, particularly night-bombers. The emphasis was on doing something quickly as losses of aircrew over Germany were soon to approach an unacceptable level. But there was also need for some long-term research in connection with the defence of a new four-engine bomber that was being built by Barnes Wallis's team at Vickers.

In the light of subsequent events, I believe that the right advice might have been to preserve radar silence[1] and to forego the attempt to defend heavy bombers with small-calibre machine guns. However, such advice would certainly not have been acceptable to the Air Staff, who were committed to a major programme of night-bombing with Lancasters carrying H_2S in a large radome in the belly and a powered rear turret with four 0.303 inch machine guns as the main defensive weapon.[2]

There were several other constraints on any proposals that might be made. One of the most serious was that in general we were not allowed to recommend the use of new components, including miniature valves, for equipment to be built in Britain. Wherever possible, such 'long term developments' were moved to the United States, often, as in the case of the proximity fuse, with research personnel as well. The argument, which appealed to many industrialists and civil servants who mistrusted scientific innovations, was that only by sticking rigidly to standard components could industry hope to satisfy the demands of the armed forces. The ban on miniaturization made it impossible for us to design the light and compact equipment needed in several gunnery applications. A more serious consequence was that it reversed the substantial industrial lead that we might otherwise have had in the post-war applications

225

of radar. Even in the short term it was unwise to standardize so rigidly, as was found when the production specifications for a television receiver valve proved unsuitable for computing circuits where we had been told to use it on a large scale.

By 1942, A. A. Hall's group at Farnborough had developed a very neat and compact predictor-gunsight for use in gun turrets and fighters. To make this work, it was necessary to feed an estimate of the range into the predictor. In the Farnborough model, range was found by setting the aircraft type, e.g. Me 109, into the control box and contracting a circular ring of dots until it just matched the wing-span of the approaching fighter. The range was then obtained from the wing-span of the fighter and the angle that its wings subtended in the gunsight. No one thought this a very satisfactory method but the gunner had no time to follow any more elaborate optical procedure. This left watching the arc of tracer bullets as the only alternative method of gun-laying without radar.

The upshot was that there was a strong operational requirement for a lightweight range-only radar which would automatically feed 'range' into the predictor-gunsights of fighters or bombers. The radar had to be fairly directional but it could be of relatively low power as the gunner didn't need a range of more than a mile or two. For this development it seemed essential to use miniature valves and other components available in America but not in wartime Britain. The right course, therefore, was to rely on American research and obtain 'range-only' radar on Lend-Lease when it was ready and we needed it. This made sense because a 'range-only' radar was well suited to the heavily armed day-bombers used by the Americans but not to our more lightly armed night-bombers. From the outset it was clear that the British Air Staff wanted a radar that would enable the guns of a night bomber to be fired blind, i.e. on radar information alone. For Barnes Wallis's new aeroplane, the Windsor, we planned a system with radar in the tail controlling guns in turrets and engine nacelles, but this could not be done in a Lancaster which was where Bomber Command wanted immediate action.

As a result of discussions along these lines, it was agreed that TRE would concentrate initially on the Lancaster bomber where we would attempt to project a cathode ray tube spot into the predictor-gunsight of the rear turret in approximately the true position of the approaching night fighter; at the same time range would automatically be fed into the predictor. I thought we could get sufficient accuracy with a conical scan of 10 centimetre radiation generated by rotating an offset aerial in a 16-inch mirror. However there definitely was not room for all this inside the gun turret. A. W. Whitaker of Nash & Thompson broke through the impasse by suggesting that we link the guns mechanically to a radar mirror housed in a perspex fairing that rotated with the gun turret. The rear gunner would search for the enemy fighter by swinging the turret to and fro in the usual way.

There were several obvious difficulties of which perhaps the most immediate was the absence of miniature components in Britain. I asked Bernard Lovell if he would mind if we stole some power from H_2S, with which it was intended all Lancasters should be equipped. He agreed at once that we should take about a quarter; as he said, that would only reduce the range by 7 per cent. This meant that we could dispense with a massive power pack and modulator and reduce our requirements to a standard transmitter–receiver box and a small control unit in the turret, together with a single fairly large box containing the receiving amplifier and computing circuits in the body of the aeroplane.

There remained other major snags. I felt that if I were a gunner in a rear turret equipped with 7.7 millimetre machine guns, I would hesitate to open fire on a fighter armed with 20 or 30 mm cannon.[3] Bomber Command's answer to such a question would have been: that is our problem, not yours. They also told me that 12.7 mm machine-gun turrets were being developed by Bristol Aircraft and Boulton Paul whom I visited during the course of the next few months. However, these developments were a long way off and there was in effect no satisfactory answer to my question. It turned out eventually that a better way of defending bombers was with long-range fighters such as the Mustang or the Mosquito. So far as I know, bombers designed after the war were not equipped with gun turrets but relied on speed, fighter cover and flying at either very low or very high altitudes. Defence by air-to-air missiles may have become important later, but it was barely thought about in 1942. Our long-term plans for Barnes Wallis's new four-engine bomber collapsed when the geodetic structure was found to be unsuitable for high-performance aeroplanes. The Warwick crashed on a test flight and work on the four-engined Windsor was abandoned at the end of the war.[4]

Another difficulty was the usual one of identification. My

contacts at Bomber Command told me not to worry about this as they had a highly secret but simple method coming along. I learnt later that the plan was to use coded infra-red at the front of all friendly bombers. I don't think this got very far and am doubtful if it would have worked for long. A simpler proposal was to weave or jink the Lancaster and see if the radar contact stuck to the Lancaster's tail. If so, it was probably hostile.

After discussion at TRE and RAE, we decided to project an artificial cathode ray tube image of the radar target into the gunsight with the usual type of spot indicator display. The rear gunner would see the image focused to infinity in more or less the correct position in his gunsight. He would then train his guns on the target and either open fire blind or wait and return fire as soon as the night-fighter opened up. In making these proposals, I was greatly helped by Andrew Huxley who worked in London for the Admiralty Gunnery Division and designed and built a neat model of a gunsight with a cathode ray tube image focused to infinity and reflected from a semitransparent glass surface. This first model was made in his small workshop at his home in Hampstead.

Another proposal which helped things along came from Whitaker who said that Nash & Thompson would provide us with a gun turret appropriately modified for gun-laying with 10 cm radar. They said they could do this within a month or two and kept their word so that by the end of 1942 we had a turret looking out over the Worcestershire plain from the roof of Preston House at Malvern School. Meanwhile the powers at TRE had told me to build up a small team[5] and make a prototype as quickly as possible.

From time to time I had serious misgivings about the whole project but these were no worse than my initial feelings about centimetre AI. On the credit side, the project had the great advantage that its merits would be tested in a clear and definite way. What had to be done was to get the prototype airborne in a Lancaster bomber and check the accuracy in a blacked-out turret by firing a camera-gun that had been fitted in place of one of the real guns. Before that, an observer could see how well the artificial image coincided with the real one, but these trials could only be done in the air.

The first lash-up AGLT was installed in the rear turret of a Lancaster early in 1943. A young graduate, W. R. Beakley, was in charge of installation and flight-testing at Defford aerodrome. This was a tough assignment as fault-finding in an aeroplane was time consuming and the RAF naturally got cross when a four-engined bomber was grounded for any length of time. In February, when Beakley had flu, I had to take over for a week and my letter speaks of working under horrible conditions of mud and cold, with everything going wrong. However, by late March the main faults had been eliminated and Beakley was soon able to take camera-gun photographs showing that in the air the guns could be trained with an accuracy of about half a degree or better.

In spite of the primitive condition of the equipment and the absence of full gunnery trials, the Ministry decided on another crash programme of 200 sets to be followed by a production of several thousand. EMI (or strictly HMV)[6] were the main electrical contractors and carried through the programme with great speed and efficiency. All this kept us very busy in 1943 but I got some relief from taking part in service trials at an RAF station on Newmarket Heath. Today all traces of the aerodrome have disappeared. I now don't know exactly where it was situated but remember that one of the runways ran through the Devil's Dyke, a section of which was demolished on the understanding that the thousand year old earth wall and ditch would be restored after the war, a promise faithfully kept.

I enjoyed the outing, partly because the trials were reasonably successful and partly because it gave me the chance to spend the weekend in Cambridge. After dinner in Hall at Trinity, I made the usual Sunday evening call at the Master's Lodge where I was warmly greeted by George Trevelyan and entertained to find Janet Trevelyan and a friend playing Word-making and Word-taking in French! This seemed very much like old times and helped to restore my confidence in the survival of the world to which I really belonged. I managed to visit the Physiological Laboratory where I talked with Rushton and Adrian and spent a few hours in the Library. There I was interested – and not altogether pleased – to find that Curtis and Cole had pushed on for another two years along the road that Huxley and I had helped to open up in our work on nerve conduction. There was nothing I could do about that but it increased my determination to get back to Cambridge as soon as I decently could.

The rest of the AGLT story can be told briefly. There was one snag, which did not show up until a relatively late stage, when it turned out that the combination of the lag in the gyrogunsight, together with the smoothing lag in the AGLT set, tended to make the system unstable and very difficult to operate. The solution we proposed was to put potentiometers on the turret axes and add voltages proportional to the rate of change of position to those controlling the cathode ray tube spot. I believe that Faulder was responsible for making this work. Otherwise the programme went off all right and a few squadrons of Lancasters were equipped early in 1944. We got a signal about that time to say that it had been used successfully over Germany, but we did not know what that meant. Had the rear gunner really opened fire on an enemy fighter? Had he shot it down? Driven it off? Or what? We were not told and I didn't try to find out.

Soon after this signal, Wing Commander Saward, our main contact with Bomber Command, told me that Lancaster bombers were being used for daylight bombing in the run-up to the invasion of France. As a result, AGLT was being ripped out of the aircraft. Saward was angry about this because he really believed that the war could be won by night-bombing if only we had persisted in it a bit longer. My reaction was one of relief even though it meant that all our hard work had been largely wasted. I had grown utterly sick of the night-bombing offensive and was relieved to hear that the invasion was on the way.

Part of my group continued with the development of a long-term system called AGLT III. They flight-tested a very nice model with a radar in the tail which searched around and automatically locked onto an approaching fighter, thus providing information which would enable guns amidship and in engine nacelles to bear on the fighter. This was a much better system than the one pushed through on the crash programme but it had to wait for Barnes Wallis's four-engined bomber. When this project collapsed at the end of the war, it brought down AGLT with it. But that was much later.

R. Willmer who worked on AGLT in the 1940s and again on AI at Malvern from 1948 to 1955 reminds me of one interesting thing that came out of AGLT. In devising a ground test for the system we coupled a highly resonant echo box to the concentric line between the transmitter–receiver box and the aerial. When properly tuned such a system could store the transmitter frequency for a millisecond or so. If this stored frequency was added to an echo from an object moving with relative velocity v, the combined signal was modulated at a frequency of approximately $2v/\lambda$. This modulation is a beat between the transmitted frequency f and the signal from the moving target which is altered by the Doppler effect to $f + 2v/\lambda$ where λ is the wavelength transmitted and v is the relative velocity of the target towards the transmitter. A similar beat frequency for a stationary target was produced by rotation of the aerial. At the time it seemed to me that these striking effects might have all sorts of applications but I could not interest anyone at TRE in these possibilities. At that stage of the war the British electronic industry was saturated with radar projects and the last thing anyone wanted was any major change in the basic technique. However, after the war Doppler Radar became an important method and the idea underlying it must have occurred independently to many people.

My last contact with AGLT involved my group in a great deal of work without having any impact on the war at all. During the period between 12 June 1944 when flying bombs began to reach London, and the end of August when their launching sites in France were overrun by Allied troops, some 8,600 V1s were launched from France, of which about 2,300 fell in the London area. About 5,500 people were killed and 16,000 seriously injured. It turned out that the most effective form of defence were the heavy (i.e. 3.7 inch) anti-aircraft guns controlled by the splendid American gun-laying equipment, SCR 584; the shells, which were usually fired over the Channel, were fitted with proximity fuses, again supplied by our American allies. In the end, this type of anti-aircraft fire brought down 60–70 per cent of flying bombs approaching over the Channel. Day- and night-fighters also accounted for 10–20 per cent of incoming missiles.[7]

With piloted aircraft, losses of more than 10 per cent would rapidly have brought the offensive to an end, but against missiles one wants something closer to 100 per cent. (Having seen how very difficult it is to achieve a high proportion of 'kills', I remain deeply suspicious of programmes like 'Star Wars' in which it is assumed that an extraordinarily high degree of success can be achieved.)

In 1944 our anti-aircraft defence consisted partly of 'heavies' and partly of Bofors guns firing clips of 40 mm shells. These were virtually useless against flying bombs as there was not enough time to lay the guns after picking up the target visually. Since there were plenty of AGLT sets around, TRE was asked whether a modified set might not be used to control a 'Kerrison' predictor and a battery of Bofors guns. From such calculations as we could make, this looked possible – but not a good bet. One argument in its favour was that the SCR 584 heavy anti-aircraft programme might be defeated if the Germans built a gentle weave into the flight path of the V1s. At all events, TRE was once more asked to launch a crash programme, and Denis Taylor, Bill Penley and I were put in charge of deciding on the substantial modifications and additions to be made to AGLT.

We did get something to work after a fashion, but guided missiles were really needed for that job and I was enormously relieved when the Allies overran the V1 launching sites and brought the programme to an end.

In reflecting over the second half of the war, I have wondered why the Air Staff gave such high priority to our primitive AGLT. Was there any real chance of defending bombers with four 7.7 mm guns in a rear turret? Would it not have been better to rely on accompanying long-range fighters, keeping H$_2$S switched off most of the time? Or use some other navigational aid? Or rely on confusion with Window and other counter-measures?

A partial answer is to be found in the scale of aircrew losses which amounted to 55,000 for Bomber Command throughout the war. In 1943 the chance of being shot down in a raid over Germany averaged about 6 per cent. This does not sound too bad until you realize that in a tour of thirty missions your chance of surviving will only be one in six. I doubt if AGLT would have made much difference, but it may be that when gunners were given a chance to fire back, they would accept greater risks than if they had to sit passively during a long flight over enemy territory. Another unpleasant possibility is that Bomber Command found it easier to encourage their senior commanders if they could say that there was a wonderful new device just round the corner. Perhaps there is no clear answer to these questions, for it seems that massive technical developments of a military nature acquire a momentum that carries them beyond the point at which they are logically justifiable. The best way of preventing such situations is to avoid war altogether – if that is possible.

Chapter 25

A wartime visit to America

TOWARDS THE end of 1940 a group of American physicists persuaded the National Defence Research Committee to found the Radiation Laboratory at the Massachusetts Institute of Technology in Cambridge, Mass. Some of those principally concerned were Vannevar Bush, a former Dean of Engineering at MIT, Karl Compton, the President of MIT; and Lee Du Bridge, the first Director of the Radiation Laboratory.

The growth of the Radiation Laboratory was stimulated in general by the Tizard–Cockcroft Mission and in particular by E. G. Bowen who helped with the detailed exchange of information about the magnetron and other developments in centimetre radar.

After America entered the war at the end of 1941, HMG arranged that a British electrical engineer, D. M. Robinson, would spend most of the year at the Radiation Laboratory where he would organize a continuous interchange of information on developments in radar between Britain and America. Robinson returned to this country every year for a visit of a month or two and during his absence from MIT it was customary to replace him with a British radar expert. Several of my colleagues had visited the Radiation Laboratory for this or some other good reason and I was mildly annoyed that one crash programme after another had made it impossible for me to do the same. However, by 1944, AGLT Mark I was clearly out of the research phase and its development was in the capable hands of a production engineer, C. J. Carter, who had been seconded to my group. Rather suddenly, I found myself posted to the British Air Commission in Washington for about six weeks from mid February 1944. The idea was that I should spend most of my time at the MIT Radiation Laboratory but that I should also visit some of the major electronic firms and perhaps watch air-gunnery trials if the opportunity arose.

I was glad to be leaving England. My relationship with Tess had come to an end and though I saw her from time to time I deeply missed our intimate friendship. It was the sort of occasion on which people advise you to change your job, travel or at least take a long holiday abroad. But none of these options was open to anyone in wartime and, after four years, TRE was becoming an increasingly depressing place. So it was a real pleasure to be looking out at the familiar islands and mountains of the west of Scotland as we sailed away from Britain on a grey winter morning. The weather was not bad, but the long Atlantic swell made the Queen Mary roll and pitch in a dignified sort of way. This didn't have much effect on me, except to keep my appetite down and to make me very sleepy. I wrote that I slept enormously: about eleven hours at night and another hour or two in the afternoon. I shared a cabin with a party of aircraft engineers who were giving information to American firms about the substantial progress we had made with jet engines and fighters. I knew something about this development from a visit made to the Gloucester Aircraft Company during the previous year so we could talk fairly freely.

The ship was not full. Most of the passengers were US Army officers returning on leave, with a sprinkling of ferry pilots and civilians travelling on jobs similar to mine. I made friends with a Norwegian army officer and a large Dutch woman with an attractive placid face. The Norwegian was an athletic, weather-beaten man with a bitter look about him which made me think his family was still in Norway. (I didn't ask.) He fought all through the Norwegian campaign and finally got away from the North Cape by crossing the Atlantic in a fishing boat. I got to know the Dutch woman on the strength of her small son who had to be carried up to boat drill. This was an arduous task since the child was wearing a lifebelt, as I was, and had to be held more or less at arm's length. The Dutch woman (whose name I never learned) came from Java but spoke perfect American. Her husband was an instructor in the US Naval Air Corps and spent most of his time with his unit somewhere else on the ship.

These notes on the transatlantic crossing are typical of a period when people came together for a brief period and then separated, like dead leaves eddying round one another for a moment before being blown apart in an autumn gale in the run-up to a long winter.

I arrived in New York on a grey wintry Sunday at the end of February. As soon as I had cleared customs I rang up Marni Rous, with whom I had corresponded intermittently through the war. By great good fortune, I found her at her flat on East 51st Street and we spent the rest of the day in a long, delightful conversation, interrupted only by a visit to a cocktail party at her parents' home. We had met seven years before, during the year that I spent at the Rockefeller Institute in New York, and again, later, in Connemara where she was staying with her Irish cousins in the summer before Munich, 1938. She had planned to come to Cambridge on a Henry Fellowship in the autumn of 1939, but the war put an end to that project. We kept in touch by occasional letters, but our correspondence grew increasingly irregular although never completely given up. Marni hit off our feelings exactly when she said that it seemed like writing to a character in a novel: someone with whom you had no direct connection but in whose welfare you were intensely interested.

I had to leave almost at once for Washington and then the Radiation Laboratory at MIT, but there were frequent opportunities to return to New York and it was not long before Marni and I decided to get married. At first we were told that there was not the slightest chance of her getting a passage to England. However we persevered, pulling such strings as were available to us.

Getting married and trying to organize Marni's passage was a major distraction, but I managed to get through the work that I had been assigned without too much difficulty. While at the Radiation Laboratory I stayed with Britton Chance whom I had met while he was working in the Physiological Laboratory, Cambridge, in 1939, and who now held a senior position in the lab. It was an odd time to be there because almost every week a senior physicist left the Laboratory to work at Los Alamos or one of the other centres of nuclear research. People were supposed to disappear without comment and you got into trouble if you asked what had happened to X or Y. In America at that time the unmentionable really was unmentionable. It was rather like *Watership Down*, where the rabbits in one particular warren never allow themselves to mention or think about the regular disappearance of members of their community as a result of trapping by their human neighbours. Before leaving for

America I had been asked whether I would join the Manhattan Project but had no hesitation in refusing. This was partly because I disliked working on such a very destructive weapon and partly because there was a reasonable chance that I might be able to return to my own field after five or six years of war service.

For security reasons I left my technical notebooks at TRE and so have only vague recollections of my visit to the Radiation Laboratory. I remember being much impressed by the magnificent anti-aircraft radar SCR 584 which was soon to play such an important part in the destruction of flying bombs, also by an airborne surveillance radar for a naval task force, a forerunner of the modern AWACS plane. More generally, I was deeply impressed by the seemingly inexhaustible supply of people with a college degree in engineering. At the top level, I don't think there was much to choose between our applied scientists and theirs, but when it came to building or servicing electrical equipment, the technicians available had a college degree in America, whereas at home you were lucky to find anyone with a school certificate in physics.

I enjoyed my visit to firms like Westinghouse, Sperry and the General Electric Company though I found occasionally that the engineers resented having to exchange information with a British scientist. I couldn't blame them as America was beginning to forge ahead in radar technology, but felt that they were also beginning to forget how much they had learnt from our early work. It was yet another example of the maxim, 'it's always the good actions you regret most in life.'

Towards the end of March I flew to a large air base, Eglin Field, in Florida to witness a remarkable gunnery trial. The mention of Florida suggests 'keys', everglades and delicious swimming off semitropical beaches. But these happy thoughts proved misleading. If you look at a map you will see that at the base of the peninsula, Florida extends several hundred miles towards Alabama. I have been to Alabama since and found it an agreeable *Gone with the Wind* type of country, but Eglin Field, which lies on the coast roughly midway between Pensacola and Tallahassee, was a really God-forsaken place: an enormous aerodrome surrounded by mile after mile of sandy scrub with only an occasional black family, scratching a living from the infertile soil, to break the monotony. However, the gunnery trial which took place over the Gulf of Mexico, was full of interest.

Someone had persuaded the Mitchell Company to fit a four inch gun into the front half of their B25 bomber. The strategic plan was that the Pacific war might have to be fought in Japanese-occupied China where small boats in rivers and coastal waters would be an important military target. Tactically, the idea was that as the aeroplane swept down onto a naval target it would be able to get off five or six rounds as the range was reduced from 15,000 to 1,000 feet. For this to work, it was essential that radar range should be fed into the gun sight – as had been done in these trials. The demonstration was dramatic, with shell splashes all round the target and the gun recoiling about six feet in the fuselage as each shot was fired. Of course this was really a situation for rockets rather than artillery but radar range would be equally valuable in either situation so the trials were of some general interest.

After a few days at Eglin Field I found that I was trapped there by a transport strike of some kind. This was blow as Marni and I were anxious to get married and had tentatively planned to do this at the end of March. However I managed to cadge a lift on a Liberator bomber that was flying to Boston, and telephoned the good news to her in New York. In those days, long-distance telephone calls, like travel, had a romance that is lacking today. As the connection went through, you heard a succession of voices, all with different regional accents, handing on the call from one exchange to the next. 'This is Tallahassee calling Atlanta for New York . . . This is Atlanta calling Richmond . . .' and so on, across a series of states. The journey too was more interesting than it would be today. I sat in the rear gun turret of the Liberator and watched the countryside unfold, in a way that is only possible today if you are flying in a small aeroplane in a remote part of the world.

Marni and I were married on 31 March, and on the following day we left New York for Washington where I had a report to write for the British Air Commission and she had to see various officials about her passage, passport and the like. One of the strings we had pulled had had the right effect because she was granted an interview with the redoubtable Mrs Shipley who, after some questioning, gave her permission to sail. It turned out later that the effective string had been Felix Frankfurter, the distinguished liberal Supreme Court Judge, who was a friend of Marni's parents. Several years later we called on him to say thank you and learned that he had done the trick by finding himself sitting next to Mrs Shipley at dinner.

Seizing the opportunity, he said to her suddenly, 'Do you believe in love?' and followed up by painting an affecting picture of Marni and me, stressing the importance of this transatlantic union of two young persons and the way that it symbolized the Anglo–American alliance at an important moment in the history of our two great countries, and how imperative it was to allow the young bride passage to Britain.

Originally I was supposed to fly back in a British plane but I knew there was no chance of getting Marni on such a flight so I asked to be sent back by sea accompanied by my wife. The British Air Commission took a dim view of this and simply struck out 'with wife'. Marni got passage slightly before me on an eight-knot convoy, taking eighteen days to cross the Atlantic, while I sailed for home in mid-April on the Queen Elizabeth which was jammed full of troops.

My colleagues at TRE were very nice about our hurried marriage, but I had a certain amount of teasing about my mind not really being on the job – which it wasn't. Joyce Lovell told Marni later that the Atkinsons, with whom I lived, raided my desk when the news came through and were baffled to find photographs of two very different ladies, Marni and Tess, and were very curious to know which one to expect. In the event I reached England at the end of April, about a week before Marni did, but there was no way, either in America or England, that I could discover where or when she was likely to arrive.

After she left New York, I spent my last few days there with her father and mother. In contrast to what one reads in novels about in-law relationships, I found both of mine enchanting people and since that time we have spent many happy weeks together both in England and America.

Marni's father, Peyton Rous, was the head of the Department of Cancer Research at the Rockefeller Institute for Medical Research (now the Rockefeller University). He was already famous for his work on a tumour-producing virus, the Rous Sarcoma virus, for which he received the Nobel Prize in 1966, long after this discovery, once suspect, had been repeated and extended over and over again. Many of us felt that he should have received the prize much earlier either for the virus research or for his pioneering work in the first World War on the preservation and transfusion of human blood. However, neither of these injustices seemed to worry Peyton and it is nice that he lived long enough to receive the prize and enjoy the visit to Stockholm.

Peyton was a man of wide interests and combined his devotion to science with a love of natural history and the countryside, both American and English. He had an optimistic outlook on his own research and frequently thought that he had found the answer to one of the many scientific puzzles that crop up in cancer research. When this happened, he would rush out and buy an antique or engraving for a few dollars which at that time might purchase a Hokusai print or even a Goya etching. Later, when the new theory collapsed, he would console himself with the thought that at least he had the tea caddy or whatever it was. Unlike many scientists, he cared passionately about literary style and wrote an American version of Mandarin prose over which he took much time and trouble. An example can be found in his biographical memoir of Simon Flexner, the pathologist and administrator who directed the Rockefeller Institute for the first twenty years of its existence. Peyton was a strong anglophile but without the snobbery which characterized so many American anglophiles like Henry James, Edith Wharton or the Cambridge historian, Gaillard Lapsley.

Marni's mother, Marion deKay, was seventeen years younger than Peyton and quite different in character, though sharing many of his interests in literature and painting. She had little use for the countryside, however, and always looked forward eagerly to Labor Day when the Rouses would return to New York after spending the hottest months in the Adirondacks or the gentler hills of Connecticut. She combined a remarkable memory with an intense interest in people and family relationships. I remember her pleasure on finding out that I belonged to a large family with several branches and a tangle of cousins, complicated still further by the many names added by my stepfather's family. It was not long before she knew far more about my relations than I did myself. All this helped her to forgive me for stealing away her beloved eldest daughter. We both looked forward to the time when we would meet again, either in England or America.

Chapter 26

The last year of the war

AFTER A week in Malvern without any news of Marni I was growing increasingly anxious. We all knew that the Allies would soon invade Europe and it seemed only too likely that during the weeks before the invasion the German Navy would make a tremendous effort with submarines in the Western Approaches.

After eighteen days at sea, Marni's convoy landed at Liverpool in the pouring rain on the morning of May 11th. She sent a telegram to me which didn't arrive until the following day and then set out on one of those interminable wartime journeys with many changes, waits at wayside stations and prolonged periods in 'those curious British carriages with two doors but no corridor or w.c.'. Somehow or other she arrived at Great Malvern and with the help of a one-eyed Station Master (who also made a pass at her) she finally got through to me on the telephone. I was then living with Jimmy and Mona Atkinson in Upper Welland, some four miles from TRE and Great Malvern and, as so often happened in wartime, the only civilian garage prepared to do repairs, appropriately called Dalleymore, had failed to finish a minor job on my car. However, Jimmy drove me to Malvern where we saved Marni from the Station Master and were soon settled in to The Chace, that being the name of the large, beautifully situated house which the Atkinsons had rented.

It turned out that Marni's voyage though exceedingly slow had been free from obvious danger. Her greatest worry was the safety of several small children, often in her care, who spent their time rushing about the top deck from which the railing had been removed. She shared a cabin with an ebullient Hungarian–American, Sophy, of about the same age and also newly wed. Sophy's family were travelling carnival-proprietors who had been grounded in New Jersey by the petrol (or more accurately, gas) shortage. Here Sophy had met and married a British rating from HMS *Asbery*, Fred, and was now sailing to the UK to live with his totally unknown parents in London. 'If they don't like me, it's their hard luck. It's Fred I'm married to!' Fred had a friend whom Sophy knew to have sailed on one of HM's corvettes and she became convinced that it was the very corvette protecting the convoy, so when it came round on its morning patrol she rushed up to the top deck and screamed, 'Is Soapy Watson aboard?' The result was that the next time the corvette came by, its entire complement was on deck whistling and hallooing and this was followed by a severe signal forbidding civilians to address protecting vessels.

The other passengers on the small cargo boat were an odd mixture of British colonials and some service personnel going home the long way round, a war correspondent or two, and wives and children who had been evacuated during the 1940–1 blitz and were now returning to Britain just in time for the buzz-bombs, had they but known. In this curious situation, suspended between existences, extraordinary confidences were imparted, in Marni's case (for she was a good listener) often culminating in the suggestion that, after all, one more infidelity wouldn't matter, and anyhow who was to know?

The Atkinsons were very hospitable and The Chace, although large, was already full to overflowing, with a TRE couple and baby in the servants' quarters, an aged aunt and a couple bombed out of London in another wing, as well as their two children. I knew that we could not stay there for more than a day or two and had put our names down for a TRE flat, but for these there was a long waiting list and little prospect of anything materializing for many months. The only possibility was a cottage in the village with rooms to let and eventually we found one, rather grandiosely called Ashdene, where Mr and Mrs Hill were prepared to let us have two rooms. We had to share a kitchen with them and were rationed to two baths a week but at least it was somewhere.

I do not think I have ever seen a house of such ugliness and pseudo-gentility. It was like a stage-set for some malignant comedy, with dark ochre walls, a dado of pink and blue flowers, and *The Monarch of the Glen* in a fumed oak frame. I remember my mother, who was not particularly fussy about her surroundings, shuddering with horror at our first matrimonial home. For a while our landlady was pleasant enough but her husband (always referred to as 'e') deeply resented our presence and insisted on our moving out at once in September when he learnt that Marni was pregnant. Fortunately the attic flat in The Chace became vacant at that moment and we were able to live there until the end of the war.

In spite of the squalor and discomfort of Ashdene, I think of that summer with a great deal of pleasure. May 1944 was beautifully fine and we spent the long summer evenings drinking cider in the local pub or listening to cuckoos and nightingales in the woods below the British Camp at the southern end of the Malvern Hills.

Almost as soon as we arrived we used up some of my leave to make a quick trip to Edinburgh to meet my mother and stepfather who had been equally astonished at our precipitous marriage. After getting back I had people to see in London, Farnborough and Oxford and we managed to combine these meetings with visits to my relations who figure entertainingly in the long chronicle that Marni wrote every week to her parents in New York. My life got busier and grimmer after 13 June when the flying bombs started to fall and I had to make several visits to London as well as trying to cope with the ill-fated crash programme using AGLT that I have described in an earlier chapter.

Marni, too, was busier. Like all wives without children, she had to do war-work and after the Ministry of Employment had digested her qualifications as an editor of children's books, they gave her a job at TRE in the film unit. She much enjoyed this though not greatly impressed by the urgency of the work which soon consisted mainly in lettering captions for a film about the Royal visit to TRE.

In mid-September when the V1 launching sites in France and Belgium were overrun, the AGLT crash programme for Bofors guns was abandoned and I started to look round for some new venture. On one of my trips to Farnborough I met two clever but somewhat dishevelled characters, Philpot and Benson, who had designed a steerable rocket small enough to go in a fighter aircraft. They wanted either radar or infra-red to go with it. As TRE was prepared to sanction this work, Rutherglen and I started thinking about beam-riding or homing projectiles and even went so far as to build a certain amount of hardware to go with the missile. This resulted in my keeping a watching brief for TRE on guided-missile control in general and I had to attend several interservice meetings in London on that subject. I was relieved to find that at last an effort was being made to get miniature valve production going on a large scale. This was a project in which W. B. Lewis (the No. 2 at TRE) took a keen interest. But although the guided missile work was intellectually interesting, I could not rouse up much enthusiasm for research which seemed unlikely to be useful in the current war.

After moving into the attic flat at The Chace in the autumn, Marni and I felt that for the first time we had a room of our own. I had been living in suitcases for five years and it was a great relief to spread out books and papers and think about the experimental work on nerve that I had been doing before the war, or make plans for what might be done in the future. As the winter drew in, we no longer wanted to wander in the woods or drink cider in the pub below the British Camp. Instead, Marni worked on a detective story, *Student Body*, about her college in America, Swarthmore, and I returned to the analysis of two sets of 1939 experiments, one with Andrew Huxley and the other with William Rushton. We had a surprising number of visitors: my brothers, Robin (from the Sudan) and Keith (an MD in the Navy), my mother from Edinburgh, Andrew Huxley from the Admiralty, Philpot from Farnborough, Heywood Broun and Bill Fleming from the US Army, are some that I remember. We had also made friends with the doctors and some of the patients at the large psychiatric hospital which the US Air Force had built in wooden huts on land adjoining the garden of The Chace. When this first got going, we were told not to worry unless we saw (as we did one day) someone dressed in purple pyjamas. In that case we were at once to ring the camp and if necessary collect the Atkinson children from the garden. But most of the patients were ordinary young men suffering from the various neurasthenias that come when casualties run at 10–20 per cent as the American losses did when they started the day-bombing of Germany.

In the summer of 1944 food rationing was very severe and I remember the extreme pleasure of drinking whisky and (simultaneously) eating luscious rich chocolate cake when we went to a dance at the American Camp.

As the war went on, it grew increasingly complicated to carry out the ordinary tasks of life. Shopping, cooking and travelling were all beset with difficulties. Marni found that the same was true of pregnancy and was soon to hear horrid tales of the local maternity home in Malvern. It seemed that you were supposed to start the birth there but that if anything went the least bit wrong, you were rushed ten miles to Worcester in a Police ambulance; Black Maria is mentioned in one of her letters home but I think that must be an exaggeration. There was the added difficulty that her father's position as an eminent medical research worker had always assured her of the best medical assistance that was available in New York. We knew that most Americans regarded British nursing homes as Dickensian institutions and felt sure that Marni's parents would be horrified at the idea of her having a baby in Malvern even if we were happy ourselves – which we weren't.

A satisfactory solution to the problem came on one of our visits to Oxford. Ever since I was a schoolboy I had had a great affection and admiration for my stepfather's youngest sister, Barbara, and her husband, Hugo Cairns, a brilliant brain-surgeon and Nuffield Professor at the Radcliffe Infirmary in Oxford. We inquired tentatively of them whether there was any possibility of Marni having the baby at the Radcliffe maternity wing. Hugo thought he could fix that up, and not only did he do so but Barbara invited Marni to stay for a week or two before the expected birth date. Thanks to their great kindness, all this worked out as planned and our first daughter was born in Oxford early in April 1945. By then it was almost time to leave Malvern and start a new life in Cambridge.

III

STARTING AGAIN

Chapter 27

Malvern and Cambridge, 1944–5

An early visit to Cambridge

By the autumn of 1944 the V1 launching sites had been overrun by the Allies, the invasion of Western Europe had gone reasonably well, if not exactly according to plan, and I was once more able to work normal hours and take such leave as I was entitled to. It also seemed likely that the European war would be over fairly soon and, having done five years war service, that I should be able to resume my Cambridge career if the University asked for my return. At about this time my boss, Philip Dee, wrote to one of the powers at the Ministry of Aircraft Production saying that I was basically a research man and that 'it would be a tragedy' if I were not allowed to continue in my own line. These were some of the reasons why Marni and I felt we should spend as much time as we could in Cambridge, something that we wanted to do in any case. Such trips were helped by the fact that I usually had business to transact with firms in Cambridge, or with the airfield at Newmarket, and so was able to use our increasingly decrepit car for the journey.

Cambridge was very pretty in the early autumn weather, which pleased me; I wanted Marni to see it at its best, as it was a place she remembered only dimly from the year 1926–7, when her father spent a sabbatical year in Cambridge, working in the Pathology Laboratory, and she went to school at Miss Tilley's in Millington Road.

We stayed in comfort at the Garden House Hotel, then a pleasant Edwardian building located in a rather romantic site close to the river. We met several of my friends and colleagues, Nora David, Andrew Gow, Adrian, William Rushton and Rawdon-Smith all of whom obviously took to Marni and urged our speedy return, though the general view was that finding a house might prove difficult and expensive.

The war must have been a hard time for Adrian, who was then Head of the Physiological Laboratory, in which I hoped once more to resume work when we got back to Cambridge. In the summer of 1942, while on a short holiday in the Lake District both he and his wife Hester were involved in a climbing accident, in which Hester's right leg was injured so badly that it had to be amputated above the knee. Hester did not allow this to interfere with her many private and public activities, and although never able to walk far was often to be seen bicycling about Cambridge. This accident was a source of great distress to both the Adrians, and although neither complained, the mechanics of life in war-time must have been difficult. Adrian took on all the shopping and could often be seen bicycling at high speed or rushing into the laboratory with a rucksack full of provisions on his back.

There was a great deal to be done in the laboratory. Medical students had to be put through a quick two-year course by a skeleton staff and there were practically no demonstrators or research students to help with experimental classes. Adrian lectured to both first and second-year students and in addition taught elementary physiology to student nurses at Addenbrooke's Hospital. He was also involved with experimental work on nerve gases and with meetings of the Chemical Board in London. On top of all this he managed to do much interesting work on the sensory cortex of various animals, some of which is summarized in his excellent book *The Physical Background of Perception* (Adrian 1947). Adrian was always reluctant to unload work onto other people but when I got back to Cambridge in August 1945 he asked me to take on the lectures to nurses, which I was glad to do, not least because it made me brush up my knowledge of physiology.

I had several things to discuss with William Rushton, with whom I had collaborated before the war on experiments on single nerve fibres from lobsters. These were almost large enough to qualify as giant nerve fibres, but unlike those from squid they could be obtained simply by buying a live lobster from a fishmonger and dissecting the nerve supplying one of the walking legs. The edible qualities of the lobster were in no way impaired by this procedure and I became very popular with my friends on the days that I was likely to be experimenting.

248

The object of our experiments was to examine the distribution of electric currents in a single nerve fibre with stimuli that were too weak to set up a nerve impulse. We hoped that the distribution and time course of the voltages we recorded would conform to the theoretical predictions of cable theory and, if they did, that we should then be able to calculate three basic constants which are important, though not in themselves sufficient, for the development of any electrical theory of nerve conduction. The three quantities we wanted were the specific conductivity of the protoplasm inside the cell, known as axoplasm, and the capacity and conductance of the cell membrane per unit area. The membrane conductance is not a constant in the simple sense since it varies with the applied voltage, but it can be treated as constant if it is measured with sufficiently small voltages, as we had done in our experiments.

The first step in the analysis was to make enlarged tracings of the experimental records and compare them with the theoretical curves whose analytical form had been worked out by Rushton and by Campbell and Foster (1931). I proposed to do this in the evenings and weekends at Malvern and for this purpose took away an enlarger and the experimental films from my room in the basement of the Cambridge Physiological Laboratory. At the same time Rushton said that he would compute the family of theoretical curves that I would need for the analysis.

I also wanted to get Rushton's opinion of a theoretical paper that I had written while crossing the Atlantic in April 1944 in the Queen Elizabeth, in theory a fast passage. In fact, although the Queen did travel at high speed she must have followed a circuitous route because the whole passage took seven or eight days and the weather changed in 24 hours from subtropical to almost arctic as if we had steamed north for a thousand miles. There were 18,000 US Army troops on board, and the ship was so crowded that there was nowhere to sit down; eating was confined to two large but very hurried meals a day and there was absolutely nothing to do except lie on your bunk reading, writing or playing cards. This leisure forced me to crystallize an idea that had been buzzing in my head for some time.

Provided one assumes that nervous conduction is brought about by local electric currents and makes some drastic simplifications, it is not difficult to calculate the conduction velocity expected in continuous 'unmyelinated' nerve fibres such as those in most invertebrates or the smallest nerve fibres in vertebrates. I tried several different models and found that in every case the predicted velocity was inversely proportional to $\sqrt{(r_i + r_o)}$ where r_i is the resistance per unit length of the core of the nerve fibre and r_o that of the fluid outside the nerve fibre. The external resistance r_o is usually negligible, but, as I had shown in my 1938 experiments, if it is raised artificially the conduction velocity can be reduced by about the amount predicted by the simple formula quoted above. A related prediction made by all the models I tried was that if fibres of different size have identical membranes and axoplasm then their conduction velocity should be proportional to the square root of the diameter. This led me to seek a general proof of the two square-root laws which depended only on the assumptions that the nerve fibre is capable of propagating an electrical wave without decrement and that conduction is brought about by local circuit current. To begin with, Rushton was doubtful about the general validity of my dimensional argument and partly for that reason, partly because I had other things to do, it did not get published until 1954. By then Rushton had accepted the argument and I was able to formulate the theory better than I had done in 1944. In this connection it should be said that whereas velocity should be proportional to the square root of the diameter in continuous unmyelinated nerve fibres, the proposition does not hold for the myelinated nerve fibres of vertebrates, where conduction is saltatory and the impulse skips from one node to the next. Here, velocity varies as the first power of the diameter rather than with its square root, the main reason being that the thickness of the myelin sheath which insulates the internodes increases with fibre diameter (Rushton 1951).

During this visit to Cambridge I met Rawdon–Smith who had designed and helped build the amplifier and associated equipment that I had used before the war, and left at Plymouth in August 1939, in the vain hope that there would be no war and that Andrew Huxley and I could continue our exciting experiments on the giant nerve fibres of the squid. In 1940 Rawdon borrowed the entire electronic rack, which was fortunate as otherwise it would have been destroyed in the air raid which demolished the part of the Plymouth laboratory in which I had worked. In Cambridge he told me that he planned shortly to move to America, but promised to return my equipment before he left. I had been a bit reluctant to lend it in the first place, but unlike most good actions this one paid off handsomely. Not only did I get my own equipment back intact instead of smashed to smithereens in an air raid, but I had it multiplied by three as neither Rawdon-Smith nor his colleague Rowan Sturdy had any further use for their own sets. Yet another incomplete set came my way, as Kenneth Craik, the brilliant neuropsychologist whose ideas are still important, was killed tragically in a traffic accident on the last day of the war as he bicycled home from his laboratory in Cambridge, and no-one seemed to have any use for his equipment. These sets proved to be extremely useful as the d.c. amplifier was much better than anything that could be bought commercially, even though designed in 1939. So far as I remember, Andrew Huxley, Richard Keynes and I each had one set, and we kept one for visitors, of whom Silvio Weidmann from Berne was one of the first. But in the autumn of 1944, with the Germans retaliating vigorously at Arnhem, one would have needed to be very optimistic to imagine any such development within a year or two.

I had one more, slightly melancholy, task in Cambridge. In my absence, my beautiful rooms in Nevile's Court had very reasonably been allocated to someone else, and I found that all my possessions, furniture, books and papers had been dumped in a gloomy room in Whewell's Court. A good many things seemed to have disappeared but Marni and I decided that what was left would be a big help if and when we could find a house.

Winter 1944–5: Malvern

Towards the end of October 1944 when Marni and I moved into the attic flat at The Chace, I at last had time for doing my own work in the evenings and at weekends, and decided that my first priority was to write up and illustrate a paper with Andrew Huxley on the work we had done together in 1939. This did not take long as our main conclusion, the same as that implied in our 1939 *Nature* note and strengthened by Curtis and Cole (1942), was that during the nerve impulse the membrane potential reversed by 40 millivolts or more, instead or falling to zero as previously supposed. I finished the first draft by mid-January 1945 and gave it to Andrew Huxley when he joined Marni and me for lunch at the Café Royal in London, looking very smart in naval uniform; afterwards we walked to the Admiralty where he worked on gunnery and I had a meeting. We arranged that he would spend the first weekend in February with us in Malvern so that we could revise and if possible finalize the paper. All this worked according to plan. The *Journal of Physiology* made a nice job of the figures and the paper appeared within about six months.

Later on we both came to regret the discussion in that paper and I have often been asked why we did not mention the simple idea that the reversal of membrane potential might be due to a selective increase in sodium permeability, or whether we had thought of this idea at that time. Without documentary evidence it is dangerous to answer such questions from memory, but I know that things looked rather black for the sodium theory both then and several years later.

In the first place there was the report by Curtis and Cole (1942), which later proved incorrect, that the action potential of *Loligo pealii* might exceed the resting potential by as much as 110 mV; on the sodium theory this required an internal concentration of less than 6 millimolar, which compared unfavourably with the value of 270 mM obtained by subtracting potassium from total base in the analyses of Bear and Schmitt (1939). There was also a preliminary report which again proved to be wrong, to the effect that the action potential and resting potential of a squid axon were unaffected by removing all ions and circulating isosmotic dextrose (Curtis and Cole 1942). This seemed to fit with the well-known observation that frog nerve would continue to conduct impulses for many hours in a salt-free isotonic sugar solution, a result now known to be due to an impermeable layer in the perineurium.

A winter visit to Cambridge, 1945

The paper with Rushton on *Electrical constants in a crustacean nerve fibre* entailed much laborious analysis and did not get finished until the autumn of 1945. During the intervening period I made several visits to Cambridge, partly to talk to Rushton, partly to look for a house and partly so that Marni could meet various people we had missed on our autumn trip in 1944.

Our next visit was made early in January in appalling weather with snow so deep that it was difficult to see the edge of the road and the surface so slippery that on one occasion, though moving quite slowly, we managed to skid right round. This was somewhat alarming as Marni was six months pregnant, but as it was our last opportunity of visiting Cambridge for some time we thought we had better take it.

To begin with the visit did not go too well. The only house we could find was unattractive and, we thought, hideously expensive, and Marni was coming to the conclusion that Cambridge was very much a man's world. However, after a day or two things cheered up largely because we spent some time with Barbara Rothschild who, as Marni wrote to her parents, was the perfect antidote to her depression on the subject of drear Cambridge female society.

Barbara's marriage to Victor Rothschild seemed to have come unstuck during the war, but for the time being she was living at Merton Hall, where we spent an entertaining evening. It turned out that Barbara's youngest daughter had been born at the Radcliffe Maternity wing in Oxford where Marni planned to have her baby in April; also that Marni was going to the gynaecologist, Mr Stallworthy, who had looked after Barbara and of whom she had a high opinion. She thought the Radcliffe would probably remember her because she cut open her mattress with nail scissors. Why? 'Because the place is all very nice, but the mattresses! The mattresses are *that* thick and stuffed with coils of black hair, the hair of Jude's first wife, you know, the pig-sticking one – Arabella, yes, stuffed with Arabella's hair, so I just cut mine open to find out what made them so hideously uncomfortable. The nurses were furious, but Mr Stallworthy was most sympathetic and said that something ought to be done, so perhaps they'll be easier now.' There was more of the same kind with much half-true gossip about people we knew or would like to know. Her society and hospitality (at Victor Rothschild's expense) were to enliven our lives for the first six months that we spent in Cambridge, but at the end of the day we were a little relieved to escape from the tangles of her private life into our own more humdrum existence.

Spring and early summer 1945: Malvern, Oxford, Cambridge

February and early March remained bitterly cold, until suddenly the Eastnor woods were full of wild daffodils and it was time for me to drive Marni to Oxford, where she was to stay with Hugo and Barbara Cairns until she had to move into the Radcliffe. Our friendship with them, already close, was strengthened by Hugo's

recent visit to the USA in which he stayed with Marni's family in
New York. We had given Hugo their address and he came to call,
but their liking for each other was so instantaneous and mutual that
he never returned to his hotel and spent the rest of his time in New
York with them. This pattern was to be repeated with several of our
friends. New York is a beautiful city and its hotels are elegant and
comfortable, if expensive. But there is something impersonal about
the whole set-up which made it most attractive to move into the
book-lined tranquillity of the Rouses' apartment at 122 East 82nd
Street,as many of our friends and relations found during the next
thirty-five years.

In addition to being England's leading brain surgeon, Hugo was
pushing two medical developments of great importance, both in the
war and after it. The first was the practice of always wearing crash
helmets on a motor bicycle, and the second was the widespread use
of penicillin and related compounds which had been developed in
Oxford by Chain and Florey during the previous ten years. I also
heard about the work of Young and Medawar on nerve regener-
ation in which Cairns was naturally very interested. All this helped
to turn me back from physics to biology which is where I really
belonged.

Our first daughter Sarah was born on 3 April and I managed to
get to Oxford a few days beforehand. In those days husbands were
kept well away from the delivery ward, and there was nothing for
me to do except give moral support and take Marni to the hospital
at the right moment. At that time it was also the custom for the
mother to spend a week in bed instead of jumping up almost
immediately after the birth as women seem to do now. As we
both had many friends and relatives in Oxford, Marni was soon
seeing a stream of visitors, including the redoubtable Mrs A. L.
Smith (my step-grandmother) who managed to gate-crash her
way to Marni's bedside almost as soon as she was out of the
delivery room.

Owing to the great kindness of Barbara Cairns when Marni came
out of the Radcliffe she spent another fortnight in Oxford enjoying
the spring sunshine in the Cairns's garden in Charlbury Road before
plunging back into the rigours of Malvern life.

In early May I spent another few days house-hunting in
Cambridge, again with no success in the city itself. However, I did
find a beautiful Queen Anne house in Linton, about ten miles away,

with which I temporarily fell in love. Fortunately Marni was dead against it from the start and I rapidly came round to her point of view. With small children at home and me working late in the lab or college, it would have been difficult at the best of times, but disastrous with petrol rationing lasting for several years.

My next journey was to Plymouth to collect the remains of my apparatus, some of which was still usable. This made me realize just how fortunate it was that Rawdon-Smith had borrowed my electronic rack as the bombs had made a sad mess of anything fragile. I took what was left back to Cambridge and made another tour of the house-agents, this time with more success for I found a medium-sized house in Bentley Road that I thought would be suitable, though 'terribly expensive' at £4,300. I rang Marni at once and we agreed to make an offer, only to find that the house was already sold. However, a few days later the house agent rang up to say that the sale had fallen through and that we could have 17 Bentley Road if we made an offer at once. This we did, and on 5 June the house was ours. It had four bedrooms and three living rooms including a small study. There was also a summer-house in the garden, which was used in a somewhat extraordinary way during the following winter. There was a large-ish garden, described by the house-agent as in need of attention, which, translated, means that it was a sea of weeds, the lawn having been dug up for potatoes in 1940 and then left to its own devices. But it was OUR HOUSE and served us very well for eight years.

I was due for release from TRE on 11 June and had no intention of spending one second longer there than I had to. But we neither of us minded staying at The Chace and it was convenient to be there until July when the Atkinsons moved out and the property was sold. We decided that Marni and Sarah would go to stay with the Barrows in Birmingham for three weeks and I would put in a little time in Cambridge, hoping thereby to do something to justify the University's request for my urgent return.

On 1 August, Marni, Sarah and I would move for six weeks into Ashton House, the small very pretty house in Newnham belonging to Dick and Nora David, but not needed in the summer holidays while Nora and her two children were in Cornwall (Dick was in the Navy). We were to get vacant possession of 17 Bentley Road on 20th August but the inside of the house had to be painted and this we proposed to do ourselves, with help from my mother, as the maximum amount you were allowed to spend on any repair or decorating job was £10!

Summer and autumn 1945: return to Cambridge

Accidents or illnesses are apt to play havoc with such complicated plans but nothing went wrong and we spent a lovely two months getting our new house ready in time for the first Michaelmas Term of peacetime.

At the beginning of August I picked up Marni and Sarah and we settled into Ashton House. I had given Marni a somewhat optimistic picture of the glittering society that we would meet in Cambridge, but to begin with no one very interesting turned up except that we had a pleasant visit from Hugo Cairns and Adrian, for whom Marni somehow managed to provide an excellent lunch. Food seemed to be growing increasingly scarce, and I remember the tragedy when a food parcel from New York was damaged and we found that a packet of Ivory Soap had burst into the rice. However thoroughly we washed the rice we could not eliminate the taste of soap.

Adrian was working on the semicircular canals of pigeons at the time and offered us the dead birds when he had finished with them. He explained that they had been anaesthetized with nembutal, but that drugs worked on a weight basis and that the nembutal would have no effect on us. We were happy to take him at his word and found the pigeons delicious, particularly when cooked in prewar Trinity port.

On 6 August the first atomic bomb was dropped on Hiroshima, at which, according to Marni's letter my immediate comment was, 'Good heavens, they've got it to work after all.'

A few days later the war was over, with VJ-Day on 15 August. I came home from the lab to a late lunch to find that Marni had had a splendid morning, sitting in the sun across the road in the pub with Dadie Rylands of King's and Maurice Bowra, later Warden of Wadham College, Oxford, both brilliant and amusing talkers. We fixed a baby-sitter for Sarah and arranged that we would all four have dinner at the Pike and Eel at Overcote, a pleasant prettyish pub on the river Ouse in the depth of the Fens. Like most restaurant meals in those days, the food was indifferent, but we drank a great deal of beer and talked and talked, ignoring as best we could the VJ-Day celebrations that went on around us.

Eventually we got access to 17 Bentley Road and began the business of painting the interior in which we were greatly helped by my mother who spent a fortnight with us in Ashton House. We walked over every day to Bentley Road with Sarah whom we parked in her pram in the garden. I am afraid that I was not all that much help, partly because I was lecturing to nurses and had many things to do in the lab, but more particularly because I was so pleased to be on my own in my old laboratory room that I seem to have been in a permanent state of absent-minded euphoria, which Marni described in the following way in her weekly letter to her parents:

> *1 September 1945.* Alan proceeds slowly with his study . . . partly because of a great inclination to return to the lab after tea. He is like a dolphin that has suddenly been released . . . into the open sea. He plunges and gambols and cavorts in pure research after so long, and spends such time as he has to think of other things in mourning that he must do so much teaching when term begins again. To be sure this plunging and gambolling in the home mostly takes the form of prolonged brown studies. He may be here in body, but in mind he is far away on planes of higher mathematics from dawn to dusk and even I am sorry to say before dawn as he wakes about five thirty and then can't go back to sleep . . . What is your cure Daddy? . . . Outside of worrying about his loss of sleep I am happy to see him so absorbed, and overflowing with new ideas and the opening out of new avenues to explore.

What Marni says here is picturesque but roughly true. I had forgotten how long it takes to make a new technique work and, to begin with, rushed madly from one wild idea to another, trying to make bimolecular films or a special diffuse electrode one day, and then dashing on to try something else when what I had thought of didn't work at once, or proved too difficult. After a bit I came to my senses and settled down to a pedestrian experiment which eventually, and with Andrew Huxley's help, led to interesting results. But that comes later.

For some time, I had been hoping that Andrew would obtain his release from the Admiralty and resume his Research Fellowship at Trinity College. In November he wrote that he hoped to return early in the New Year and we spent a happy weekend at Bentley Road discussing research possibilities.

By a curious coincidence, Barbara Rothschild had fallen in love with Islay, the island in the south-west of Scotland where my family used to have a house. She spent a few weeks there in the summer of 1945 and returned in September with two Islay girls aged 15, one of whom, Ella, she proposed to apprentice to the Rothschilds' senior nurse, Peggy, who as it happens was also from Islay. At the last minute Ella's friend Jean Currie, who like Ella had just left school, decided she would like to come too, so Barbara brought both girls south, thinking that Jean might work for us – a proposal that we naturally accepted at once, as Jean, though completely untrained, was extremely nice and rapidly became devoted to Sarah.

Both girls came from tiny crofts where all the cooking was done on an open peat hearth. They normally spoke Gaelic in the home and their English had that attractive admixture of half-French words that one used to meet in Scotland – for example, a plate is called an ashet (assiette) and a leg of lamb a jiggot (gigot).

Barbara was extremely funny about the journey south with the two girls who, except for a day excursion to Glasgow, had never been off the island before and were thrilled to see a steam engine for the first time. According to her account, the first man Jean saw had a wooden leg, whereupon Jean said, 'But Mistress Rothschild, Islay men have two legs.' Later on we found that Jean sometimes referred to papples or papple-trees. On enquiry it turned out that Jean must have seen a quince tree and that Barbara out of mischief said, 'Oh that's a papple – halfway between a pear and an apple you know.'

Jean stayed with us for nearly two years and was a tremendous help. Nowadays a mother's help seems an unnecessary and expensive luxury, but at that time unless the husband had time to spare it was a virtual necessity – no washing machines or disposables of any kind, petrol rationing, food shortages and the necessity of bottling or preserving are a few of the many reasons that come to mind. In many ways the extreme simplicity of Jean's upbringing was an advantage, but it did have drawbacks. We couldn't help being secretly amused when the Rothschildren were found to have caught lice from Ella, but it was a different matter when Marni who had hair three feet long, found that she suffered from the same complaint. 'Do you think,' she said gently to Jean, 'that I might have caught them from you?' 'Och no, Mistress Hodgkin,' was the reply 'I have no lice.' And then, reflectively fluffing out her hair, 'But I have tons of nits.'

Early in October we went to lunch with the Master of Trinity, George Trevelyan and his wife Janet. Just as we had started, another couple, Harry and Mary Sandbach, arrived. 'Your responsibility, my dear,' said the Master. 'I always put these things down in my little book,' and we proceeded calmly with an enlarged party. Afterwards we wandered round the portraits in the Lodge and stopping in front of Marvell, Marni said, 'I almost wrote my undergraduate thesis on him.' 'Did you, by gad!' said Trevelyan, brightening. 'Finally I chose Henry Vaughan, but it was on the social history side and somewhat disapproved of by the English Faculty.' 'Ah,' said Trevelyan, 'I, too, am interested in Social History.' Before we left, Trevvy drew me on one side and said in his gruff warm-hearted way, 'Alan, I don't think your wife is exactly beautiful, but she has a very *nice* face.'

A few days later we had tea with Dennis Robertson, the Professor of Economics, described by Marni as 'a most charming shy lovely-looking man', and with him were Roger Mynors and Mounia and Cynthia Postan, with all of whom we quickly established close and lasting friendships.

At the end of October Marni mentions that we have just had Michael and Anne Grant to dinner and that today, Sunday, we had Dennis Robertson and David Hill to lunch, but that the occasion was complicated by the unexpected arrival for midday drinks of Barbara Rothschild and her party of house guests which included Ludo Kennedy, Peggy Ashcroft and Cecil Beaton. Marni and I speculated for a while as to why we should have received this high-powered visit and came to the conclusion that it was 'to see how the poor live'.

We saw a lot of Barbara Rothschild, and Victor came to stay from time to time but the two were not supposed to meet as they were getting a divorce, Barbara's infidelities both before and during the war having proved too much for Victor. My old flame Tess Mayor, whom Marni liked very much, also came occasionally, though not with Victor. This annoyed Barbara who felt it disloyal of us to invite Tess, because she thought, rightly as it turned out, that Victor and Tess might be planning to get married.

Barbara liked talking to Marni, partly because she was a good listener and also, we thought, because she was the only one of Barbara's friends who didn't know her in the old days and so could credit her picture of herself as a reformed character who might have resumed the marriage with Victor successfully if he had been willing.

Victor himself was most anxious to spend some continuous time in Cambridge so that he could restart his experiments on fertilizing trout eggs, of which he had a batch coming early in January. For the time being Merton Hall was clearly impossible and, as we knew he loathed the idea of staying in a Cambridge hotel, we suggested he move in with us, using the summer-house as a bedroom for the times when we needed the small guest room. Somewhat to our surprise, we found him an easy person to have in the house and, as he owned a farm nearby, he was a great help on the food front, which was growing increasingly difficult with the ending of Lend-Lease. He took to sleeping in the summer-house with calm, and even with the appearance of enjoyment, in spite of the cold and damp to which it was subject. We kept the electric fire going all the time , but this 'made almost no difference to the cold of last week'. From time to time Victor invited us for a weekend in the pampering comfort of Ranger's Cottage in Tring park – a very pleasant rest and change for all of us. We came to the conclusion that although Victor liked to be as comfortable as possible it was not essential for him, and that if necessary he would happily put up with conditions even as primitive as those to which he was subjected that winter.

Chapter 28

Research and teaching: Cambridge, Plymouth, 1945–7

Rebuilding my equipment; teaching

Although I was very keen to start research it was as difficult to get going in the Laboratory as it was to set up house. In six months' time the universities were to be flooded with a mass of war-surplus equipment, but in August 1945 there was nothing in the Laboratory and very little in the shops. I remember hunting for a piece of insulated sleeving. Eventually I found what I thought was a suitable piece in a drawer of miscellaneous electrical components. I was surprised to find that the object I held in my hand was tapered and seemed stiffer than I expected. I looked at it with a binocular microscope and was amazed at the intricate design that I saw. Why on earth, I thought, should anyone go to so much trouble to make a flexible tube out of those beautifully articulated joints, and how difficult it must be to mass-produce such an object. Then I realized that what I was looking at was a lobster feeler that had fallen into the drawer and lain there since the days when I had worked on lobsters before the war. There were many other difficulties. But by cannibalizing equipment and other expedients I managed to get my apparatus working well enough to start experiments on crab nerve fibres by the early autumn.

One of the teaching jobs I would have to do would be to help look after the advanced nerve–muscle class in Physiology which then involved only about ten students. I thought it would be nice to provide enough oscilloscopes for the class to look at nerve action potentials for themselves, and after ordering some relatively cheap oscilloscopes, I had fun designing a very simple stimulator and amplifier to go with them.

Adrian had obtained my release from military service on the

261

grounds that he needed help with teaching. One of my first jobs was to lecture during the summer vacation to student nurses on Human Physiology. This was good practice but not enjoyable. The nurses were in the charge of a fearsome-looking matron and, however hard I tried, I couldn't get a flicker of a smile, or other sign of interest, out of them. I felt better when Adrian, who had been giving the lectures before, told me that he'd had the same experience.

Adrian let me off with a relatively light teaching load, but I found it much harder to give tutorials in Trinity College than before the war. This was partly because I had forgotten a good deal and partly because I had ceased to believe in some of the principles that had once seemed to hold physiology together. The constancy of the internal environment remained as important as ever, but the ways in which constancy was achieved had become more complicated. It was also clear that much that I had read and taught before the war had been wildly oversimplified, if not downright wrong. An example is the hierarchical arrangement of respiratory centres postulated by Lumsden in the 1920s. I suppose that after five years working as a physicist I had little use for biological generalizations and always wanted to concentrate on the physicochemical approach to physiology. This didn't go down well with most medical students.

I gave tutorials to about twelve undergraduates, usually taken in pairs and did this between 5.00 and 8.00 p.m. before dinner in Hall on two days a week, preferably those on which I had to demonstrate or lecture in the laboratory. The idea of this arrangement, which I found hard work, was to keep three days free for my own experiments, but visitors, seminars and committees made it difficult to stick to this plan. Like a good college man and to be honest, like one strongly influenced by his old tutor, Gow, I usually dined in Hall on Sundays and went on to the Master's Lodge afterwards. The only drawback to this otherwise pleasant arrangement was that the Master, Trevelyan, who drank little himself, was liable to fill up your glass with neat whisky. After claret and port at dinner, this was not a good idea if you had to give a 9.00 a.m. lecture on the following Monday morning.

I have mixed feelings about the tutorials, or supervisions as they are called in Cambridge. With some of my brightest pupils, like Richard Adrian or Ian Glynn, it was enormously stimulating and enjoyable, particularly at the third-year level. But for the bulk of my undergraduates, who were mostly hard-pressed medical students, it could be very hard going and I was relieved to give it up in 1952 when I became Foulerton Research Professor of the Royal Society. Still, there were compensating features about even the less bright undergraduates. I remember on one cold evening getting my pupil to add up the calories of food he had eaten at lunch. I found that he had taken the College lunch in Hall, at that time a solid but unimaginative meal probably consisting of potatoes, veg., meat and gravy followed by a stodgy pudding of some kind. After adding up these calories I asked if he had eaten anything more. 'Well, Sir,' he said 'it has been a very cold day and as I was still hungry I had a second lunch at the British Restaurant.' So we added in those calories. 'And that was it, I suppose?' said I, to which he replied, 'When I got back to my rooms I found I was still hungry so I ate a loaf of bread.' I remembered this episode thirty years later when a student committee of mixed sex argued very reasonably in favour of a low-calorie option at lunch.

My plan to keep three days clear for experiments ran into difficulties early in 1946 when I was asked if I would serve on the College Council which in those days met for two or three hours before, and often after, lunch on Fridays in term-time. I asked Adrian what I should do, hoping he would advise me to say no, which would have been unlikely as he rarely gave unequivocal advice if he could avoid it. Sure enough I got the delphic reply that it would be a rest and I would get a good lunch, as well as learning a lot about human nature. I decided that I had better serve and was duly elected for four years, for which I have no regrets as there was something very splendid about the way in which Trevelyan conducted business. He was not a particularly efficient chairman and indeed would sometimes doze off if business dragged on to the afternoon. I remember one afternoon meeting of a Council Committee, in which we were interviewing candidates for a Travelling Scholarship to Italy, when Trevvy, who had dozed throughout the interview, rounded off the business by drawing the candidate's attention to the importance of the siesta in Mediterranean countries. For the most part, he presided in a calm and kindly way, but this benevolent mood was liable to be interrupted by a sudden flash of anger if he thought the College was proposing to act in a shabby or ungenerous fashion, for instance: 'Is this a college or a bloody institution?', when it was proposed to deny tenure to someone who had served the College well, on the grounds that he had not produced enough scientific papers. He would occasionally open an item of business with some reflection on life in general, sometimes obscure in nature, for example, 'The worst of living in England is that one has to keep up these bally chancels': a reference to the requirement that the patron of a living, as Trinity College being a landowner often is, must maintain the chancel in the churches on its property. This is not something that would occur to the ordinary person as a disadvantage of living in this country.

As someone who had spent much of his life walking through Europe, for example along the routes followed by Marlborough's or Garibaldi's armies, and who had a strong feeling for natural beauty, Trevelyan deplored the increasing mechanization of modern life. He felt that scientists were too pushing and had collared too much priority in Cambridge. But I don't think he allowed these feelings to affect his conduct of business or his personal relations with Trinity scientists.

In 1946 Trinity was due to celebrate its Quatercentenary, and I was put on a small committee to suggest names of distinguished Trinity scientists who should be invited to the celebrations, which had been postponed to the summer of 1947. I argued, without any difficulty, that one of these must be Professor D'Arcy Thompson, the Edinburgh zoologist best known, then and now, for his great book *On Growth and Form*, but also familiar to Greek and Latin scholars as the author of books on birds and fishes in classical literature. I am glad to say that he was invited to the lunch and evidently enjoyed himself. In 1948 he died and on the day after his death Trevelyan showed me a letter from D'Arcy Thompson which he said he proposed to destroy after I had read it. The letter had clearly affected him deeply, which was not surprising, for it began 'Master, I am about to enter a greater gate than yours' and went on to express both the writer's affection for the College and his sorrow at never having become a Fellow or Honorary Fellow. It was full of affecting memories – how his rooms looked out over the Bowling Green where he remembered seeing a redstart, and other moving detail. In retrospect, D'Arcy Thompson certainly should have been made an Honorary Fellow, but in those days very few were elected and the importance of his biological work was not as widely appreciated as it is now.

The Rockefeller Foundation's Grant for Neurophysiology

After the war, Professor E. D. Adrian, then Head of the Department of Physiology, obtained a grant from the Rockefeller Foundation which helped to support a group working primarily on the biophysics of nerve and muscle. In the first instance the grant was for £3,000 per annum for five years from 1 February 1946. We got into the way of referring to the group as the Rockefeller Unit, but this was an unofficial name, as it was the Foundation's policy to initiate rather than maintain projects.

A convenient way of summarizing the status of the unit is to quote from a memorandum written by Adrian in October 1949 for the Faculty Board at a time when the University was considering its Needs for the next quinquennium which started in 1951.

> At the end of the war, the return to the Department of several research workers of great promise encouraged me to apply to the Rockefeller Foundation, who agreed to a grant of £3000 per annum for five years. The Unit has been in charge of A. L. Hodgkin, who is at present an Assistant Director of Research in the Laboratory; the other members were originally A. F. Huxley (now University Demonstrator) and D. K. Hill. Since then, two more research workers have joined the Unit: R. D. Keynes and P. R. Lewis. In addition to these a number of scientists from various countries have come for periods ranging from a few months to two years. The work of the Unit has been of quite exceptional value and has earned it an international reputation. It has dealt mainly with the physical and chemical changes involved in the conduction of impulses in nerve fibres and has depended on the development of techniques for studying isolated nerve fibres. A special aquarium has been built to keep marine invertebrates (which have the most suitable nerves) and equipment for radio-active tracers and for precise chemical and physical estimations has been installed and is in full use. Much of the apparatus is built by the radio technician attached to the Unit.
>
> The Rockefeller grant has been used for apparatus and equipment, for paying salaries (whole or in part), and for maintenance grants to various workers who have been invited here. In accordance with their general policy the Rockefeller Foundation made the grant to encourage a new development and did not contemplate financing it after the initial five year period, but the Unit has already justified its existence and the research programme of the Physiological Laboratory would suffer very seriously if the work in biophysics had to be curtailed.

265

Adrian suggested that the best arrangement would be the creation of a sub-department for Research in Biophysics in charge of a Reader with various University posts associated with it. This particular proposal did not get far, but the Unit continued, essentally as envisaged by Adrian, with continued support from the Rockeller Foundation, as well as new help from the Nuffield Foundation and Royal Society.

Restarting experiments. Effects of potassium on membrane conductance

By the autumn of 1945 my apparatus was working well enough to allow me to start experiments on single nerve fibres from the shore crab *Carcinus maenas*. After some ambitious trials which didn't come off, I decided to examine the effect of potassium and other monovalent cations on membrane conductance. According to the membrane theory of nerve conduction, the surface of a resting nerve or muscle fibre is permeable to potassium ions but relatively impermeable to sodium ions, so that a rise in K^+ but not Na^+ should increase the membrane conductance: a point which I thought it would be useful to check and quantify.

The method of determining the effect of ions on membrane conductance depended on determining how much local circuit current spreads over a length of 2 millimetres which contained either normal saline or the test solution. The rest of the fibre was immersed in oil so that currents could be applied, and voltages recorded from the fibre without using the internal electrode technique, which had not then been adapted to anything as small as a nerve fibre only 30 micrometres in diameter.

Although the method was somewhat indirect, the results were clear-cut. Small changes in external potassium concentration caused large and rapidly reversible changes in membrane conductance, about forty times greater than those produced by adding similar molar quantities of sodium or lithium.

Trebling the sea water concentration of potassium increased the membrane conductance roughly threefold, and removing it halved the conductance.

Similar increments were produced by adding salts in the following molar ratios: $RbCl, 0.8$; $KCl, 1.0$; $CsCl, 2.2$; $NaCl, 40$; $LiCl, 40$.

An interesting point emerged when the external potassium

266

concentration was raised by gently stroking the fibre with a potassium-rich drop of solution, and then measuring membrane conductance with the whole fibre immersed in oil, thus reducing the volume of external fluid to the thin film which clings to the fibre in oil. Under these conditions the rise in membrane conductance was not maintained, but declined towards its resting value with a half-time of a minute or so. This recovery was almost certainly due to absorption of excess potassium by the axon against a concentration gradient, which must have depended on the activity of the sodium pump, about which we knew very little at that time. The average rate of absorption of potassium ions was 100 picomoles $cm^{-2} sec^{-1}$ which agrees well with estimates made later in other nerves by more direct methods.

From the discussion in this paper (Hodgkin 1947) it is evident that I had been much influenced by the work of Conway (1946) and Krogh (1946) and their colleagues, particularly Ussing, about whose work I learned soon after the war.

Potassium leakage during activity, 1946

Andrew Huxley returned to Cambridge at the end of 1945 and we worked together intensively for the next ten years, sometimes with different partners but always in close intellectual contact. It made a great difference to have someone of such penetrating intelligence and experimental skill with whom to discuss new ideas or apply difficult methods.

One of the first experiments we started was to try out an indirect, but very sensitive, method of measuring the quantity of potassium which leaks out of a nerve fibre during the passage of a nerve impulse. This measurement was important because it seemed possible that movement of ions down concentration gradients, for example an entry of sodium ions followed by the outflow of potassium ions, might be responsible for altering the charge on the membrane capacity and generating the nerve impulse.

At that time there were reports that prolonged stimulation might cause leakage of potassium ions from nerve. But there was no evidence as to the magnitude or the time course of the leakage; nor was there any certainty that potassium leakage was a normal and invariable accompaniment of activity.

An isolated axon from *Carcinus* is surrounded by a layer of saline which is only a few micrometres in thickness when it is immersed in oil. Electrical and optical methods indicated that the volume of external fluid which surrounded one centimetre length of axon was about 3×10^{-9} litres. The potassium concentration in this layer was normally the same as in sea water, 10 millimoles per litre, so the total amount of external potassium was only about 3×10^{-11} moles per centimetre. This means that the membrane conductance between impulses should increase markedly when a train of impulses passes down the axon, if there is any appreciable leakage of potassium.

We found that the cumulative effect of a train of impulses on membrane conductance was striking if the fibre was in oil but negligible in a large volume of saline. Immediately after a train of a few thousand impulses the conductance was doubled and then returned to its resting value with a half-time of a minute or two, presumably because potassium ions were being absorbed by the axon. Recovery of membrane conductance could be greatly accelerated by dipping the axon in sea water, presumably because this washed away the excess potassium.

On the assumption that the effects observed were due to leakage of potassium we obtained a value of 1.7×10^{-12} for the number of moles of potassium which leak through one square centimetre of membrane in one impulse (Hodgkin and Huxley 1947). Another way of expressing this result is to say that in one impulse roughly 10,000 potassium ions escape through 1 cm^2, or that one potassium ion escapes through an area occupied by about 500 fatty acid molecules in a monolayer. This last calculation brings home the extremely small magnitude of the potassium leakage, as does the statement that in a single impulse a $30 \mu m$ *Carcinus* axon loses only one hundred thousandth of its internal potassium.

Although very small, the quantity of potassium lost was more than the theoretical minimum that it had to exceed if entry of external sodium followed by exit of internal potassium was the basic ionic mechanism underlying the action potential. Thus 1.7×10^{-12} moles of monovalent cation carries a charge of nearly 0.17 microcoulomb which was about twice the charge on the resting membrane. The actual loss of potassium would be expected to be larger because the action potential is greater than the resting potential and because there is bound to be some simultaneous exchange of Na^+ for K^+. All this became clearer and easier to explain after we had shown that there was an initial selective increase in sodium permeability followed by a delayed increase in potassium permeability.

Analysing and writing up the potassium leakage experiments occupied much of the latter part of 1946 and made us think about the possible directions that future research on nerve ought to take.

On the experimental side it was clearly urgent to check our figure for potassium leakage during nerve activity by some more direct method. Radioactive tracers were beginning to be available and it seemed likely that they might be helpful in measuring how much sodium was gained and potassium lost in a nerve impulse. Richard Keynes, who had spent much of the war working on Admiralty radar, was keen to have a go at these ambitious projects which he started in 1946 and developed into a highly successful line of research.

In the winter of 1946–7 Andrew Huxley and I spent much time speculating about the kind of system which might generate an action potential. Our initial hypothesis, was that sodium ions were transferred across the membrane by negatively charged carrier molecules or dipoles. In the resting state these were held in one position by electrostatic forces and unable to ferry sodium ions across the membrane. These carriers could be prevented from transporting sodium by reacting slowly with some substance in the axoplasm: this process being known to us as 'inactivation'. A propagated action potential calculated by Huxley in 1947 incorporated the main features that emerged two years later from the voltage-clamp experiments, i.e. a rapid rise in sodium permeability followed by a slower decay, and a slow rise in potassium permeability. However, one of the central features of our scheme was shot down by the excellent data that we obtained in the 1949 squid season, which showed that the sodium and potassium ions that cross the membrane during the impulse are not ferried across by carriers, but move through highly selective channels controlled by voltage-dependent 'gates'. In spite of this defect, I feel that these early theoretical studies were important in helping us to decide on the right experimental approach.

In these theoretical action potentials the reversed potential difference at the crest of the spike depended on a selective increase in sodium permeability and a low internal concentration of sodium ions. Huxley felt all along that this was a likely mechanism, but I

269

was more doubtful, partly because there seemed to be quantitative discrepancies, and partly because I hankered after a mechanism which would give a transient reversal, so accounting for repolarization, oscillations and the transient nature of the action potential. We tried various mechanisms that I thought might operate in this way, but Huxley shot them all down, leaving a rise in permeability to sodium ions, or perhaps to an internal anion, as the most likely cause of the reversed potential.

Evidence for sodium entry during activity, 1947

Towards the end of 1946 Bernard Katz sent me a manuscript in which he showed, among other things, that crab axons became inexcitable in salt-free sugar solutions (Katz 1947). As this agreed with my own experience I began to think that Curtis and Cole's (1942) result must have been wrong and that there was hope for the sodium theory.

In January 1947 I decided to test the theory by measuring the effect of sodium-deficient dextrose solutions on the action potential recorded externally from single crab nerve fibres, and at the same time recording the longitudinal resistance of external and internal fluids in parallel. The second measurement was needed because if sodium chloride is replaced with dextrose the external resistance rises; this reduces short-circuiting and partly counteracts any true effect of sodium deficiency. However, if the external action potential is divided by the longitudinal resistance per unit length, one obtains a quantity equal to $\dfrac{V}{r_i}$ where V is the membrane potential and r_i is the internal resistance per unit length, which should remain constant in different solutions. By this method I found that lowering the external concentration of sodium reduced the action potential by about the right amount – for example lowering $[Na]_o$ to $\frac{1}{5}$ reduced the action potential by 40 per cent : from 120 to 72 mV if one assumed initial equality of internal and external resistances. The reduction of 48 mV was not far from 58log5 and seemed reasonable.

These experiments were brought to an end by the most serious of the energy crises which beset Britain after the war, in this case precipitated by an exceptionally prolonged cold spell which lasted until the end of March 1947. It was soon found that National coal stocks were exhausted and the central heating was switched off in many buildings, including university laboratories. We then had no Cold Room in our part of the laboratory and I remember that David Hill took the opportunity of carrying out a series of experiments at 4 °C. But you can't dissect single fibres at such temperatures and I spent the time writing at home or talking with Andrew Huxley in Trinity where he could be found cranking a Brunsviga calculating machine with mitten-covered hands. As the freeze intensified, life at home became increasingly difficult and much of the day went in trying to restore the central heating, dealing with burst pipes and trying to obtain food, which was growing increasingly scarce, particularly vegetables – on one day Marni wrote that there was not a single cabbage to be found in Cambridge. Instead of the thaw predicted by the BBC, the freeze was followed by a blizzard, which left two feet of loose snow on the ground, and finally by a flood that put inches of water into many Cambridge houses but fortunately did not reach ours.

I started experimenting again in April but by then I had heard that the Plymouth Laboratory was recovering its trawler, and that I should soon be able to do the sodium experiments properly using the squid axon and an internal electrode. I planned to go to Plymouth in mid-June as soon as term was over, but meanwhile there were many other things to occupy me, both at home and in the laboratory. I had to get the new equipment ready for Plymouth and there was a good deal to do in Trinity in connection with the Quatercentenary celebrations, which took place in glorious weather on 3 June. However, I did manage to fit in some experiments with *Carcinus* axons at the end of May which showed that the action potential disappeared if the sodium chloride in the external solution was replaced with choline chloride. An interesting point was that although the fibres were inexcitable in the absence of sodium there was still a delayed increase in conductivity in response to depolarization which we later showed was due to an increase in potassium permeability.

First tests of sodium theory at Plymouth, summer 1947

I got my equipment to Plymouth on about 18 June, as soon after term as possible. Life was not easy. Much of the laboratory had been destroyed in the great air raids of 1941 and was being rebuilt, squid were in short supply and I'd forgotten much of the technique.

Worst of all I was short of a partner. Andrew Huxley was just getting married and wouldn't be free until the autumn; Bernard Katz would be able to join me in September but not in July when I should be on my own. But I was terribly pleased to be back where we had left off in 1939, and very excited when the sodium experiments started to come out so well.

I found that the nets used by the trawler were too coarse and that the few squid that were caught were so mangled that they did not survive overnight. It seemed that the only way to get results was to start dissecting in the late afternoon, when the ship came in, and carry on until 2.00 or 3.00 a.m.

However, the experiments that I had to do were not difficult by modern standards and I quickly became convinced that the sodium hypothesis was basically correct. In a letter to Victor Rothschild, written from Plymouth on 8 July, I say that I think I have proved why the nerve membrane potential reverses during activity. 'The reason is simply that the active membrane, instead of becoming freely permeable to all ions (as supposed by classical membrane theory) becomes much more permeable to Na than to K.' A further letter on 14 July goes into more detail.

By the time of the International Congress of Physiology, which was held in Oxford from 21 to 25 July 1947, I had obtained three main lines of evidence, all of which supported the sodium theory, particularly after they had been strengthened and extended by subsequent work with Bernard Katz in September. In the first place the experiments showed that the action potential was rapidly and reversibly abolished by sodium-free solutions like isotonic dextrose or choline chloride, as would be expected if entry of sodium ions were essential for generating the rising phase of the action potential. This agreed with the conclusion drawn by Overton in 1902 on the basis of his classical experiments on frog muscle.

Another simple consequence of the sodium hypothesis is that the magnitude of the action potential should be greatly influenced by the concentration of sodium ions in the external fluid. Thus the active membrane should no longer be capable of giving a reversed electromotive force if the concentrations of sodium on the two sides of the membrane are equalized. In the extreme case where the membrane is much more permeable to sodium than to any other ion the potential should approach that given by the Nernst equation:

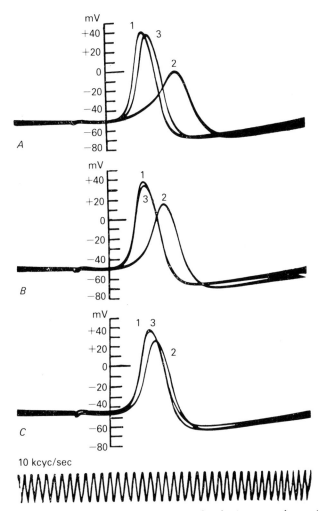

Figure 28.1. Effect of sodium-deficient external solutions on the action potential. Records labelled 1 and 3 were with the axon in sea water. *A*2 with ⅓ sea water, ⅔ isotonic dextrose; *B*2 with ½ sea water, ½ isotonic dextrose; *C*2 with 0.7 sea water. 0.3 isotonic dextrose. (From Hodgkin and Katz 1949a)

$$V_{Na} = \frac{RT}{F} \ln \frac{[Na]_o}{[Na]_i} \qquad (1)$$

This gives a limiting value of +58 mV for a ten-fold concentration ratio and accounts satisfactorily for the reversal of 50 mV commonly seen in fresh axons.

Figure 28.1, which was obtained in September with Katz, illustrates the effect of reducing sodium concentration on the action

273

potential and shows that, in the physiological region, the reversed potential varies in the manner predicted by the Nernst equation.

It was also shown that a solution containing extra sodium increased the reversed potential by about the amount predicted by equation (1). This was a particularly satisfactory result, because it seemed unlikely that an increase beyond the normal would be brought about by an abnormal solution.

Another strong argument for the sodium theory was that the maximum rate of rise of the action potential was approximately proportional to the sodium concentration. This is illustrated by Figure 28.2 which was obtained by electrical differentiation of the action potential recorded in three solutions containing the following relative concentrations of sodium: a, 0.5; b, 1.0; c, 1.56.

The maximum rate of rise is an interesting quantity because it provides information about the magnitude of the sodium current during the rising phase of the action potential. The membrane current during a propagated action potential is proportional to the second derivative of potential with respect to time, and is therefore zero when the first derivative is at a maximum or minimum. The current through the membrane consists of capacity current, $\dfrac{C dV}{dt}$ and an ionic current, I_i, due to movement of ions through the membrane. These two components must be equal and opposite at the point of inflexion on the rising phase. From the maximum rate of rise, which averaged 600 volts per second, and the membrane capacity of 1.5 microfarads per cm^2, the inward ionic current was found to be 0.9 milliamps per cm^2; allowance for movements of potassium and chloride ions gave the inward sodium current as about 1 mA/cm^2 at the point of inflexion on the rising phase.

Huxley and I were going to the International Physiological Congress in Oxford at the end of July and we had sent in an abstract of a communication on *Potassium leakage and absorption by an active nerve fibre*. Rather greedily I had also submitted an abstract on repetitive firing in a single nerve fibre. With the permission of the chairman, I withdrew the communication on repetitive firing and extended the talk with Huxley to include the recent Plymouth work on the effects of sodium on the action potential. This had a good reception, but we also collected useful criticisms, some of which Katz and I were able to answer by the experiments that we did in September.

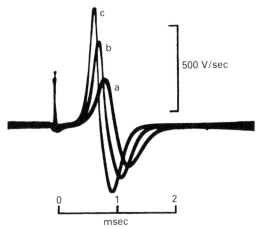

Figure 28.2. Rate of change of membrane voltage in solutions containing various concentrations of sodium: *a*, in 50% sea water, 50% isotonic dextrose; *b*, in sea water; *c*, in sodium-rich solution containing 1.56 times the sea-water concentration of sodium. (From Hodgkin & Katz 1949a)

A characteristic feature of isolated squid axons is that the main part of the action potential is followed by an underswing in which the internal potential is more negative than in the resting condition. Katz and I found that the underswing was markedly increased by potassium-free solutions and decreased by potassium-rich ones: a result expected if the potassium permeability increased during the action potential and took a few milliseconds to return to normal.

To begin with we avoided making any quantitative assumptions about the relative permeabilities of the membrane to sodium and potassium ions. The resting membrane was considered more permeable to K^+ than Na^+ and this condition was assumed to reverse during activity. However, as time went on it became increasingly clear that we needed a theory which would predict the potential difference that would arise across a thin membrane, permeable to K^+, Na^+ and Cl^-, and separating different concentrations of these ions. For this purpose we used a simple equation derived by Goldman (1943) who assumed that the voltage gradient through the membrane was constant, that ions move under the influence of diffusion and the electric field, and that the concentration at the edges of the membrane are directly proportional to those in the aqueous solutions.

These assumptions led to the following relation for the internal potential in the steady state and in the absence of applied current:

275

$$V = \frac{RT}{F} \ln \left[\frac{P_K[K]_o + P_{Na}[Na]_o + P_{Cl}[Cl]_i}{P_K[K]_i + P_{Na}[Na]_i + P_{Cl}[Cl]_o} \right] \qquad (2)$$

where P_K, P_{Na} and P_{Cl} are the permeabilities to K^+, Na^+ and Cl^- respectively. In calculating steady potential differences in the absence of applied current only the relative values of the permeability coefficient need be known.

From an analysis of our data in terms of the constant field theory and likely values of the ionic concentrations, we concluded that in the resting state the relative permeability coefficients were approximately

$$P_K : P_{Na} : P_{Cl} = 1:0.04:0.45$$

which gave a resting potential of about -60 mV for the resting potential.

On activation, the permeability to sodium was assumed to increase some 500 times giving

$$P_K : P_{Na} : P_{Cl} = 1:20:0.45$$

corresponding to an active membrane potential of $+40$ mV and an action potential amplitude of 100 mV.

At the bottom of the underswing, sodium permeability was taken to be completely inactivated and potassium permeability enhanced giving

$$P_K : P_{Na} : P_{Cl} = 1.8:0:0.45$$

corresponding to a membrane potential of -74 mV and an underswing amplitude of 14 mV.

Doubts about the size of the junction potential between the sea water in the microelectrode and the axoplasm inevitably introduced some uncertainty into the values for membrane potential, but the changes in potential difference produced by altering the ionic concentrations agreed well with the theory.

There was also satisfactory agreement between the value of the sodium current obtained from the maximum rate of rise of the action potential multiplied by the membrane capacity and that predicted by the constant field theory with $P_{Na} = 20P_K$: the sodium current being close to 1 mA/cm^2 in both cases. In this calculation we assumed that P_K remained at its resting value, which was estimated from the resting membrane conductance as 1.8×10^{-6} cm/sec.

Analysing and writing up the Plymouth experiments occupied me for several months and we did not get the typescript sent in to the *Journal of Physiology* until mid-January 1948. This delay seemed bad enough but, owing to some printing difficulty, the paper did not appear for a further fifteen months (Hodgkin and Katz 1949a). An equally long delay occurred with the manuscript dealing with the effects of temperature on electrical activity, on which Katz had done most of the work (Hodgkin and Katz 1949b).

I was so fed up with these long delays, as well as being lectured at by a senior colleague on the dangers of over-publishing, that I temporarily deserted the *Journal of Physiology* and sent my next paper with Nastuk to the *Journal of Cellular and Comparative Physiology* (Nastuk and Hodgkin 1950).

Chapter 29

New York and Chicago, spring 1948

General

One of the reasons for embarking on this trip was to enable Marni to visit her parents in New York, and this we managed to do in a month's visit lasting from mid-March to mid-April 1948, just between two Cambridge terms. My mother in Edinburgh had very kindly said that she would look after our two children, with the assistance of our new mother's help, who turned out to be no help at all and had to be fired for throwing shoes at my stepfather in response to a polite request. This, firing the new help, was something we had both longed to do, but lacked the experience to take the step that my mama saw at once was inevitable. Transporting two children aged one and three to Edinburgh and back by train under near-wartime conditions was a somewhat daunting experience for Marni. However, she enormously enjoyed the month with her parents in New York and for three weeks was able to resume her old job as an editor of children's books at the Viking Press in exactly the position that she had left four years before. Apparently, not only was the job the same, but she occupied her old desk, with the result that former colleagues at first passed her with a casual greeting and then turned in surprise, feeling that they had been transported four years backward in some kind of time machine.

The other main reason for our American visit was for me to give a lecture in Chicago and exchange information with Cole and Marmont; this occupied the middle fortnight of our month in America.

Victor Rothschild, who was a Director of BOAC, a forerunner of British Airways, had helped us to obtain tickets at short notice.

In return he asked me to let him know what we thought of the service and of the trip in general. I hadn't anything much to suggest but filled up the letters with scientific gossip, some of which is given below:

21 March 1948
New York City

Dear Victor,

This letter is long overdue but we have lived in such a whirl that there has been no time to settle down and write the promised account of our trip. I am afraid that this account will not be a racy affair like Venetia's since everything was without exception most efficient and comfortable. No whipped rabbit brains but frequent sandwiches (which were good if a trifle dull) and an excellent breakfast, including haddock, on the aircraft and a still better one with ham and eggs, cream and grapefruit in Gander. The general arrangements at Heathrow, customs, passports, etc. struck us as considerably pleasanter and more efficient than those at La Guardia but of this we shall be able to judge better on our return. We liked the little speech made by the steward before take-off, thought the service good and appreciated the fact that we were left alone and not pestered by solicitous inquiries about our comfort. We liked the maps and the information given about the trip but thought that further information might have been given on two points.

(1) Whether or not to tip: we gathered not, but were left with a slightly uneasy feeling. (2) The nervous traveller might be glad to know that when a screaming noise is heard just before landing or take off it is not because the propellor has flown off or the engine seized up but merely the normal operation of lowering the flaps. We were delayed an hour at Gander – which seems to me one of the most desolate places in the world – and thought the airline showed good psychology in telling us that we should probably be delayed for at least two hours. We were slightly annoyed to find that our copies of Time magazine disappeared at Shannon, but discovered that the steward had taken them into safe-keeping, in order to avoid the depredations of the Irish, who are starved for literature of this kind. So all ended well. I asked several passengers whether there was anything to choose between BOAC and the American lines but gathered that the general standards of comfort and efficiency were almost identical. We were rather surprised to find that the aircraft was only about half full but I suppose transatlantic travel has been greatly reduced by financial restrictions.

On arrival we were met by Marion and Peyton and had soon settled down to a long gossip which has continued with occasional

279

interruptions until the present time. On the day after we arrived I had to go down to the annual conference of experimental biologists and physiologists at Atlantic City, where I spent two days. This was a somewhat exhausting affair as several hundred papers are read and the congress is attended by several thousand people. However it was useful because I managed to meet a great many people including Curtis whom I should not otherwise have seen. I think the most interesting piece of work which I came across were some experiments done by a Chinese working with Gerard in Chicago. It looks as though it really is possible to measure membrane potentials without injury by the use of a micro electrode pushed through the cell surface. But I shall find out more about the technique in Chicago.

Have you ever been to Atlantic City? It must be one of the most hideous and fantastic resorts in the world. It is a sort of monstrous compound of Blackpool and Bournemouth with everything magnified about 500 times.

. . . I go to Chicago on the 24th and shall be there until April 6th. I hope all goes well with you and that you are having a successful time at Millport.

With love from all to you and Tess,
Alan

26 April 1948
17 Bentley Road, Cambridge

Dear Victor,

I have been meaning to write ever since I got back but we have both been in a tearing rush and I have been catching up with my medical students and other chores. And now there are only a few days before you return so I will not write more than a line. We are both looking forward to your return with great pleasure: Cambridge does not seem at all as it should without you and Tess.

We had an excellent time in America though it now seems rather remote. The old folks were in fine form and of course send love and all sorts of messages to you both. Peyton's work is going tremendously well and he is full of energy and enthusiasm. So much so that he took almost no notice of an episode which would have flattened me out completely. Too complicated to explain, but all about some purple tumours – or were they papillomas after all – which had been induced by extracting yellow scurf from other tumours. All discussed at great length over dinner, to the disgust of the squeamish Ellen. Anyhow one day they were tumours and cancer was practically solved and the next they were papillomas and we were back at the beginning. But Peyton didn't seem to mind a bit and I suspect Marion knew they were

280

papillomas all along. However, apart from minor ups and downs of this kind I gather everything is going extremely well and the results sound very interesting in so far as I am able to understand them.

The scientific part of my trip was worthwhile but not overwhelmingly exciting. Kacy is doing some nice things but his Institute of Radiobiology and Biophysics is a curious place because almost the only living things in it are physicists. (All their nerve work is done at Woods Hole.) I met a great many scientists one way and another and have picked up a certain amount of technical information about microelectrodes and the like.

When not working, our time was spent in sociabilities, going to plays, looking at pictures, a weekend in the country and most important in endless gossip of a delightful kind with Marni's family.

We had a successful trip home in company with an elderly Glaswegian who assured us that 'BOAC es the best', a sentiment which I am sure everyone echoed after we landed safely in almost thick fog at Heathrow.

Marni is back with the children who are in fine form but exhausting. Cambridge is looking very nice in brilliantly sunny weather.

Love from all,
Alan

Chicago, March 1948: K. S. Cole, G. Marmont

Since the end of the war I had corresponded intermittently with Kacy Cole and on 26 August 1947 I had written to tell him about the sodium results and to discuss future joint research. We had made a tentative plan, which never came off, to join forces at Woods Hole in 1948 and were starting to discuss research possibilities. I wrote:

> I should rather like to have a shot at perfusing the inside of the axon with potassium or sodium salts and have some ideas about the best method of doing this. I am also interested in the possibility of stimulating an axon with a diffuse electrode in such a way that the axon is excited uniformly over a length of one or two centimetres. This might give useful information about the nature of the active process uncomplicated by propagation and local circuits. What are your plans and views?

In his reply of 7 October 1947. Cole said:

> . . . I am sure that you will be excited to hear that we spent the whole summer with an internal electrode 15 millimetres long and

281

about 100 microns in diameter . . . The two principal ideas are first the use of the central outside region with a guard region on each side, and second the use of a feedback circuit to control either the current flow in the central region or the potential difference in that region to the desired value . . .

In Chicago I spent several days with Cole and Marmont in the Institute of Radiobiology and Biophysics, where they showed me the results they had obtained the previous summer at Woods Hole with the membrane current or potential of a giant nerve fibre under the control of electronic feedback. I gathered that Marmont was more enthusiastic about current control and that, perhaps for this reason, they had not done many experiments with voltage control. However, the results which Cole showed me clearly illustrated the essential features of records obtained with the 'voltage-clamp' technique, this being the name subsequently used for the method in which the membrane potential is suddenly displaced and then held at the new value by electronic feedback. Thus when the membrane was suddenly depolarized by 50 millivolts, corresponding approximately to an abrupt short-circuit of the membrane, there was first an initial surge of capacity current, then a transient phase of inward current and finally an outward current which also declined, perhaps owing to polarization of the axial electrode.

As Cole recognized, the most interesting feature of these records was the transient inward current, seen with depolarizations of 20–100 mV, which was in the opposite direction to that expected in a simple physical system. A possible explanation, proved by later work, was that the depolarization increased the permeability to sodium ions, thus allowing them to move down their concentration gradient. The inward current varied smoothly with depolarization and there was no sudden discontinuity or threshold, a result explained if negative feedback had stabilized a regenerative loop.

Andrew Huxley and I were anxious to test our carrier theory and when we heard about Cole and Marmont's experiments we felt that voltage control, with current applied through a long metal wire, might be a good way to prove or disprove the theory. But I was worried about electrode polarization and decided to use two fine silver wires, one for current and the other for voltage. Before leaving for America, in March 1948, I made a short double-spiral electrode out of two 20 micrometre wires wound round a $60\,\mu$m glass rod. I believe that I also drew the circuit diagram of the feedback amplifier that I thought might be suitable for voltage-clamp experiments. After my return to Cambridge in mid-April, our new instrument maker, R. H. Cook, built the feedback amplifier that we used, with some modifications, later in the summer.

However, this was neither used nor tested until mid-August, when Huxley joined Katz and myself at Plymouth. Our apparatus differed in several respects from that of Cole and Marmont, but it owed much to the experiments which they started in 1947 and to the information which they generously provided in 1948. The reason I did not try out the feedback amplifier earlier was partly that I had other interesting experiments to do, and partly that I did not want to tread too closely on the heels of Cole and Marmont.

Chicago, April 1948: G. Ling, R. W. Gerard

At the big conference at Atlantic City, in mid-March, I heard about a very interesting experimental method that was being developed by Gilbert Ling in Professor Gerard's laboratory in Chicago. He had found that if he drew a 1 millimetre glass tube down to a really fine tip, it would penetrate a frog muscle fibre easily, giving a stable resting potential and apparently causing no damage or local contracture. The trick was that the tip had to be so fine that it 'disappeared' when looked at with visible light under a high-power objective. Suitable electrodes had to have a sharp taper, and a resistance of 100 megohm when filled with isotonic potassium chloride solution. Examination with an electron microscope of samples which satisfied these criteria indicated that the tip diameter of a suitable electrode was about $0.3\,\mu$m.

Gilbert Ling taught me how to pull these electrodes by hand, which is tedious but not difficult, how to fill them, and minor tricks such as the importance of using the inner rather than the outer surface of the sartorius muscle. He and Gerard were using an electrometer, which was unsuitable for recording rapid changes, and had not attempted to measure action potentials. I asked them if they would mind if I had a shot at this when I got back to Cambridge. They replied that they had no immediate plans in that direction and I should feel free to go ahead. This delighted me as it provided an opportunity for seeing whether some of the results I had obtained on squid nerve could be duplicated in a vertebrate tissue.

283

Chapter 30

The electrical activity of muscle, Cambridge 1948

SOON AFTER I got back from America in May 1948 I tried recording muscle resting and action potentials with an electrode made in the way that Ling had shown me. I used a special type of input stage in which the effective input capacity was reduced by a metal screen connected to the cathode of a conventional type of cathode follower. With this arrangement I recorded resting potentials of about 80 millivolts, as Ling had done, and action potentials of 100 mV or more. I was pleased that the action potential should exceed the resting potential, as this was the first point I wanted to check. However, it was quite clear that the action potential was being cut by the input time constant which might be as high as 500 microseconds with electrodes filled with isotonic (120 millimolar) potassium chloride. As the squid season was approaching I decided to leave this problem until the autumn when I should have an American visitor, Dr W. L. Nastuk, working with me.

However, before switching off the muscle problem I was able to demonstrate the technique and preliminary results to the Physiological Society at their meeting in Cambridge on 22 May. I remember the occasion because every time that Sir Henry Dale rested on the bench it pulled the electrode out of the muscle and I had to start again.

After a gruelling two months at Plymouth and a short holiday with Marni in the Isle of Barra, I returned to the question of recording action potentials with a microelectrode when Bill Nastuk arrived at the beginning of October. This was a heavy teaching term and I relied on Nastuk to keep the experiments going. Our initial problem was to reduce the lag in recording action potentials. As a first step I thought we might try filling the electrodes with a stronger solution of potassium chloride. Nastuk asked me how strong and I quickly said 3 molar, from a vague and erroneous recollection that such a solution was sometimes used by electro-chemists, instead of the traditional saturated KCl bridges, to eliminate junction potentials. We didn't want crystals appearing in the electrode and I knew 3 M wasn't saturated, but otherwise the decision was a pure guess made in a hurry as I was rushing off to a teaching assignment. I have often been asked why we chose 3 M and wish there was some good reason for a somewhat arbitrary choice that has been followed in a large number of scientific papers.

When dipping into Ringer solution, electrodes with a tip dia-meter of 0.4 micrometre had a resistance of 10–30 megohm if filled with 3 M-KCl as against 50–100 megohm if filled with 120 mM-KCl. Calculations and various controls indicated that the leakage of potassium chloride from an electrode filled with 3M-KCl into the muscle fibre had a negligible effect on its electrical properties. Their lower resistance made them much easier to work with than the original Ling electrode and enabled them to record the action potential without appreciable distortion or loss. They also appeared to abolish, or at any rate greatly reduce, the junction potential between solutions of different ionic composition.

Having established that the resting potential was close to 90 mV and the action potential to 120 mV, we went on to study the effect of replacing sodium with choline, or of adding extra sodium chloride and found that the changes in active membrane potential agreed almost perfectly with that expected in a sodium electrode.

All this sounds very smooth and tidy but we did not finish the experiments without a last-minute struggle. Nastuk had to leave at the beginning of January and just before Christmas we discovered a trivial but tiresome mistake in one of the solutions. The lab was locked and unheated for a week, but we switched on as much electric heating as we could and managed to repeat the crucial experiments with the solutions as we wanted them. This little episode accounts for the room temperature of 13 °C recorded in one of the experiments in Table 2 of Nastuk and Hodgkin (1950).

I was pleased with this research for several reasons. In the first place, the experiments helped to answer the criticisms of physiolo-gists who felt that work on crab or squid nerve might not be relevant to our own excitable tissues, and that much more needed to be done closer to home before one could accept our ionic explana-tions of the action potential as general. It was also pleasant to be working in Cambridge on a tissue as accessible as the frog's sartorius muscle, and doing experiments which could easily be demonstrated to students, or even repeated by them. Another advantage of frog muscle was that its ionic composition was better known and easier to measure than that of crab or squid nerve.

The paper dealing with these experiments (Nastuk and Hodgkin 1950) is interesting because it introduces, possibly for the first time, the concept of the equilibrium potential for each ion. By this is meant the potential difference at which no net transfer of the ion would occur in the absence of metabolic work or secretory activity: this p.d. being given by the usual expression for a concentration cell:

$$E = \frac{RT}{zF} \ln \frac{c_o}{c_i}$$

where c_o and c_i are the concentrations of the ion outside and inside the cell and z is the algebraic valency of the ion. R, T and F have their usual significance; on the modern convention E is measured in the sense inside minus outside.

For frog muscle E_K was calculated as -99 mV, E_{Cl} as -80 to -120 mV and E_{Na} as $+52$ mV. A potential difference of about -100mV would then be expected for a membrane permeable to potassium and chloride but impermeable to sodium. A very slight permeability to sodium would be sufficient to reduce the resting potential to the observed value near -90 mV. On the assumption that the membrane becomes highly permeable to sodium during activity, the membrane potential should tend to approach E_{Na} at the peak of the action potential. It could only reach that value if the permeability to sodium became infinitely large compared to that to potassium and chloride ions. This condition is unlikely to be realized so it was reassuring to find that the potential at the peak averaged $+31$ mV which was well below the limit allowed by the hypothesis.

The *Journal of Physiology* had taken so long with my last two papers that we sent this one to the *Journal of Cellular and Comparative Physiology*, who produced proofs so quickly that it seemed likely that our paper would appear before those of Ling and Gerard in the same Journal. This seemed a poor way of repaying them for their generously given information about the method, so we asked that our paper should be delayed until after theirs – which is what happened.

Microelectrodes filled with 3 M KCl, but otherwise of the type introduced by Ling, were soon being used in many laboratories. Early in 1949, Bernard Katz and Paul Fatt came to see the technique and applied it in an illuminating way to the neuromuscular junction. In Cambridge, Weidmann and his colleagues started the elegant studies of Purkinje fibres in the mammalian heart; Richard Keynes used them on the electroplates of the electric eel and Jack Eccles launched a massive attack on the central nervous system from his laboratories in Otago and Canberra. Naturally there were minor changes and improvements, and it was sometimes necessary to use a solution other than 3 M KCl, for example 4 M potassium acetate was frequently used in the 1970s. Life was made easier by the introduction of automatic electrode pullers which could be set to apply the right heat and pull at the appropriate moment in a reproducible way. By using improved methods of pulling and filling electrodes and tricky electronics, it eventually became possible to record from cells as small as a retinal rod or cone with electrodes that tapered down to 0.1 μm or less. However, that was a long way in the future.

Chapter 31

Excitation and conduction in nerve

Voltage-clamp experiments at Plymouth

For some time Huxley and I had been interested in the idea of avoiding cable complications by applying electric currents uniformly to a length of nerve fibre; in particular we wanted to know what shape the nerve impulse would be if it arose simultaneously over a length of nerve fibre, instead of propagating along the fibre as it normally does. To achieve this, Andrew made an elegant perspex cell in which ten electrodes communicated through small holes, at 1 millimetre intervals, with a slot containing a single crab nerve fibre; this was early in 1946. We were slightly disappointed to find that the action potential set up by a brief shock applied uniformly was not strikingly different from that recorded from an ordinary point electrode. However, we did not pursue the matter at that stage, partly because we had other easier experiments to do and partly because axons did not survive well in the perspex cell, perhaps because of the release of toxic plasticizer from the perspex.

When we heard from Cole in October 1947 about his and Marmont's success with a long metal electrode it was clear that this was the way forward, and we were keen to try out voltage control on the squid axon as soon as we reasonably could. In acknowledging our indebtedness to Cole and Marmont I may sometimes have given the impression that we hadn't discussed the possibility of using feedback before I heard from Cole in 1947. I evidently did this in the first draft of the lecture *Chance and Design* that I gave in 1976, on which Huxley made the following comment in a letter dated 6 June 1976:

> This reads as if you hadn't thought of using feedback till it was suggested in Kacy's letter of October '47. I think I have a clear memory

of your talking about a feedback system to do a voltage clamp before the end of the war (or at any rate before I came back to Cambridge at the beginning of '46). My memory is that I was naive enough about polarization to say, 'Why bother about feedback and why not just apply the required potential to a metal electrode inside the fibre.'

I don't remember this exact conversation but it would be surprising if we didn't have a discussion along these lines as the equations for our carrier models had simple analytical solutions at constant membrane potential but not for constant, or zero, current.

With the help of our instrument maker, R. H. Cook, who proved to be a treasure, I got all the equipment ready for Plymouth, loaded most of it into a trailer and set out at 5.00 a.m. on 5 July 1948. The journey there and back used up my entire petrol ration for the rest of the year, as my application for a supplementary ration of 18 gallons had been turned down on the grounds that I *could* do the journey without exhausting my ration for the rest of the year. Katz soon joined me at Plymouth and we spent several weeks trying to perfuse nerve fibres, with virtually no success, except that we learnt that calcium ions would liquefy the normally solid protoplasm inside the nerve fibre. Having failed here, we started to make and insert double spiral electrodes of the type shown in Figure 31.1. This didn't work either until we realized that one should first pre-drill the axon with a smooth glass capillary, which left a track down which the double spiral would slide without buckling. Then things started to move and by using short shocks and constant currents with different external solutions we obtained much indirect information about the permeability changes to sodium and potassium during the action potential.

Andrew Huxley arrived in mid–August and settled down to make the feedback amplifier work. We managed to do a few voltage-clamp experiments which we wrote up for the conference held in Paris in 1949, but realized that we needed a proper system of guard electrodes and would do better to work at low temperatures. We didn't shoot down the carrier hypothesis until the next year and initially had no clear evidence about the inactivation of the sodium channel.

During the next year Huxley and I spent some time improving the equipment and we returned to the attack at Plymouth in June 1949. At first squid were in poor supply and we took a few weeks to get going. But by mid-July 1949 Katz had joined us, there was a

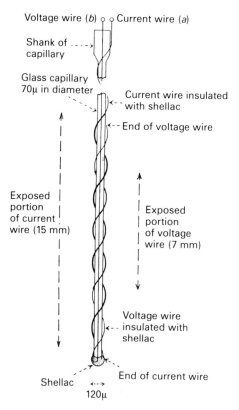

Figure 31.1. Diagram of internal electrode (not to scale). The pitch of each spiral was 0.5 mm. The exposed portions of the wires are shown by heavier lines. (From Hodgkin, Huxley and Katz 1952.)

In the early electrodes both wires were wound from the outside of the shank.

good supply of living squid and in the next month we obtained virtually all the voltage-clamp records that we used to illustrate the papers published in 1952. I believe we were able to do this quickly and without leaving too many gaps because we had spent so long thinking about the kind of system which might produce an action potential similar to that in nerve. We also knew what we had to measure in order to reconstruct an action potential.

Analysis of voltage-clamp results: Cambridge, 1949–52

There were several reasons why it took us over two years to analyse and write up the results. In the first place we both continued with some experimental research and a mild amount of teaching. The second reason, which now seems surprising, was that although we had obtained much new information, the overall conclusion was initially a disappointment. We had started off to test a carrier hypothesis and believed that even if that hypothesis was incorrect, we should nevertheless be able to deduce a mechanism from the electrical data that we had collected. We soon realized that the carrier model could not be made to fit certain results, for instance the nearly linear instantaneous current–voltage relationship, and that it had to be replaced by some kind of voltage-dependent gating system. As soon as we began to think about molecular mechanisms, it became clear that the electrical data would by itself yield only very general information about the class of system likely to be involved. So we settled for the more limited aim of finding a simple set of mathematical equations which might plausibly represent the movement of electrically charged gating particles. But even that was not easy, as the kinetics of the conductance changes were unlike anything we had come across before, particularly the S-shaped 'on' and exponential 'off' of the conductance curves. I think we both appreciated the need to involve several particles, but it was Andrew who eventually came up with the ideas that led to the m^3h and n^4 formulation.

Finally there was the difficulty of computing the action potentials from the equations we had developed. We had settled all the equations and constants by March 1951 and hoped to get these solved on the Cambridge University computer. However, before anything could be done we heard that the computer would be off the air for six months or so, while it underwent a major modification. Andrew Huxley got us out of that difficulty by solving the differential equations numerically using a hand-operated Brunsviga calculating machine. The propagated action potential took about three weeks to compute and must have been an enormous labour for Andrew. But it was exciting to see it come out with the right shape and velocity and we began to feel that we had not wasted the many months we had spent in analysing records.

Separation of membrane current into its components carried by sodium and potassium. Our records showed that when the membrane potential was suddenly depolarized by a potential step of between 20 and 110 millivolts, the membrane current consisted of a brief surge of outward capacity current, followed by a transient phase of inward current, which lasted about a millisecond and was then replaced by

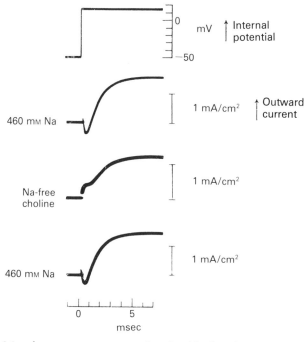

Figure 31.2. Membrane currents associated with depolarization of 65 mV in the presence and absence of external sodium ions. The change in membrane potential is shown at the top; the lower three records give the membrane current density; temperature 11°C; outward current and internal potential shown upward. (From Hodgkin & Huxley 1952a.)

a maintained phase of outward current. There was strong evidence that the early inward current was carried by external sodium ions moving into the axon and the late outward current by internal potassium ions moving out. Thus the inward current could be reversed in sign if the external sodium concentration was made low enough or the internal potential raised beyond the sodium equilibrium potential.(Figures 31.2 and 31.3) It was also shown that the potential at which the early current changed sign varied with external sodium concentration in the manner predicted by the Nernst relation. For potassium, the clearest evidence was obtained in studies with radioactive tracers which identified potassium ions as the carriers of maintained current in cuttlefish axons.

Having established that the early current was carried by sodium ions and the delayed current by potassium ions, the next step was to separate the total ionic current into its two components. Figure 31.4 illustrates the method. Here curve A shows the ionic

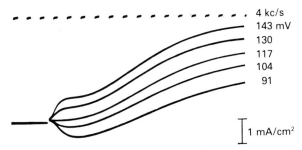

Figure 31.3. Membrane currents for different displacements of the membrane potential at a temperature of 3.5°C; outward current upwards. The figures at the right give the change in internal potential. (From Hodgkin, Huxley and Katz 1952.)

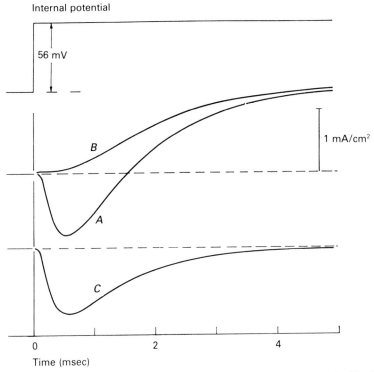

Figure 31.4. Separation of membrane current into components carried by Na and K; outward current upwards. A, Current with axon in sea water $= I_{Na} + I_{K}.B$, Current with most of external Na replaced by choline $= I_{K}.C$, Difference between A and $B = I_{Na}$. Temperature 8.5°C. (From Hodgkin & Huxley 1952a.)

current when an axon in sea water was depolarized by 56 mV. In principle the sodium current could have been eliminated by repeating the measurement with the external sodium concentration reduced to the point at which the initial hump of current was neither

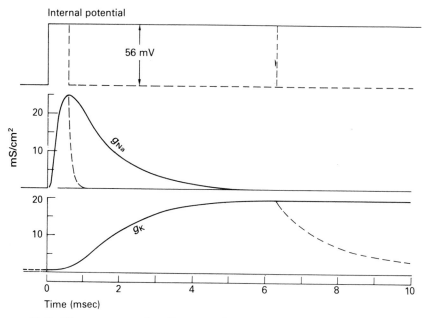

Figure 31.5. Time course of sodium conductance (g_{Na}) and potassium conductance (g_K) associated with depolarization of 56 mV; vertical scale in mS/cm². The continuous curves, which were derived from those in Figure 31.4, are for a maintained depolarization; broken curves give the effect of repolarizing the membrane after 0.6 or 6.3 msec. (From Hodgkin 1957, based on Hodgkin and Huxley 1952a, b.)

inward nor outward. In practice it would have been difficult to hit off exactly the right sodium concentration, so the procedure was to interpolate between curves in high and low sodium. Curve *B* in Figure 31.4 which was in fact very close to the record with external sodium reduced tenfold, was obtained in this way and is taken as the potassium current. On the assumption that the potassium current was independent of the sodium and choline concentration, the sodium current in record *A* can be obtained by subtracting curve *B* from curve *A* and is shown as I_{Na} in *C*.

The sodium and potassium conductances. From the individual currents it was not difficult to estimate the conductivity of the membrane to sodium and potassium ions. Figure 31.5 illustrates the changes that occurred when the inside of the fibre, initially negative to the outside by 50–60 mV, was suddenly made positive by 56 mV, a change corresponding to suddenly short-circuiting the membrane. The sodium conductance, g_{Na}, started at an exceedingly low value

294

and rose rapidly to about 25 milliSiemens per cm²; it then declined exponentially. The potassium conductance, g_K, started at a small but finite value; it did not change at once but rose along an S-shaped curve to a steady level. Both changes were graded and reversible; if the membrane potential was restored to its resting level the speed with which the conductance declined was about ten times greater for sodium than potassium.

It is important to realize that the sodium conductance may be reduced in two different ways. If, as in the example shown by the dotted line in Figure 31.5, the resting potential is restored after a short time, the system controlling sodium permeability reverts rapidly to its resting state; in this kind of experiment a second pulse, applied immediately after the first, leads to a second rise in sodium conductance. On the other hand, if the depolarization is maintained, the sodium current is reduced more slowly by a process known as inactivation. After the sodium channel has been inactivated the membrane potential must be repolarized for a few milliseconds before a second pulse is again effective. The system controlling potassium permeability in squid nerve does not show any appreciable inactivation and the potassium conductance remains at a high level as long as the nerve is depolarized.

The sodium conductance, g_{Na}, is defined by

$$I_{Na} = g_{Na}(V - V_{Na}) \tag{1}$$

where I_{Na} is the component of current carried by sodium ions, V is the membrane potential and V_{Na} is the equilibrium potential for sodium ions. A similar relation applies to the potassium conductance.

Equation (1) is essentially a definition of g_{Na} and would apply whatever the relation between I_{Na} and $V - V_{Na}$. However, its usefulness was greatly increased by the observation that under normal conditions the instantaneous value of the sodium current was found to be directly proportional to the driving force $V - V_{Na}$. It is essential to include the word instantaneous in the statement because the conductance moves towards a new value when the potential is altered, and the current is proportional to the voltage only if the time interval is very small.

Figure 31.6 shows a family of curves defining the changes in conductance associated with steps of different magnitude. The amplitude and time course of the two conductances vary greatly

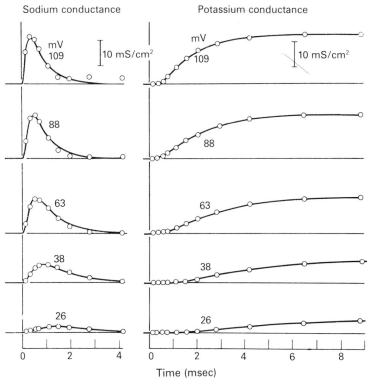

Figure 31.6. Time course of sodium and potassium conductance for different displacements at 6°C; the numbers give the depolarization used. The circles are experimental estimates and the smooth curves are solutions of equations (1) to (4) in Chapter 31. (From Hodgkin and Huxley 1952d.)

with membrane potential, but there is no sudden break or discontinuity in the relation between conductance and potential.

The normal all-or-nothing threshold of nerve arises because sodium permeability and membrane potential are linked regeneratively in the following manner:

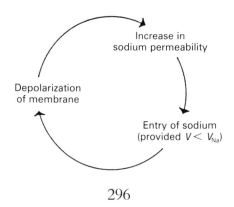

A quantitative description of membrane current in nerve. The circles in Figure 31.6 are from experimental measurements of sodium and potassium conductance and the smooth curves have been drawn from the quantitative theory which Huxley and I developed (Hodgkin and Huxley 1952d). Fitting theoretical equations to biological processes is not always particularly helpful, but in this case Huxley and I had a strong reason for carrying out such an analysis. A nerve fibre undergoes all sorts of complicated electrical changes under different experimental conditions and it is not obvious that these can be explained by relatively simple permeability changes of the kind seen in Figure 31.6. To answer such questions we needed a theory and preferably one that could be given a physical basis of some kind.

The main features to be incorporated in the theory are shown in Figure 31.6. A striking point which caused us some difficulty was that each of the conductances rose with an initial S-shaped delay, but fell along an approximately exponential curve when the membrane potential was restored to its resting value. A useful simplification was achieved by assuming that each conductance was proportional to the third or fourth power of a variable which obeyed a first-order equation. A fourth power was used for potassium and in this case the rise of conductance from zero to a finite value was described by $[1 - e^{-t}]^4$ and rose with a marked inflexion, whereas the fall was given by e^{-4t} and remained exponential but with a faster rate constant.

To account for the change in potassium conductance we assumed that a path for potassium ions was formed when four charged particles had moved to a certain region of the membrane under the influence of the electric field. If n is the probability that a single particle is in the right place, then $g_K = \bar{g}_K n^4$ where $\bar{g}_K$ is the maximum potassium conductance and n obeys first order kinetics, i.e.

$$\frac{dn}{dt} = \alpha_n(1 - n) - \beta_n n \tag{2}$$

where α_n and β_n are rate constants which depend on the membrane potential V; α_n increases and β_n decreases as the inside of the fibre becomes more positive.

For the sodium channel we assumed that three simultaneous events, each of probability m opened the channel to Na^+ and that a

297

single event of probability $(1 - h)$ blocked it. These events were not specified, but might be thought of as the movements of three activating particles and one blocking particle to a certain region of the membrane. The probability that there will be three activating particles and no blocking particle is therefore m^3h so $g_{Na} = \bar{g}_{Na}m^3h$ where $\bar{g}_{Na}$ is the maximum sodium conductance. The values of m and h are given by relations similar to equation (2):

$$\frac{dm}{dt} = \alpha_m(1 - m) - \beta_m m \qquad (3)$$

$$\frac{dh}{dt} = \alpha_h(1 - h) - \beta_h h \qquad (4)$$

The effect of making the inside of the nerve fibre more positive is to increase α_m and β_h and to decrease β_m and α_h.

It is relatively simple to apply equations (2), (3) and (4) to the voltage-clamp data. At a fixed voltage the αs and βs are constant so the equations lead to exponential expressions for n, m and h; conductances can then be calculated from n^4 and m^3h. This is how the smooth curves in Fig 31.6 were obtained. Further information about the way in which membrane potential influences the rate constants was obtained by double step experiments in which, after the conductance had been turned on by the first step, the potential was switched to a new level before restoring the resting potential. The inactivation mechanism was investigated by preceding the test step with a conditioning step of variable amplitude that lasted long enough (about 40 msec) for inactivation to reach a steady level at all voltages.

A striking property of the nerve membrane is the extreme steepness of the relation between ionic conductance and membrane potential. Thus the sodium conductance may be increased e-fold by a change of only 4 mV and the corresponding figure for the potassium conductance was 5 mV. In a physical device, such as a vacuum tube, crystal rectifier or transistor, an e-fold change in conductance is usually associated with a potential change of the order of kT/e . At room temperature kT/e is 25 mV and, from the Boltzmann equation, one would expect that a similar quantity would apply to any system in which the conductance is controlled by the presence or absence of a single electronic charge.

Since the ionic conductances of the membrane change *e*-fold in 4–5 mV it would seem that they must be brought about by the simultaneous movement of 5 or 6 charges. These charges might all be located on one particle or, more probably, several particles with a smaller charge might be involved at each site. The assumptions that Huxley and I used to describe the changes in conductance were a compromise between the ideas of several singly charged particles and one bearing multiple charges. Thus the assumption that the potassium conductance is proportional to n^4 implies that four particles are involved at each site, whereas the expressions used for the variation of α_n and β_n with membrane potential were roughly consistent with each particle being divalent.

The steepness of the relation between conductance and membrane potential must be of value to the animal since it enables the nervous systems to work with voltages that are much lower than those normally used in man-made computers. On the other hand, although they are efficient in this respect, ionic gating systems in membranes are many orders of magnitude slower than their electronic counterparts in any ordinary computer.

Reconstruction of electrical behaviour of nerve

Although the quantitative theory which Huxley and I developed was partly empirical it accounts satisfactorily for many aspects of a nerve's behaviour.

The expression for the membrane current density, I, is

$$I = \frac{CdV}{dt} + (V - V_{\text{K}})\bar{g}_{\text{K}}n^4 + (V - V_{\text{Na}})\bar{g}_{\text{Na}}m^3h + (V - V_{\text{L}})\bar{g}_{\text{L}} \tag{5}$$

The first term on the right-hand side is the capacity current, C being the membrane capacity per unit area. The second and third terms give the potassium and sodium current, while the last term which is relatively unimportant gives the current carried by ions other than sodium and potassium through a constant leak conductance $\bar{g}_{\text{L}}$.

The simplest case to deal with is the membrane action potential in which all parts of the membrane are activated synchronously by a brief shock applied to a length of nerve; here, I in equation (5) is zero after the initial displacement. In Figure 31.7 the lower part illustrates what happens in an actual nerve following the displace-

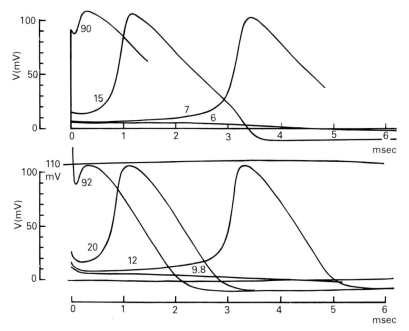

Figure 31.7. Upper curves, theoretical solution for different initial depolarizations of a uniform area of membrane. Lower curves, tracings of membrane action potential at 6°C obtained on same axon as that which gave Figure 31.6. The numbers attached to the curve give the strength of the shock in nanocoulomb/cm². (From Hodgkin and Huxley 1952d.)

ment of the potential by a short shock. The upper curves are the numerical solutions obtained by Huxley from equations (2)–(5) which describe the changes in sodium and potassium conductance. The agreement between real and model nerves is clearly satisfactory.

Calculation of the form and velocity of the propagated impulse is more difficult but of greater general interest. Here equations (2)–(5) are used with the well-known relation for the current density in a continuous nerve fibre surrounded by a large volume of external fluid, that is

$$I = \frac{a}{2R} \frac{d^2V}{dx^2} \tag{6}$$

where a is the radius of the nerve fibre, R is the resistivity of the axoplasm inside it, and x is distance along the nerve. In the case of a fibre propagating at constant velocity, θ, x may be replaced by $-\theta t$. I in equation (5) is then replaced by

300

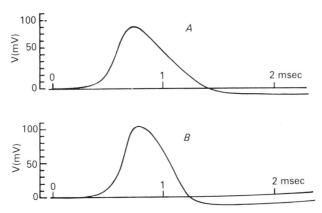

Figure 31.8. Propagated action potentials in *A*, theoretical model and *B*, squid axon at 18.5°C. The calculated velocity was 18.8 m/sec and the experimental velocity 21.2 m/sec. (From Hodgkin and Huxley 1952d.)

$$I = \frac{a}{2R\theta^2} \frac{d^2 V}{dt^2} \tag{7}$$

In the resulting second-order equation, the conduction velocity, θ, is constant but its value is unknown at the beginning of the computation. The procedure is to guess a value for θ and start a trial solution. It is found that V goes to $\pm \infty$ according to whether θ has been chosen too high or too low. The correct value of θ, which corresponds to the natural velocity of propagation, brings the potential back to its resting value at the end of the run.

A numerical solution along these lines was worked out by Huxley and was found to agree with the behaviour of a real nerve in the following respects: the form, amplitude and velocity of the action potential (Figure 31.8); the time course and amplitude of the conductance change; and the movements of sodium and potassium during the action potential.

Having carried out the reconstruction of the propagated action potential, it is possible to give a clear picture of the sequence of events during the nerve impulse. Figure 31.9 shows the calculated variation of sodium and potassium conductance during the theoretical action potential. As the impulse advances, the potential difference across the membrane just ahead of the active region is altered by electric currents flowing in a local circuit through the axoplasm and external fluid; this causes a rise in the conductance to sodium ions which enter, making the inside positive and giving the current

301

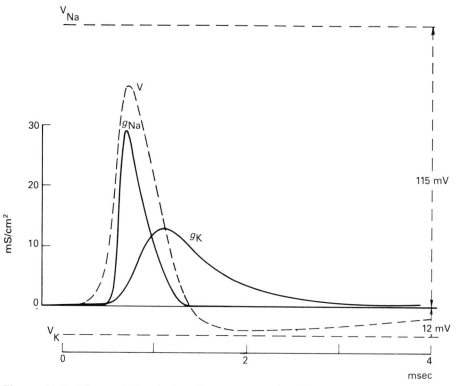

Figure 31.9. Theoretical solution for propagated action potential and conductances at 18.5°C. (From Hodgkin and Huxley 1952d.) Total entry of sodium = 4.33 pmole/cm²; total exit of potassium = 4.26 pmole/cm².

required to activate the next section of nerve. At the crest of the impulse the slower changes which result from depolarization begin to take effect. The sodium conductance declines and the potassium conductance rises so that the rate at which potassium ions leave the fibre exceeds the rate at which sodium ions enter; this makes the potential swing towards the equilibrium potential of the potassium ion. As the potential approaches the resting level any sodium conductance which has not been inactivated is cut off, so the rate of repolarization may be accelerated. This last effect is conspicuous in frog myelinated nerve fibres and although it is not obvious in a squid nerve fibre at room temperature, at 6°C both the experimental and theoretical action potentials show it quite plainly, as can be seen in Figure 31.7. (Temperature has a large effect on the rate at which the permeabilities change and in this way alters the form of the action potential.)

The slow effects of depolarization, raised potassium conductance and inactivation of the sodium-carrying system, persist for a few milliseconds and give rise to the refractory period. About 10 msec after the spike, the fibre is back in its original condition and can conduct another impulse of the same form as the first. The only difference is that it has gained a small amount of sodium and lost a similar amount of potassium ions. These quantities can be calculated theoretically and are in good agreement with those found experimentally. The number of sodium and potassium ions which cross the membrane during the impulse is small compared with the number inside a giant nerve fibre and, except in very small non-myelinated fibres, it takes many impulses to make an appreciable change in the internal concentration. However, the fibre must obviously pay off the debt incurred and it does this by a slow process which requires metabolic energy, during the period which follows a burst of electrical activity. Experiments dealing with this aspect are considered in Chapter 34.

In addition to the membrane and propagated action potentials, the equations accounted satisfactorily for recovery during the relative refractory period and a wide range of subthreshold behaviour including accommodation, anode break excitation and the oscillatory response of the membrane to a rectangular pulse of current.

The 1952 analysis in retrospect

Huxley and I were very pleased that our equations were good at reconstructing, and to some extent explaining, the mechanism of conduction and excitation in nerve, but we knew that we had made several fairly drastic assumptions, and would not have been surprised if new experimental evidence called for substantial revision of our theory.

Although it is rather like skipping to the end of a novel to see whether the heroine marries the hero, I shall devote a paragraph or two to consideration of some of the very interesting results that have appeared since our paper was published in 1952. As I left the nerve field fifteen years ago, my account will not be that of an expert, nor will it be particularly up-to-date. The scientific reader who would like to pursue these matters further will find good general accounts in the books by Hille (1984) and Aidley (1989).

Separation of ionic currents by drugs. First, it is a relief to learn that in most excitable tissues there are separate channels for sodium and potassium, and that their density is relatively low, as we thought might be the case. Life has been made much easier by the discovery that the virulent nerve poisons, tetrodotoxin and saxitoxin, block the sodium channel when applied to the outer surface of the nerve membrane but have no detectable effect on the potassium channel, or, when applied internally, on the sodium channel. At the node of Ranvier in vertebrate nerve the potassium current is blocked by tetra-ethylammonium ions without affecting the sodium channel. With these agents Hille was able to obtain beautifully clear separations of the ionic current into sodium and potassium currents which look remarkably like those that we calculated so laboriously from records taken at different sodium concentrations in the squid giant axon. Considering that the squid is a very distant relative of man – our last common ancestor, of whose form we are totally ignorant, died several hundred million years ago – this similarity in behaviour points to the survival value of the sodium channel in the animal kingdom.

Another advantage of tetrodotoxin is that it can be used to label sodium channels, which helps in estimating their density or in determinations of their chemical composition. Incidentally tetrodotoxin, one of the most poisonous substances known, is obtained from the gonads of the Japanese puffer fish, which is considered a great delicacy in that country, but should only be eaten in restaurants licensed to prepare it properly. In 1774 Captain Cook nearly died from eating puffer fish from which the gonads had not been properly removed.

Studies of tetrodotoxin binding, or gating currents, indicate that the density of sodium channels is about 300 per square micrometre in the squid giant axon and some ten times higher at frog nodes. The density of potassium channels in the giant axon is perhaps 70 per square micrometre but this measurement is harder to make.

Other evidence confirming the essential correctness of the separation into sodium and potassium conductances during the action potential comes from the work of Atwater et al (1969) who used radioactive tracers to prove that the inward current is carried exclusively by sodium ions. The work on internally perfused and dialysed axons, which is considered further in Chapter 34, was another important element in establishing the ionic basis of the action potential.

Gating currents. Our equations were based on the idea that the ionic channels are opened by the movement of charged particles within the membrane. A necessary consequence of this idea is that the conductance changes should be preceded by a small current associated with the opening of the sodium or potassium channel. Huxley and I looked for such 'gating' currents, but failed to see anything when pulsing to the sodium equilibrium potential, which is where we felt we had the best chance of seeing such currents. It was partly for this reason that we concluded that the density of sodium channels must be rather low. Eventually such currents were detected by authors on both sides of the Atlantic and are proving helpful in elucidating the tricky mechanism underlying the opening and closure of the sodium channel (Armstrong and Bezanilla, 1973; Keynes and Rojas 1974). However, it has to be said that although the existence of gating currents agrees with our theory, their exact form does not and there seems to be disagreement as to what this means.

Patch clamping. This is an important new technique introduced by Neher and Sakmann in 1976, that enables a very small area of membrane to be voltage-clamped and so provides information about the current flow through individual channels. The method was first applied to the acetylcholine-sensitive channels of denervated muscle, but a few years later Sigworth and Neher (1980) extended it to a tissue culture preparation of embryonic rat muscle containing sodium channels. The records obtained consisted of unitary square pulses of current of constant amplitude, but varying considerably in duration and frequency. A single channel is either open or closed, and there are no obvious transitional stages. The sodium channels were identified as such from the blocking action of tetrodotoxin and the effects of sodium concentration and membrane potential on single channel current.

When the membrane is depolarized the probability of a channel opening is increased so the conductance of a large number of channels increases smoothly. After inactivation or repolarization, the number of open channels falls so the overall conductance decreases. All this seems to agree reasonably well with a restatement of our 1952 equations in molecular terms (Clay and de Felice 1939).

Molecular structure of sodium channel. Using recombinant DNA techniques Numa and his colleagues in Kyoto have carried out a remarkable study of the amino-acid sequence in sodium channels from the electric organ of the electric eel. This channel has a molecular weight of about 260 kiloDaltons and consists essentially of a single polypeptide chain 1,820 amino acids long. The Kyoto workers consider that their evidence is consistent with a bead-like structure in which sodium ions pass though a central hole lined by negative charges, and the walls of the bead are composed of four nearly identical quadrants each containing six rod-like α-helices, Noda *et al.* 1984; Aidley 1989. Catterall (1986) has discussed the way in which such a structure might give a total gating current of six charges per channel, this being the valency required by our analysis.

Chapter 32

Everyday life, holidays, conferences, 1946–53

THE LAST three chapters have been largely technical because I did not want to interrupt the continuous scientific thread that runs through the experiments and analysis between 1946 and 1952. But although I worked very hard, it would be quite wrong to suppose that I did nothing except work, or that we remained lashed to the mast in Cambridge. So I will switch back to early 1946 and pick up our Cambridge life where I left it at the end of Chapter 27.

By January 1946 life in Cambridge was beginning to assume a fairly normal aspect, as can be seen from the synthetic 'diary notes' that I made when reading through my wife's weekly letters to her parents in New York.

13 January 1946. Our daughter Sarah unwell; met Mark and Sophie Pryor. My brother Keith, a doctor on a destroyer in the Adriatic, has been 'mentioned in despatches' for looking after casualties incurred when his sister ship was blown up by a mine.

20 January 1946. Skating on Grantchester meadows. My attempts to get the typical photographic record of one of my experiments. (Typical scientific records are hard to get because everything has to work perfectly, which it rarely does). Dick David, a great friend, is back from the Navy.

27 January 1946. In Barbara Rothschild's black books. Met Arthur Marshall and Ann Barnes.

4 February 1946. I am to go on Trinity Council. I find that I have made a mistake in my experiments. I receive a medal from the Société Philomathique in Paris, my first, but unfortunately it is inscribed to the wrong person! Keith is coming to stay; he has bought an island in the

West of Scotland for about £500; tribute will have to be paid to the Duke of Argyll!

10 February 1946. Gow to lunch; acid about everyone including Trevelyan.

18 February 1946. Pampering weekend at Tring with Victor, who plans to stay with Marni's parents in New York. Met Peggy Ashcroft and Jeremy Hutchinson (her husband), also Stuart Hampshire and C. Day Lewis.

24 February 1946. Both Tess and Victor going to New York for entirely different reasons. Tess might stay with Marni's parents. Victor too might stay. Our suspicions are hardening. Barbara talks to Marni about her marriage to Victor: 'I never thought he cared. But he remembers every man – some I'd forgotten myself.'

3 March 1946. Crocuses drowned in snow. Jean is delighted as they never see snow, except on the mountains, in Islay. Bernard Katz staying overnight, David Hill and Andrew Huxley to dinner. M. working on her detective story.

Victor has left a flock of goodbye notes; says he is in a state of neuroticism that makes him unbearable. Peyton and Marion will soon see him.

M. has seen Little Tackley, Barbara Rothschild's new home; it's been done up for £17000 on black market by Crumbs Crowther who will keep Squire and Vicar quiet. M. doesn't mind Barbara's sexual immorality but is deeply shocked by this.

With Barbara she met John Betjeman, who thought M. was the governess and was greatly taken with her.

10 March 1946. News that Peyton and Marion have visas and are coming to stay with us in mid-June.

18 March 1946. 'Glad that you liked Tess and Victor' whom the Rouses saw separately.

Pat and Richard Llewellyn Davies to dinner and then to *Antony and Cleopatra* with us. Marni has a part-time job – agenting for Scott books.

24 March 1946. 'Your food packages are a life line'. As a result of which Marni is getting a reputation as a good cook. The Vice Master, Winstanley, to lunch; we gave him shrimp which were much

appreciated. Gow, in the Evelyn Nursing Home, complains that their fish is like steamed blotting paper so we have promised him tuna.

Marni electrified to see a green woodpecker on our lawn – such a large and brilliant bird for an English garden.

Islay, April 1946

In early April we spent ten days in Islay, this being our first real holiday since we were married. We took Sarah (aged one) and Jean to Edinburgh and left them there with my mother. Jean didn't at all mind not going home, partly because she had never seen Edinburgh and partly because she liked having Sarah to herself.

In her letter to New York, Marni makes light of an awful night journey to Edinburgh, with Sarah wakeful in a crowded carriage and concentrates on the lovely flight from Glasgow to Islay, with six passengers in a tiny plane.

Sadly, our house, Newton, was no longer available, as the Morrison estate needed it for the vet, but we were comfortable enough in the Bridgend Hotel, who provided us with massive amounts of eggs, cream and butter. We had no car and so visited the sacred spots that I wanted Marni to see, on foot or bicycle. Her letters to New York say that one day we bicycled twenty-two miles and walked six, seeing seals and great flocks of barnacle geese; and on another we got ferried across to Jura where we visited a heronry and climbed the nearest of the three Paps, on the way watching an eagle patrolling the valley, and many deer keeping an eye on us over the brow of the hill.

As we had arranged with Jean, we visited her mother and found her in a croft so small that it was difficult to believe that it could ever have contained two parents and four daughters.

By way of a change from the beauties of Nature we called on Barbara Rothschild's friend, Mr Hunter, who ran the distillery at Port Ellen, where they make Islay Mist. As we had heard from Barbara, Mr Hunter, who gave us a warm welcome, had had what he described as a 'wee shock', i.e. a slight stroke, which I imagine must be an occupational hazard for a professional whisky taster. But it didn't seem to cramp his style and we greatly enjoyed our visit. Marni wrote that I got quite tight which is probably true, as we both like Islay Mist and she has a stronger head than I.

For our journey home I had the good sense to order sleepers and wonder now why on earth I had not done this before for the northward journey.

Shortly before we left Islay we got a letter from Marion in New York confirming what we had already guessed, that Victor and Tess were definitely going to get married as soon as the divorce became absolute. We were very pleased – though saddened for Barbara whom we liked very much and who had done a great deal for us.

Pilar, Arcachon, August 1947

Our next holiday was not for more than a year but was worth waiting for. In the meantime life in Cambridge proceeded tranquilly. With Andrew back, my work flourished and, during the summer, life at home was enlivened by the presence of Marni's parents and sister Ellen. They greatly enjoyed meeting our many friends and relations, whose connections were sufficiently complicated to defeat even Marion's wonderful memory, so that she was reduced to the expedient of keeping a small notebook of genealogical detail. The dreadful winter of 1946–7 has found its way into an earlier chapter, since the power-cuts forced me to abandon experimental work and concentrate on theory – and eventually to spend all my energy in domestic tasks such as trying to keep our house warm or, when the thaw finally came, in preventing the burst pipes from flooding us with water.

Our second daughter Deborah was born on 2 May in the Radcliffe Maternity Wing in Oxford, as Sarah had been. Marni, who had liked the Radcliffe, had accepted a warm invitation from Hugo and Barbara Cairns to stay with them in Oxford. I drove her over in mid-April and brought her back a month later. With help from Jean and my mother, who joined us for some of the time, we managed to look after Sarah and keep things going at home in a reasonably satisfactory manner. But I was very glad to have Marni home in mid-May.

At the end of the bitter winter of 1947, Victor and Tess, who were married the previous August, invited us to stay with them for a few weeks in a house near Arcachon that they had taken jointly with their friends the de Gunzbourgs. As one can see, there were considerable logistic difficulties. What about our Sarah and the new baby? the Oxford Congress? my planned visit to Plymouth? and what about Marni's mother and sister who were to come to us for two months in the summer? A further complication was that the Rothschilds hoped that we might bring out Sarah and Jacob, the two elder children of the first marriage. At first the idea looked impossible but, after the bitter cold and hardship of the previous winter, we were reluctant to give it up. So with the help of others the following plan was concocted.

Marni's mother and sister, who were with us from mid-July to mid-September, would look after our Sarah (with help from Jean) for three weeks in August while we took Deborah to Arcachon, where the de Gunzbourgs' English Nanny would welcome another baby and could give Marni any help that she needed. Victor arranged that a small plane would fly us and his two elder children from Cambridge to Bordeaux. All of which worked perfectly and gave us a delightful journey in which we could really see the French towns and countryside; the plane flew so low that we could even see people playing tennis – something that would be impossible in a modern journey by air. Apart from the Loire châteaux the only details that I remember are the La Rochelle salt pans, looking like a gigantic greenhouse in the way that they caught the sun, and the innumerable bomb-craters in every Normandy field, clearly marked after being filled in with concrete, for what reason I now cannot imagine.

Deborah was no bother at all as she lay tranquilly in her carry-cot and was fed at regular intervals. Indeed on the way home she was a positive help as the customs men all made the same joke, 'No duty on that!' as they waved Deborah's cot through without looking at the wine, which we intended to declare, but for safety had packed in her bedding.

At Bordeaux we were met by Victor, looking brown and very French in a beret, who drove us rapidly to Pilar beyond Arcachon in his magnificent open Bentley. On arrival we were greeted warmly by Philippe and Antoinette de Gunzbourg, to both of whom we became very much attached and they, I think, to us.

Philippe was more like Pierre Bezhukov in *War and Peace* than any one else I have met. He had spent much of the war with the Resistance, often in discomfort and danger but also with enjoyment, as he felt that this was the only worthwhile part of his life. Now, he tried to generate some of the same feeling by hard work and simple living on one of his country estates. But I think happiness escaped him and after a while his marriage to Antoinette

311

broke up. Probably the first signs of trouble with the marriage were there in 1947, but we were too happy and comfortable with them to want to look beneath the surface.

Antoinette too had had an adventurous war. After the Germans had invaded southern France she led a party of children, including her own, across the frontier into Switzerland. We found it hard to think of this elegant Parisian lady wading rivers and climbing mountains at night, with a crowd of frightened children and only an English Nanny to help her. But she managed this, apparently without turning a hair, and had made the transformation back to peacetime life more successfully than her husband.

Deborah aged three months was very popular with the English Nanny, Joyce, who was also looking after Hélène, the youngest de Gunzbourg and a month or two older than Deborah. I remember that the two babies were wheeled out at opposite ends of a gigantic pram. They were much admired, but Nanny, or 'Mees', as she was generally known, was mortified by the contrast between the jewellery which covered Hélène, in the manner of French babies, and Deborah's unadorned nun's veiling. On returning from an expedition we were amused to find that parity and English honour had been restored by the attachment of a conspicuous diamond brooch (borrowed) to Deborah's baby clothes.

The food was marvellous particularly after the chef had discovered that we had good appetites. He had heard much of food shortages in Britain during and after the war and thought that it would be unwise to subject our constricted English stomachs to too much initial strain. But we soon removed that notion and enjoyed unlimited quantities of delicious French food, especially vegetables.

When tired of swimming or sunbathing we sat on the veranda reading or watching the tide sweeping in or out. Arcachon bay is a vast, nearly land-locked lagoon which fills or empties through a relatively narrow channel of which our villa had a good view. We knew that you had to be careful not to swim too far from the shore, and that sailing dinghies had to be moored securely or they would be carried out to sea and pounded to bits by the Atlantic breakers on the sand bar at the mouth of the bay. Sure enough, our lunch was interrupted one day by the sight of an empty sailing dinghy drifting rapidly towards the ocean. Marni and I jumped to our feet planning to take the family canoe and rescue the dinghy. The de Gunzbourg butler, Henri, strongly advised against such action which he regarded as officious and liable to land us in trouble. When we persisted he shrugged his shoulders, muttered, 'C'est la scoutism' and gave up the argument. Without much difficulty we managed to board the dinghy, beach it and tie it up reasonably securely. But Henri was to some extent justified because the only thanks we got from the owner was abuse for not having returned the dinghy to its proper mooring.

I don't care for indefinite sunbathing but happily we were lent a small sail-boat which I was just able to manage, though we had a nasty moment when all the occupants of a crowded beach rose to their feet shrieking, 'Attention au tuyau!' as we were about to run into one of those submerged pipes, which rather unhygienically carried sewage into the bay. But here it was my French rather than my knowledge of sailing which let us down, as I hadn't the faintest idea what a tuyau was or why everyone should be so agitated. By sheer luck we just cleared it.

Arcachon and Les Landes are not a particularly beautiful part of France, but we made one or two pleasant expeditions, including one to Chateau Lafite, where we were told that the house was not yet open so they could give us only a very light lunch. Bearing this in mind, we did ourselves well on what we took to be the main course, only to find that it was followed by a string of courses of increasing grandeur. To make matters worse, we soon remembered that Victor had ordered an enormous and delicious dinner to be eaten in Bordeaux on the way home. I am afraid this was an occasion when the French comment about the constricted stomachs of the English had some justification, and we were forced to toy with the splendid dinner in Bordeaux. It had been a dry summer and on the way south from Bordeaux the night sky was brilliantly lit up by forest fires burning over the horizon.

Another memory of that holiday is of playing beach games with the four children, Patrice and Jacques de Gunzbourg and the two Rothschild children. This brought home to me the fact that a good athlete usually cannot bear to lose even a children's game. This was certainly true of Victor who ran about so swiftly and vigorously in one game that he blistered the soles of his feet badly enough to bring that particular game to an end.

Towards the end of the holiday, I wrote a long letter to Cole telling him about the sodium theory and the experiments that I had done to test it in July. This got me in the mood for the gruelling but exciting month of experimental work with Katz that was coming up when I got back to Plymouth at the end of August 1947.

Isle of Barra, September 1948

Apart from the American trip which, though enormously enjoyable, was really work for both of us, our next genuine holiday, in the Hebridean island of Barra, was in every way a complete and absolute contrast to the one at Arcachon. We went there partly because I like islands and had been told that this was a particularly nice one, and partly because it was sufficiently small for us to be able to explore the entire island without needing a car. This was important as I had been forced to use six months of my petrol ration in transporting equipment to Plymouth.

The flight to Barra was one of the most remarkable that I have ever done. Again, we flew in a tiny plane and this time landed on the great sandy beach at the north end of Barra, which has wonderful views of the mountains in Rhum and Skye as well as those in the neighbouring island of South Uist. On the way we flew past a golden eagle and for some time appeared to be advancing into a completely circular rainbow. But views are not everything and we approached the Castle Bay Hotel, where we were to stay, with some curiosity. We had heard two things about it which gave us pause. The first was that the food was not good, which proved true in the sense that it was no better and no worse than the average for the West Highlands or Eire, though why Celts should be poor cooks is to me a mystery.

The second potential drawback was that the hotel was said to be full of actors playing in the film *Whisky Galore*. This also was true but was in no way a drawback as the people we saw most of, namely Jean Cadell and James Woodburn, with whom we shared a table in the dining room, were exceptionally nice and full of interesting information.

The first thing we learnt about the film was that shooting it should have been finished several weeks earlier, but had been delayed for the opening shot of the MacBrayne steamer coming into the harbour. That didn't sound very difficult until we learnt that the steamer only called twice a week and *the harbour had to be in sunlight when it arrived*, an unlikely coincidence, as any one who knows the Hebrides will tell you.

At first we were puzzled by the selection of people in the hotel and realized how lucky we were to have succumbed to my obsession for long-range planning by booking us in many months in advance. For there evidently was keen competition to get into the hotel rather than be boarded in a croft, which though romantic was not particularly comfortable. It seemed reasonable that an elderly distinguished actress like Jean Cadell should be in the hotel, but how was it that James Woodburn, a relatively minor character-actor, was there when a star like Joan Greenwood was in a croft?

James Woodburn, who played the postmaster in the film and knew all about the Highlands, supplied the answer. 'When the agent was drawing up my contract,' he said, 'I asked him to put in an additional clause saying "No E.C." Never mind what it means. Just put it in and they will understand.' Now E.C. then stood for Earth Closet and Woodburn had guessed, correctly as it turned out, that acceptance of that condition would guarantee him a place in the hotel.

We got very friendly with the cast of *Whisky Galore* and my children have never forgiven me for declining to be an extra in one of the scenes they still had to film. But the part was that of one of the hated excise men and would have involved a great deal of waiting around. Besides, I am a rotten actor, as I discovered at school where initially I was in demand for female parts because of my somewhat girlish looks and ability to memorize, but turned out to have no real talent at all.

On fine days we did the usual things, like walking all round the island, or climbing the highest mountain, which was not very high but gave us a wonderful view from St Kilda far to the west, to the Paps of Jura, just appearing above the horizon, seventy miles to the south east. One of the fishermen ferried us over to the neighbouring island of Vatersay, which is quite different in character, with lovely flower-strewn machair and white shell beaches on the Atlantic side. Unlike Barra it is largely Roman Catholic in faith, but although tempting to do so I think it is probably fanciful to attempt to correlate the islands of Catholicism with the existence of sandy beaches.

Autumn and winter 1948–9

The six months after our holiday in Barra passed very quickly, probably because I was exceptionally busy. In Trinity I was still on

315

the Council and found myself directing the studies of some sixty medical students. This exceptionally large number arose because Rushton was spending a well-earned and highly productive sabbatical year in Stockholm, and Bailey, who was to take over from Roughton, was to be let off lightly for his first year. I am afraid that I was reduced to taking supervisions in groups of three or four, which is something I neither like nor approve of. However, the microelectrode experiments with Nastuk went smoothly so that I think of this period as happy and successful.

That autumn our technician R. H. Cook helped us to make an unexpected advance. It turned out that he was a keen marine biologist and had been able to keep cuttlefish alive for a few weeks in a small aquarium in his home near Cambridge. I have always been very keen on extending the range of animals studied by physiologists and had used some of our Rockefeller money to get an aquarium built in the basement. Cook wanted to take charge of the aquarium, which he did most successfully, and from the autumn of 1948 we had cuttlefish, scallops and other marine animals living there quite happily. Not squid, unfortunately, as these active but fragile creatures invariably damage themselves in transit. However, the cuttlefish which are much more robust do have giant nerve fibres which, though smaller than those from the squid are much bigger than anything we could get from crabs and lobsters. Richard Keynes found them to be good material for his measurements of the entry of radioactive sodium associated with the nerve impulse.

The aquarium was a pleasant port of call for the children of scientists working in the Laboratory, who often looked in there to see whether there was anything new in the aquarium or how an old friend was getting on (Figure 32.1).

The Paris Colloquium, April 1949

In the early spring of 1949 Alfred Fessard and Ali Monnier organised an excellent International Colloquium on recent advances in electrophysiology. The meetings which were supported by the CNRS were held in the Sorbonne or in the Institut Marey. They tended to start early and last late which knocked out the idea that I might do some sightseeing with Marni who had come with me. However, Richard Keynes had brought his wife, Anne, who joined forces with Marni in exploring Paris and visiting picture galleries.

316

Figure 32.1. With daughter Sarah, spring 1949.

There was a good selection of people from overseas, including Lorente de Nó, Kuffler and Cole from the USA, Katz, Gray, Huxley, Keynes and me from Britain, Wyss and Stämpfli from Switzerland and Bernhard, Skoglund and T. Teorell from Sweden. We were very well looked after by the Monniers and Fessards who entertained every night and thereby managed to have the entire conference to private dinner parties. The papers were followed by much discussion, some of it hard-hitting, but all reasonably good-natured – something that I attribute mainly to the friendly atmosphere created by our hosts.

This conference saw the opening rounds of a battle between Lorente de Nó and his critics over the properties of the sheath, or perineurium, which surrounds vertebrate nerve trunks. The background to the controversy explains why it was eventually fought so fiercely by Lorente. Between 1937 and 1946 Lorente carried out a massive investigation of the effect of electric currents on frog nerve under different experimental conditions. The preparations employed were usually the sciatic nerves of frogs. These nerves were not desheathed, as they would be nowadays by anyone wishing to investigate the effects of salts or other chemicals on nerve fibres. Like countless biologists before him, including for a time the present writer, Lorente assumed that a stocking made of a web of connective tissue threads would be unlikely to constitute much of a diffusion barrier. He satisfied himself that it didn't by some rather flimsy arguments and went ahead with his experimental analysis, which was published in 1947 in two volumes each about 500 pages long. This made it difficult for him to accept what presently emerged, which is that Nature, who had so generously provided experimentalists with nerves wrapped in neat connective tissue stockings, had laid a devilish trap for them by including in those stockings a very thin layer of flattened, pavement-like cells, which is only very sparingly permeable to salts. (This layer forms part of the blood–brain barrier and helps to ensure that neurones have a somewhat different environment from ordinary cells.) The existence of a layer of flattened endothelial cells in the perineurium was known to classical microscopists like Ranvier (1875), Key and Retzius (1876) and Sharpey-Schafer (1929), but its importance as a diffusion barrier was not recognized by Lorente who insisted (1950: p.238) that 'frog nerve fibres may conduct impulses for several hours after the concentration of sodium ions in their external medium has become utterly negligible.' This was after Feng and Liu (1949a, b) and Huxley and Stämpfli (1949, 1951b) had shown convincingly that the action of potassium ions or of removing sodium ions was greatly accelerated by slitting or removing the sheath.

I did not like this controversy, but there was no way of avoiding it as Lorente's writings were taken seriously by many electophysiologists and silence at a meeting would have been taken as consent. Fortunately, however, we did not have too much of this at the Paris meeting, and the Cambridge group concentrated on recent experiments. Huxley talked about our early voltage-clamp experiments with Katz and the carrier theory that we at first used to explain them, but subsequently abandoned; Keynes described his radioactive tracer experiments, measuring sodium entry and potassium loss during the electrical activity of crab and cuttle fish nerve fibres, and Stämpfli talked about his work with Huxley on saltatory conduction in vertebrate nerve fibres. I saw a good deal of Stephen Kuffler, and we arranged that he would stay with us in Cambridge and that we would try out a few experiments with microelectrodes on a frog muscle containing slow fibres on which he had read an interesting paper. I very much enjoyed this brief collaboration and think we were a little unlucky in that all the muscle fibres we impaled were fast fibres with normal action potentials. But perhaps this was just as well as I already had quite enough on my plate.

After the Colloquium, Marni and I spent a few days in Vézelay, where we had lovely spring weather, and divided our time beween the splendid mediaeval sculpture in the cathedral and gentle walks in the valley of the River Cure which runs below the town.

Italy, April 1950

Paris and Vézelay had whetted our appetite for travel in Europe and we were determined to get to Italy, where Marni had much-loved cousins, as soon as possible. However, this could not be for some time as she was expecting another child in late August 1949 and I had arranged another squid season with Huxley and Katz from mid-June to mid-August. All this worked out, more or less as planned. Our son Jonathan was born in a Cambridge nursing home on 24 August and we acquired a new and reliable mother's-help who went with the younger two children to the Rothschilds while Sarah, now five, stayed with Nora David with whom we had worked out a happy child-sharing arrangement.

At that time each person was allowed to take out of the country only £35 of money, either English or foreign. So we planned to fly to Rome and then make our way as cheaply as possible via Orvieto, Assisi and Arezzo to Florence where we would stay with Marni's Cousin Edith Rucellai who lived at the top of the Palazzo Rucellai in the heart of Florence. We flew by Alitalia to Rome over the Alps, breathing oxygen from a not particularly efficient arrangement, which gave Marni the typical headache associated with oxygen-lack. However, she recovered quickly and after catching the train we wanted we soon found ourselves in a romantic hotel in Orvieto, as we had decided to be just a little bit extravagant to begin with.

319

By the time we got to Arezzo, having stayed briefly in Assisi, we had made a large hole in our £70 and were reduced to the noise and squalor of the Albergo Autostazione in Arezzo. However, I am very glad we did make the effort and spent a long time looking at the wonderful Piero della Francesca *Legend of the True Cross* because those particular frescoes have deteriorated rapidly and seem to have resisted the best efforts of the highly skilful Italian experts to conserve or restore them.

By that time, we had discovered that travelling by bus in Italy was a cheap and pleasant way of seeing the countryside and managed to fit in a visit to San Sepolcro in order to see the Piero *Resurrection*, a painting that Aldous Huxley helped to popularize by describing it as the greatest picture in the world.

Still, although we enjoyed Arezzo we were glad to be translated from the honking, bustling, bathless milieu of the Albergo Autostazione into the pages of a novel by Henry James. It was indeed a translation from rags to riches and you had but to change your shirt to find it whisked away and returned in no time, beautifully washed and ironed. Breakfast was brought to our room but, as we had hoped, we ate our other meals with Cousin Edith at the top of the Palazzo where we were waited on by her ex-coachman, Alfredo, wearing white gloves and a smart uniform.

The food was out of this world as, in a different way, was the conversation. Edith (née Bronson), who was then 89 but didn't look it, was of wholly American descent, but had lived all her life in Italy, having been brought up partly in Venice and partly in Asolo, until she married Cosimo Rucellai in 1895. After Cosimo's death, her son Bernardo and his wife Christina took over the main part of the Palazzo, while Edith lived in the charming apartment at the top. She wasn't a great talker but conversation covered a wide range of time and place.

On our first evening, I mentioned that I had been reading Iris Origo's book *The Last Attachment*, based on recently discovered letters between Byron and his last flame, the much younger Teresa Guiccioli. I said how much I had enjoyed this, to which Cousin Edith replied, 'I do think it a pity to rake over the love-affairs of persons so recently dead. I discouraged Iris Origo as much as I could, but of course it's the fashion now.' And strange as it seemed then, and even stranger now, Byron and Teresa Guiccioli were to Cousin Edith 'persons recently dead' as Teresa lived to a great age and was well known to Edith and her mother.

We were relieved to find that Florence had not been badly affected by the war. All the bridges except the Ponte Vecchio were down and had been replaced by temporary pontoon affairs; but the bridges would be repaired and restored to their original form. We were pleasantly surprised that Florence was rather empty, which is not something you would think nowadays in mid-April.

The trouble about Florence is that there is too much to see, and unless you are very strong-minded you can easily crowd too much into a few days. We decided that we must come back in a year or two and I think managed to avoid the trap of 'over-sightseeing'. Our time was indeed nearly up and we soon had to catch the train to Rome and plane home.

Copenhagen, XVIII International Physiological Congress, 15–18 August 1950

This was a well-organized conference with many interesting papers. Richard Keynes and I went by sea together, leaving our wives and families to have a joint holiday in the Adrian house on the Norfolk coast, where we joined them after the Congress had finished. This was the first of several such holidays, mostly in Norfolk but one at Bamburgh in Northumberland, all of which were much enjoyed by both grown-ups and children.

Huxley and I were still in the middle of analysing our voltage-clamp records, but the qualitative results were by then pretty clear and made a nice story. Keynes presented his tracer studies of sodium gain and potassium loss during the impulse; and he and I also gave a short paper which showed that radioactive potassium ions inside a cuttlefish nerve fibre had the same mobility and diffusion coefficient as those in an ordinary aqueous solution. This helped to answer the objections of those who felt it incorrect to treat protoplasm as an aqueous gel in which ions could move about reasonably freely. Huxley and Stämpfli described a new method which gave the absolute values of the action potential and resting potential in myelinated nerve fibres, and allowed them to determine the effect of sodium and potassium ions on these values: all of which turned out to be very similar to those in squid nerve or frog muscle.

An agreeable feature of this conference was that the organizers had provided us with a large hall, where you could sit around drinking beer or coffee with your friends and 'enemies'. As far as I am concerned, one of the main advantages of these big conferences is that, after I have met and had a discussion with a colleague from overseas, then I hear his voice and see his face when I subsequently read his papers, which otherwise would be quite impersonal. There were in fact many interesting people at this congress' Hans Ussing and Arthur Solomon being two that I remember particularly.

Although the congress was small by modern standards, I was depressed by its size, and also by the extent to which time was wasted by people with maverick views. This appears from the following letter that I wrote to Victor Rothschild who was then in North America and had been staying with Marion and Peyton Rous in New York, where he had retired to bed with virus pneumonia.

21 September 1950
Cambridge.

Dear Victor,

We were dreadfully sorry to hear about your pneumonia and hope you are now entirely recovered and enjoying Mexico and California. How is the golf? and can you drive further at 7000 ft or isn't the effect statistically significant? Which reminds me to confess that I haven't opened Weatherburn. Perhaps I shall do so before you return, but I rather doubt it as it looks as though next term is going to be busier than ever.

How was the conference? I gather from Tess's letter (or perhaps Marion's) that your paper went well – which probably means very well – and that you met a lot of people; but that the formal proceedings were on the dull side. However I am sure they couldn't have been worse than those at Copenhagen at which conference 1600 people attended and something like 900 papers were read! Light relief was provided by the crazy members, of which each major country managed to produce at least one. Needless to say, the most eccentric was contributed by England. A man called Burridge who invented a thing called Burridge's Law which explains everything in biology. Someone once wrote a poem beginning:

> When Schmelling comes down with a sock on the jaw
> It's all an example of Burridge's Law

(Ask Tess if you don't know who Schmelling is.) Anyway Burridge has been on the physiological scene for a good many years, but has fortunately spent most of his life as Professor at Cawnpore, which appointment he owes to an excellent testimonial from Oxford (where he worked before). He has now retired to England, but for some reason is usually escorted by a retinue of rather seedy looking Indian women with dirty saris and smudged caste marks. He always carries a heavy stick which he waves about as though he were going to bang you on the head if you disagree with him. Great excitement was created when a Danish newspaper said that Burridge had brought a bomb to the congress – but it turns out that it was only a verbal bomb not an atomic one. Then there was Barnes, of Beutner and Barnes, who launched a big attack which included the happy phrase – re Richard's paper – 'hitched to the Hodgkin hallucination'. And a crazy Swede called Ekehorn who read out of the manuscript of his book for an hour or so. My suggestion is that we should in future have a section entitled 'Wider Implications of Physiology' which would in effect be a dump for the loonies. Talking of which Lorente didn't show up but went to Spain instead. Gasser says that he needs a long holiday.

I have not done anything at all exciting although I've spent a lot of time on my review. But it's going to take much longer than I thought.

Everything at Merton seems to be going smoothly to judge from the confident tone of your secretaries. We shall be delighted when you return. You've been away too long.

With love from us both,
Alan

Cold Spring Harbor symposium on the neuron, June 1952

The next major international meeting that I attended was one of the Cold Spring Harbor Symposia on Quantitative Biology, a well-known series which started in 1933 and has continued to the present day. The meetings were held at the Biological Laboratory, Cold Spring Harbor, in a pleasant part of Long Island; they lasted a week and were attended by roughly a hundred scientists. The contributions covered a wide range of topics, including some excellent papers on visual mechanisms by Kuffler, and by Hartline and his colleagues which I remembered many years later when I decided that I wanted to switch from peripheral nerve to vision.

So far as I was concerned this conference differed from the discussions in Paris and Copenhagen in that, after several years of experimenting, my joint papers with Huxley on membrane currents in nerve were now either in press or just published. This made it much easier to give a coherent account of our work. Huxley and I submitted a joint contribution to the Symposium so that his name appears as a participant, but I was the only person invited from Cambridge, which in retrospect seems a shame – I suppose the organizers of the conference were short of money.

At all events I had to work hard for the privilege of being there. After my talk, someone, possibly Ralph Gerard, organized an evening session with Cole and perhaps a dozen nerve people there and cross-questioned me step by step on the details of our five papers; this took several hours.

The Cold Spring Harbor meeting covered much the same ground, but in a more detailed way, as a Royal Society Discussion on Excitation and Inhibition held on 21 February 1952, under the leadership of J. C. Eccles. Both meetings marked a complete turn around in the ideas of Eccles about the role of chemical transmitters such as acetylcholine at synapses. At the Paris conference in 1949, Eccles was convinced of the part played by acetylcholine at the neuromuscular junction but still thought of electric currents as being responsible for excitation and inhibition in the central nervous system. By 1952 he had changed his view completely and had started his excellent experiments on the chemical basis of excitation and inhibition.

This change of front, which did Eccles no harm, was in striking contrast to that of Lorente de Nó, who stuck resolutely to his view that the perineurium of the sciatic nerve did not constitute a barrier to diffusion. He did not submit an abstract of his talk and I do not remember what he said but I expect it was similar to the paper he gave at the Cold Spring Harbor Symposium a few months later.

Richard Adrian reminds me of a comment made by his father, who, as President of the Royal Society, was in the chair for the discussion meeting on 21 February. 'It reminded me,' said Lord Adrian, 'of being in Trinity Chapel. The seat was just as hard and hearing Lorente making the best of a bad case was remarkably like listening to a sermon on a difficult text by a preacher in Chapel.'

After the Cold Spring Harbor conference I went for a few days to Woods Hole. Early in 1949 Cole had accepted an offer to become Scientific Director at the Naval Medical Research Institute at Bethesda, Maryland. This was an important position but administration clearly took up much of his time and it was a while before Cole had trained a small group to carry on the squid work in the way that he wanted. Curtis, too, had taken up a major administrative post in atomic energy at Brookhaven and had abandoned his nerve work. I found it a little sad to remember the summer of 1938 when Cole and Curtis had been on top of the world.

In a letter to Victor Rothschild I wrote, 'The most interesting person I saw at Woods Hole was Osterhout [the plant physiologist], who is practically blind and can hardly move but is still getting experiments done, with the help of his new wife Icky [née Irwin, a former pupil]. Samples of Icky's conversation: "I was up till 3.00 a.m. last night but I must finish the old man's work before he dies," or "When I first started this year I thought the old man had cheated." They are doing interesting work on *Nereis* [a marine worm] – artificial fertilization from very small concentrations of detergents or from sperm extracts. I don't know enough to tell whether it's any good.'

Chapter 33

The Rockefeller and Nuffield Units, 1946–56

THE ROCKEFELLER grant to Professor Adrian for Neuro-physiology, on which we depended for many of our activities, ran for five years from 1 February 1946. By the autumn of 1949 it was time to think about the future of the Rockefeller Unit. However, before decribing what Adrian first proposed and what eventually happened to the unit, I should consider some of the varied activities of the unit during the first ten years of its life.

The original members of the unit were David Hill, Andrew Huxley and myself. David Hill left in 1947 for a permanent post as Physiologist at Plymouth, where we continued to see a good deal of him, and we had two new members, Richard Keynes and Peter Lewis. Adrian left the day-to-day running of the unit to me, which in practice meant writing reports and letters for Adrian to sign – it was good practice trying to write in his style – dividing up money between people, and all the rest of the activities which go with running a scientific team. At that time we had a rather dour head lab assistant who didn't much approve of newfangled notions like Units and External Grants. The result was that the only way of knowing where we stood financially was for me to keep the Unit's accounts. Fortunately, for the first time in the history of the Physiological Laboratory, its Head, Professor Adrian, had taken on a secretary, Miss Elton, who besides running the Physiology Library and typing for the Professor, had an immense capacity for work and was a big help to me.

Naturally after a while different members of the Unit specialized in different techniques on which they advised other members of the team: Huxley on microscopy and optical matters in general, Keynes on radioactive tracers and Lewis on chemistry and biochemistry.

Radioactive tracers: R. D. Keynes

After spending five years of the war designing and testing sonar and radar equipment for the Navy, Richard Keynes returned to Cambridge to read Part II Physiology, an admirable way of qualifying him both for his degree and to take a PhD. To my great satisfaction, he decided that he wanted to use radioactive tracers with the general aim of complementing the electrical studies which Huxley and I had been doing. In describing the development of Huxley's and my ideas on nerve, I have not so far said much about Keynes's work, but I doubt if we would have got on so fast, and certainly would not have carried other people with us, without Richard's pioneering work with radioactive tracers, activation analysis and flame photometry.

In the autumn of 1946, when Keynes started his experiments, the neutron pile at Harwell had not yet been built, but the Cavendish Laboratory in Cambridge still had a cyclotron, on which Keynes was allowed running time, provided he prepared his own metaborate targets for deuteron bombardment and carried out the subsequent chemical purification. In this I could give him no help, but he did all right on his own and soon had ^{42}K and ^{24}Na of higher specific activity than those supplied later by Harwell. He then devised a neat method of mounting nerves in a thin-bottomed chamber above a Geiger counter, and soon succeeded in measuring the leakage of potassium during the nerve impulse (Keynes 1948, 1989) confirming the accuracy of Huxley's and my indirect measurements. For this purpose he used little bundles of about six 30 micrometre nerve fibres from *Carcinus*.

These bundles of crab nerve fibres were not suitable for measuring the entry of sodium with ^{24}Na, but Keynes found that the experiment could be done with the $200\,\mu m$ axons of the cuttlefish *Sepia*, which we could now keep successfully in our aquarium. With this preparation and later with the squid giant axon, Keynes and Lewis were able to show that the net gain in sodium and loss of potassium per impulse were 3.8 and 3.6 picomoles per cm^2 in good agreement with the values calculated by Huxley and myself from purely electrical measurements.

Besides developing the tracer side of our work Keynes had familiarized himself with the microelectrode technique and put this to good effect in a visit to Professor Chagas's laboratory in Rio de

Janeiro. Here, he and Martins-Ferreira investigated the way in which the electric eel *Electrophorus* generates a massive electric shock. They found a basically simple mechanism. When the eel produces a shock, each electroplate gives a sodium-dependent action potential but only on one side, the innervated side, while the noninnervated side is not excitable and remains potassium-permeable. Hence each flat plate developes a voltage of 0.12 volts, and 5,000 plates in series, as in a large electric eel, would give 600 V. One by-product of this exciting piece of work is that the electric organ is an excellent source of sodium channels and provided the material which enabled Numa and the Kyoto school to determine the amino-acid sequence of the channel. The other by-product was the development of Keynes's interest in South America, his excellent writings on the voyage of the Beagle and his discovery of the drawings by Charles Darwin's artist Conrad Martens (Keynes 1979).

The Berne Connection; saltatory conduction, 1947–9

Partly through E. D. Adrian's connection with Alex von Muralt we had heard about some of the work that was being done in the latter's laboratory in Berne. Briefly what had happened was that the Berne workers, notably Robert Stämpfli, had succeeded in breaking the Japanese monopoly by dissecting myelinated fibres from vertebrates – by which I mean frogs in the first instance. This opened up the possibility of seeing whether the hypothesis of saltatory conduction was correct and, if it was, whether our work on the entry of sodium and exit of potassium applied to the discontinuous, myelinated nerves of vertebrates as well as to the continuous unmyelinated nerves of invertebrates.

At this point I should explain that the myelinated nerve fibres of vertebrates, which range in diameter from about 2–18 μm, are covered with a fatty sheath which is interrupted at regular intervals by the nodes of Ranvier where the plasma membrane is exposed to the external medium (Figure 33.1)

On the saltatory theory, the inward current generated by the entry of sodium during the rising phase of the action potential, which is what drives the impulse along, should be confined to the nodes. The internodes act as passive sections of cable which allow the change in membrane potential to spread from one node to the

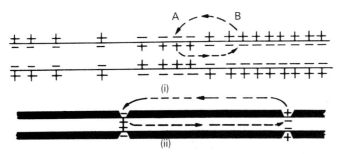

Figure 33.1. Diagrams illustrating the local circuit theory; the upper sketch represents an unmyelinated nerve fibre, the lower a myelinated nerve fibre.

next. The evolutionary point of saltatory conduction is that it allows the nerve impulse to be conducted with greater speed and with greater economy, i.e. with less ionic exchange, than a continuous mechanism. After the war it seemed to us that the saltatory theory was very attractive but that it was not supported by any strong positive evidence. This was the position in 1939 but not after 1942 when Tasaki and Takeuchi (1941, 1942) obtained strong evidence that the inward current originated at the node. However, these important papers, which appeared in *Pflüger's Archiv*, were not available in British or American Libraries during the war or for some years after it.

This was the general background when I wrote the following letter to Robert Stämpfli on 2 April 1947:

> Professor Adrian tells me that you are planning to visit Cambridge for about six months and I am writing at his suggestion to discuss details of your visit. First of all I would like to say how extremely pleased I am about this plan.
>
> I have a tremendous admiration for Professor von Muralt's work and your own and look forward very much to your visit. I hope that we may be able to collaborate in some experiments and you are welcome to use any of my apparatus while you are here. But I expect that we shall be able to let you have a complete rack of cathode-ray and amplifier equipment, which is now nearing completion and which should be in working order by the time you arrive. We should also be able to provide you with a Zeiss Binocular Microscope, dissecting stand etc. although I fear these may not be of such excellent quality as those you are used to. If there is any special equipment which you will require please let me know and we will try to obtain it for you.
>
> [Paragraph about timing of visit, housing, etc., adds that I shall be at

Plymouth till end of September with Katz. Perhaps Stämpfli will join us there?]

Professor Adrian asks me to say that the Rockefeller Unit will make you a grant of £250 to cover a stay of six months. I hope this will be sufficient for you.

I should like to discuss with you a possible joint research program on medullated axons. I am rather diffident in making suggestions in this field and you must not hesitate to tell me if I make foolish suggestions. Do you think it would be worth trying an experiment which I started some time ago but abandoned because of the difficulty of isolating a single myelinated axon. The experiment was designed to test the hypothesis that the myelin sheath is a passive insulator and that the action potential is generated only at nodes of Ranvier. According to this hypothesis the longitudinal current flowing during activity should be constant over the whole of one internode at any given instant of time. This means that the time relations of an action potential recorded with a pair of electrodes spaced about 0.5 mm apart should not alter when the electrodes are slid along an internode. Hence instead of getting a straight line relation between conduction time and conduction distance one should get a series of steps like a staircase.

Another line of research in which we are interested is the measurement of the electrical resistance and capacity of the myelin sheath and nodes of Ranvier . . .

What are your own ideas and plans for experimental work during your stay in Cambridge?

Robert Stämpfli did visit us at Plymouth and we did a preliminary experiment together after I got back to Cambridge in October. He brought with him some beautifully made forceps and scissors which greatly facilitated the dissection of single nerve fibres. Soon after that, Andrew completed the setting up of his new home and immediately made a helpful contribution by producing a very nice dark-ground optical arrangement for dissecting single myelinated axons. He also had excellent ideas about the right way to do the experiment I had been discussing with Stämpfli. As I had a mass of Plymouth data to analyse, it became clear to me that the right thing to do was to abandon experimental work in my room for a few months, and leave Huxley and Stämpfli to continue without me. Which they did, most successfully.

The main conclusions of Huxley and Stämpfli's experiments can be stated briefly. In their method the distribution of current during an impulse was measured by drawing a single fibre through a short

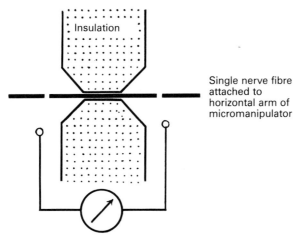

Figure 33.2. Arrangement used by Huxley and Stämpfli (1949a) for measuring longitudinal current during activity of single myelinated fibre.

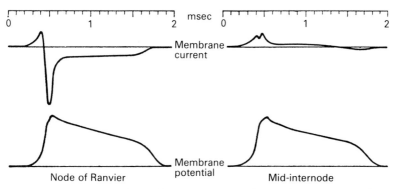

Figure 33.3. Time course of membrane current and membrane potential at a node and in the middle of an internode. Outward current is plotted upwards. (From Huxley and Stämpfli 1949b.)

length of fine capillary filled with Ringer's solution (Figure 33.2). By sliding the fibre through the capillary and recording the potential difference across the capillary at different points the longitudinal current was found as a function of time and distance. The radial current through the myelin or node of Ranvier was obtained by differentiating the longitudinal current with respect to distance, and the potential difference across the surface by integrating the potential gradient with respect to distance.

The position in a myelinated nerve fibre is well summarized by Figure 33.3 which Huxley and Stämpfli derived from one of the experiments which they published in 1949; it shows the membrane potential and membrane current at the node and in the middle of the internode. The current through the myelin is of the kind expected in a system consisting of a capacity and resistance in parallel; that through the node can only be explained by postulating an active generation of inward current, such as might occur if the nodal membrane underwent a large increase in the permeability to sodium.

From their experimental results and a further analysis, published in 1951, which gave the values of the resting potential and action potential as 71 and 116 millivolts respectively, Huxley and Stämpfli were able to calculate a number of electrical constants of which one of the most important was the capacity of the myelin sheath, estimated as about 10 picoFarads per centimetre or 0.0025 microfarads per cm^2 in the 14 μm diameter fibres on which they worked.

From electron microscope studies of developing nerve fibres Geren (1954) showed that the Schwann cells laid down myelin by spiralling many times round a nerve fibre, and then losing their protoplasm, leaving the cell membranes piled up on top of one another. In the fibres used by Huxley and Stampfli there would have been about 240 membranes in the myelin sheath, so their electrical capacity of 0.0025 μF/cm^2 indicated that each membrane was about 0.6 μF/cm^2, which seemed about right for a cell membrane.

Stämpfli's visit was followed by one from another Berne physiologist, Silvio Weidmann, who spent two years with us, mainly studying Purkinje fibres in the mammalian heart with microelectrodes. With M. H. Draper he showed that the rising phase of the action potential is carried by sodium ions moving through voltage-dependent channels that seem very like those in nerve. Purkinje fibres have only slight contractile activity and form a specialized conducting system which carries excitation through the ventricle. In most types of heart muscle there are voltage-dependent calcium channels as well as the sodium channels responsible for the initial spread of activity.

In the decade after 1950 we usually had two or three postgraduate visitors working with the group. These included I. Tasaki (Japan), J. E. De Smedt (Belgium), B. Frankenhaueser (Sweden), P. Horowitz (USA) and R. Niedergerke (Germany). In addition we had a limited number of home-grown people working for a PhD., sometimes before, but more often after completing their clinical studies. Some of these were T. I. Shaw, I. M. Glynn, R. H. Adrian and P. F. Baker. The reasons there were not more were partly that there was very little room, and partly that the natural course of action for a medical student who had just completed the Part II Physiology course in Cambridge was to go straight on to his clinical work at a London hospital.

Andrew Huxley and I never worked together after 1952, although we continued to discuss the problems we encountered in our experiments. In addition to theoretical work on nerve, Huxley's main interest was in the mechanical and optical properties of striated muscle, particularly in relation to the mechanisms of activation and contraction. This work was greatly assisted by an interference microscope which Huxley designed and built himself. In experiments done at about the same time as those of Hugh Huxley and Jean Hanson (1954), Andrew Huxley and R. Niedergerke provided strong independent evidence for the sliding filament theory of muscular contraction (1954). Later, he and his colleagues made equally good use of the new microscope in showing that there were sensitive spots at which one could obtain local activation of the contractile mechanism (Huxley and Taylor 1958). This laid the basis of modern ideas about the way in which the contractile machinery inside the muscle is turned on by the action potential of the surface membrane.

In the 1950s my interests were in the electrical and ionic properties of giant axons, the effects of calcium and the mechanism of the sodium pump. But I shared Andrew Huxley's interest in the activation of muscle, though approaching the subject from a different kind of experiment.

Planning for the future of the Rockefeller Unit

At the end of October 1949 Adrian submitted a memorandum to the Faculty Board on the need for a sub-department of Biophysical Research. After describing the work done with the help of the Rockefeller Grant and explaining that I was responsible for running the Unit, Adrian continued:

> The Rockefeller grant comes to an end shortly before the beginning of the new quinquennium, and I hope that the University will agree to take over responsibility for the research unit as soon as the new period begins. The best arrangement would be the creation of a sub-

department for Research in Biophysics in charge of a Reader, with one Assistant Director of Research and two members of the standing of Lecturers or Demonstrators.

In retrospect these proposals seem optimistic, though University College London provided a model of the kind of system Adrian had proposed. I do not know whether Adrian really thought his proposals for a sub-department of Biophysics would go through, but I think he hoped to get me made a Reader and was relieved when I didn't mind staying an Assistant Director of Research (ADR). This title will sound odd to non-Cambridge ears; it arose because the Professor, as Head of the Department, was responsible for directing all its research, so the term director had to be qualified by the addition of assistant. It is true that the position was not well paid and that it had to be renewed every five years, but the formal teaching duties were small and I knew how lucky I was to have time for my own research and to share facilities with brilliant and congenial people like Andrew Huxley and Richard Keynes.

Of course we all wondered what would happen when, as we expected, Adrian became President of the Royal Society and then Master of Trinity College, Cambridge. The first of these events took place in December 1950; the second, in October 1951, was accompanied by Adrian's resignation as Head of the Department of Physiology.

When it appoints a new Professor, Cambridge University can act fast, if the Vice-Chancellor and Electors so wish. But more often than not, particularly if the Head of a large department is being considered, many months elapse before a meeting can be arranged. On this occasion no meeting could be arranged until half-way through 1952, when Bryan Matthews was appointed, and I never knew whether I would have been offered the post if I had not already accepted the Foulerton Professorship – nor whether I would have accepted it if asked.

Adrian, at all events, was quite clear that I should not succeed him, though whether because my research was good or my administrative ability bad, or both, I do not know. During the autumn he said that, if I agreed, he would propose me for the Foulerton Research Chair of the Royal Society. I do not remember the conversation but must have given a provisional acceptance, because on 3 December 1951 Marni wrote to her parents that

'Alan's Foulerton is through and Adrian will announce it on Friday'. She added that we were 'tremendously pleased, salary £2,000 per annum; lecturing to Part II, but no more supervisions; means that A. cannot be head of the lab.'

Application to Rockefeller and Nuffield Foundation for 1951–6

After the readership proposal collapsed, Adrian had turned over to me the question of replacing part of the Rockefeller Grant as we knew that the Foundation wanted to taper off their existing grant of £3,000 per annum The upshot was that we put in a successful application to the Rockefeller Foundation for £1,000 per annum for five years and a similar one to the Nuffield Foundation which was also successful. These grants helped to pay our technician's salary, but were mainly used for purchasing electronic, optical and chemical equipment and for radioactive tracer experiments. The main problems studied were conduction in nerve, muscle and heart; contraction and activation in skeletal muscle; and the active transport or secretion of ions in various tissues, but primarily nerve, muscle and red blood cells. The grants were initially to Adrian but there was an understanding that I would take them over when he retired.

Chapter 34

Research, 1951–63

The sodium pump; Na/K ATPase; experiments with Keynes, 1951–55

Our experimental and theoretical work had provided good evidence that the immediate source of energy for nervous conduction came from the movements of ions down their concentration gradients: an entry of sodium during the rising phase of the action potential followed by an exit of potassium during the falling phase. These movements amounted to only a minute fraction of the potassium or sodium present in the nerve fibre, which should be able to conduct many impulses without the intervention of metabolism. However, it was clear that if the ionic exchange were to be of any permanent use to the animal then nerve fibres would have to have evolved a mechanism for extruding the sodium ions that entered and reabsorbing the potassium ions lost during activity. We guessed that nerve, like other tissues, would be found to contain a system for using metabolic energy to pump out sodium from the cell interior and felt that it would be interesting to characterize such a system using radioactive tracers.

To begin with, Keynes and I used cuttlefish axons, with which we were both familiar; these were large enough to give us the necessary sensitivity and had the great advantage that we could do the experiments in Cambridge. Later we moved to Plymouth where we could extend the work to squid axons for measurements with internal electrodes, or inject substances if necessary.

Working together had several advantages as we planned to measure the influx and efflux of both sodium and potassium under different conditions with the short-lived isotopes ^{42}K and ^{24}Na

336

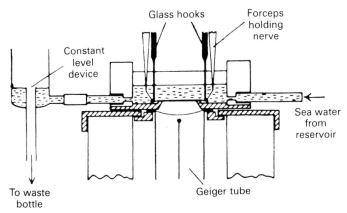

Figure 34.1. Apparatus for measuring radioactivity of nerve fibre developed by Keynes (1951). The perspex chamber and the lead shield on which it was supported were circular. The view shown is a vertical section through the centre.

which had to be used within a day or two after their arrival from Harwell on Monday afternoon.

Even with two people it was difficult to do all we wanted in the time available, particularly as the supply of cuttlefish from Plymouth was somewhat precarious. I remember Marni and I taking a taxi to the station to collect a last-minute consignment in large flat bins labelled 'tropical fish'. Perhaps foolishly, we accepted the taxi-driver's help and I remember his horror when he dropped a tin, releasing a quantity of sea water and two large cuttlefish on Cambridge station. Another hazard was the *Sepia* ink with which they were liable to drench you unless handled cautiously. Still they are fascinating creatures and I was sorry when the cuttlefish season came to an end and life returned to normal.

For influx measurements we used the method developed by Keynes (Figure 34.1) though special equipment was sometimes needed to restrict the influx of tracer to a short length of nerve. For effluxes the fibre was first loaded with tracer and then, after washing off external radioactivity, samples were collected every ten minutes or so, again employing special methods to restrict the collection length if necessary. In the simplest case, tracer was collected by looping the fibre in a narrow V and transferring it from one sampling tube to the next. It was, however, absolutely necessary to avoid collecting tracer from anywhere near the cut end, preferably excluding a length of at least 10 millimetres at each end.

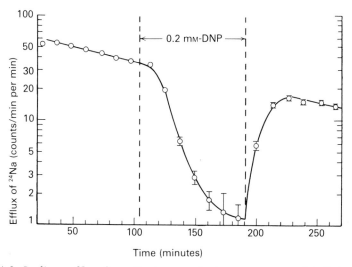

Figure 34.2. Sodium efflux from *Sepia* axon during treatment with dinitrophenol. At the beginning and end of the experiment the axon was in artificial sea water. Abscissa: time after end of stimulation. Ordinate rate at which ^{24}Na leaves axon. Vertical lines are $\pm 2 \times$ SE. (From Hodgkin and Keynes 1955a.)

The specific activity at the end of the experiment was obtained from the ratio of the final radioactivity to the Na or K determined from activation analysis by the method of Keynes and Lewis (1951).

With this technique it was easy to show that metabolic poisons like cyanide, azide or dinitrophenol caused a large and reversible decrease in the outflow of labelled sodium from the nerve fibre (Figure 34.2). Now dinitrophenol acts by uncoupling the formation of the energy-rich compound adenosine triphosphate (ATP) from oxidative metabolism so this experiment suggests that the pumping out of sodium may depend on the supply of ATP. More direct evidence for this conclusion was obtained later, when it was shown that after the sodium pump had been blocked with cyanide it could be restored by the injection of ATP or by substances which generate ATP.

The sodium outflow could also be reduced to about one quarter by removing external potassium ions which fitted with the idea that metabolism drives a cycle in which potassium ions are taken up on one limb and sodium ions are extruded on the other (Figure 34.3). This idea was strengthened when we found that much of the normal potassium uptake was blocked by cyanide and dinitrophenol.

There was a striking contrast between the action of metabolic

338

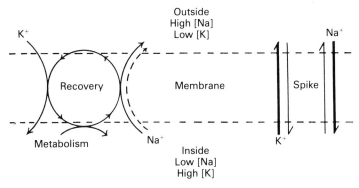

Figure 34.3. Diagram illustrating movement of ions through the nerve membrane. The downhill movements which occur during the impulse are shown on the right; uphill movements during recovery are shown on the left. The broken line represents the component of the sodium efflux which is not abolished by removing external potassium ions. (From Hodgkin and Keynes 1955a.)

inhibitors on the sodium pump and their effect, or rather lack of effect, on the downhill movements of sodium and potassium which generate the action potential. For example, when a giant axon was poisoned with dinitrophenol the sodium outflow dropped to a low value, but the action potential and resting potential changed very little, and careful measurement was needed in order to show that whereas the the action potential remained steady in the unpoisoned fibre, it did undergo a slow decline in the poisoned one.

The conclusion from the tracer experiments, summarized in Figure 34.3, is that there are two systems in the membrane: one driven by metabolism and used for building up concentration differences; the other, relatively independent of metabolism and responsible for controlling downhill movements of sodium and potassium during the action potential.

The two systems have quite different properties and it is possible to differentiate between them in many ways. Thus the cardiac glycoside, g-strophanthin, which inhibits the sodium pump by combining with an ATP-splitting enzyme in the membrane, has no effect on the action potential, whereas tetrodotoxin and other poisons which block the action potential do not inhibit the pump. The systems also differ in their ionic selectivity, for lithium which can replace sodium in the action potential mechanism is not moved at all effectively by the metabolically-driven sodium pump. Another important difference is that the maximum rate at which the

339

secretory system can move ions is about 50 picomoles cm^{-2} sec^{-1} in the squid axon, whereas peak movements of up to 10,000 pmol cm^{-2} sec^{-1} occur during the action potential. Many other distinctions could now be made, but these are some of the more obvious points which struck us in the 1950s.

Movements of potassium ions in single file

The experiments on the effect of metabolic inhibitors on the outflow of sodium and uptake of potassium all seemed to fit well with the idea of a cycle, driven by metabolism, in which the efflux of sodium was coupled to potassium influx. However, there was one aspect which bothered us.

In 1949 the Danish scientist Hans Ussing showed that if an ion moves through the cell membrane solely under the influence of the electrical and chemical gradients, then, provided that its movement is independent of other ions, the ratio between the efflux and influx of the ion should be given by a simple equation, which in the case of potassium ions would be:

$$\frac{\text{Efflux}}{\text{Influx}} = \frac{[\text{K}]_i}{[\text{K}]_o} \exp \frac{VF}{RT} \tag{1}$$

or

$$\frac{\text{Efflux}}{\text{Influx}} = \exp \left[\frac{(V - V_{\text{K}})F)}{RT} \right] \tag{2}$$

where V_{K}, the potassium equilibrium potential, is given by

$$V_{\text{K}} = \frac{RT}{F} \ln \frac{[\text{K}]_o}{[\text{K}]_i} \tag{3}$$

and R, T and F have their usual significance.

Keynes and I found that after active transport had been eliminated with dinitrophenol, the efflux and influx of potassium were equal, as predicted by equation (2), and as indeed is thermodynamically necessary if ions are moving solely under the influence of the electrical and chemical gradients across the membrane. However, something very interesting happened when the flux ratio was measured at potentials other than the potassium equilibrium potential, for we found that the flux ratio altered much more than predicted by Ussing's equation, and that the data could be described approximately by:

$$\frac{\text{Efflux}}{\text{Influx}} = \exp\frac{n(V - V_K)F}{RT} \qquad (4)$$

where n was about 2.5 as opposed to unity in equation (2). This makes a very large difference, since equation (2) predicts a ten-fold change in flux ratio for a 58 millivolt change in membrane potential whereas equation (4) with $n = 2.5$ predicts that the flux ratio will change by $10^{2.5} = 316$.

We thought that this meant that potassium ions moved through a long pore which was narrow enough to constrain these ions to move in single file. In this system, if there are n ions in a pore then an individual ion can pass through the membrane only if it is hit n times in succession from the same side. This will happen much more often if the ion is moving in the same direction as the majority of other ions. To a first approximation n is the average number of ions in the pore. This explanation has now been generally accepted and is strongly supported by the work of Lev and his colleagues in Leningrad on the passage of ions through end-to-end pairs of gramicidin molecules, which are known to promote single-file diffusion through lipid bilayers and have the right molecular structure to do so – namely a long bead-like molecule with a hole of the right size down the middle of it.

As we felt that we should have difficulty in getting across the concept of single-file diffusion, we built a mechanical model to illustrate the principle and its implications for tracer studies. This consisted of two flat compartments which could be joined by a short gap as in Figure 34.4A or by a long gap as in Figure 34.4B. One hundred small steel balls, which had been blued by heating, to represent labelled ions, were put in the left-hand compartment and fifty balls, identical except for being silver in colour, were put in the right-hand compartment. The model was then shaken vigorously by a motor for fifteen seconds. This caused all the balls to rattle about with a random 'Brownian' movement, and a certain number passed through the gap. The number of blue balls which moved from left to right was counted and compared with the number of silver balls which had moved in the opposite direction during the same time interval. In order to avoid possible errors which might have been introduced by a slight tilt of the apparatus, or by small

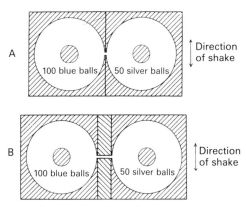

Figure 34.4. Diagram of mechanical model. In *A* two flat compartments were separated by a narrow gap; in *B* the gap width was increased by spacers. The sides and bottom were made of aluminium, the top of perspex. The diameter of the circular compartents was 74 mm and that of the balls 3 mm. The central discs helped to randomize the motion of the balls. (From Hodgkin and Keynes 1955b.)

mechanical differences between the blue and silver balls, the experiment was repeated with 100 silver balls in the right and 50 blue balls in the left compartment. Both experiments were performed many times and the results averaged.

With a short gap separating the two sides one would expect the number of balls leaving the side with 100 to be twice as great as the number leaving the side with 50. This prediction was approximately but not exactly fulfilled, the ratio in 19 trials averaging 2.7 (standard error of mean 0.2). A much greater ratio was found when the long gap was employed, for in this case the number leaving the 100-ball compartment was 18 times the number travelling in the opposite direction (34 trials, s.e.m. 4). Use of the long gap therefore greatly exaggerated the ratio of downstream to upstream movement. The reason for this is obvious.

A single collision with the short gap is sufficient to transfer a ball across the gap. This is not so with the long gap. The number of balls in the gap fluctuated but on average there were about three balls in the gap at any particular moment. This means that at least four collisions were needed to transfer a ball right through the gap. Starting with three blue balls in the gap, the first collision of a silver ball moves it into the right-hand position, the second moves it into the middle, the third moves it into the left-hand position and the fourth finally transfers it into the other compartment. On this highly simplified argument, the frequency with which four collisions occur in succession will be 2^4, that is, 16 times greater on the 100-ball than on the 50-ball side. Hence a very large flux ratio is obtained with the long gap but not with the short one. A residuum of the same effect probably occurred with the short gap because the balls were of finite size and the gap was not infinitesimal as was tacitly assumed in the initial argument.

Philosophers of science sometimes refer with slight scorn to Newtonian billiard-ball models of gases or other systems in which scientists have used thought-experiments to illustrate physical laws. So far as I know, Keynes and Cook (our assistant) and I are the only people who have ever made and tested such a model – as opposed to talking about it. Sadly the model no longer exists as it disappeared in one of those orgies of spring-cleaning to which all laboratories are subjected from time to time.

Restoration of the Na/K pump by injection of ATP and other compounds. During the 1950s evidence accumulated to show first that a metabolically driven system, which extruded sodium and absorbed potassium, was widely distributed throughout the animal kingdom; and secondly that it was driven by the splitting of the terminal phosphate group from the key compound, adenosine triphosphate, the overall reaction being:

$$ATP + H_2O = ADP + P_i + H^+$$

where ATP and ADP stand for adenosine tri- and diphosphate respectively, P_i for inorganic phosphate and H^+ for hydrogen ions. Under standard conditions this reaction liberates 7.3 kilocalories per mole of free energy which may be used for biosynthesis or for movement, secretion and all the other energy-consuming tasks that animals have to perform. ADP is rapidly rephosphorylated to ATP by oxidative metabolism or glycolysis and sometimes by another substance, phosphagen, which acts as a labile store of high-energy phosphate. In vertebrates the phosphagen is creatine phosphate, but in the squid the substance acting as a store of high-energy phosphate is arginine phosphate. These ideas, which are of great importance in biology, were first proposed by F. Lipmann in 1941.

After Keynes and I had written up our work on active transport and the single-file effect, our next move was to try to test the idea that ATP supplied the energy for pumping out sodium ions and reabsorbing potassium ions. For this purpose we designed a gadget which enabled us to inject a length of giant axon first with ^{22}Na and then with different substances, of which ATP was one. What we wanted to do was to inhibit the sodium pump with cyanide or dinitrophenol and then restore it by injecting ATP. The first experiments of this kind were inconclusive, partly because there was some doubt about the purity of the ATP (Hodgkin and Keynes 1956). It turned out later that we had good reasons for our doubts because this sample, which we had obtained from a highly reputable manufacturer, gave an absolute sea of spots, a sign of impurities, when subjected to two-dimensional partition chromatography. A repetition of the experiment by Caldwell and Keynes (1957) with purer samples gave a positive result but by then several new points were beginning to emerge.

The first step, which was taken by P. C. Caldwell who worked at Plymouth between 1955 and 1960, was to study the effect of metabolic inhibitors on the phosphate compounds in the giant axons of *Loligo*. We were fortunate that Caldwell (who died in 1979) should have been at Plymouth at this time; his training as an Oxford chemist, and subsequent research in Hinshelwood's and A. V. Hill's laboratories made him a key member of the team working at Plymouth with Keynes, Shaw and myself in the late 1950s.

Caldwell first showed that the three main forms of phosphate in the protoplasm of a squid nerve are inorganic phosphate, adenosine triphosphate (ATP) and arginine phosphate. When cyanide or dinitrophenol was applied in concentrations which inhibited sodium extrusion, arginine phosphate broke down to arginine and inorganic phosphate; and ATP broke down to adenylic acid (AMP) and inorganic phosphate. On removing cyanide, the ATP and phosphagen were resynthesized.

Our injection experiments showed that the sodium efflux in an axon poisoned with cyanide could be restored by both ATP and arginine phosphate and that this action was shared to varying extents by certain other compounds containing high-energy phosphate, denoted by ~P. Seven compounds containing ~P were tested, namely arginine phosphate, phosphopyruvate, ATP, ADP, guanosine triphosphate, inosine triphosphate and creatine phosphate. Inosine triphosphate had a very slight action and guanosine triphosphate a moderate one. In contrast to arginine phosphate, creatine phosphate was totally without effect. This was highly satisfactory since it was known that creatine phosphate was not handled by the enzyme arginine phosphokinase. I remember that as we did the experiment Peter Caldwell remarked reflectively, 'This must be the first time for 500 million years that any living cell on the squid side of the animal kingdom has had creatine phosphate inside it.'

ATP had no effect when hydrolysed to AMP and inorganic phosphate, nor did arginine phosphate when broken down to arginine and inorganic phosphate. Both ATP and arginine phosphate were ineffective when applied externally.

All the compounds which promoted an outflow of sodium ions almost certainly did so by regenerating ATP by reactions such as:

ADP + arginine P = ATP + arginine   (Arginine phosphokinase)
ADP + pyruvate P = ATP + pyruvate   (Pyruvate kinase)
2ADP = ATP + AMP   (Adenylate kinase)

Evidence for such reactions was provided by chemical experiments which showed a virtually complete conversion of AMP into ATP within five minutes of injecting arginine phosphate or phosphopyruvate into axons poisoned with cyanide.

The action of injected phosphate compounds was transient as would be expected if they were used up in driving ion transport, or in side-reactions. The quantity of sodium ions extruded was roughly proportional to the quantity of ATP injected – the ratio being about 0.7 ions per high-energy phosphate bond. This was less than the value approaching 3 found for sodium transport in frog skin by Leaf and Renshaw (1957) and would correspond to a rather inefficient system. Later, Baker and Shaw (1965) showed that after an ATP injection into a squid axon, about three-quarters of the ATP breakdown occurred in the protoplasm and only one quarter at the membrane – indicating that the true ratio in our experiments might have been about three sodium ions extruded for each ATP split. The figure now generally accepted is that two potassium ions are absorbed and three sodium ions are extruded for each ATP split.

Although we considered that phosphoarginine or phosphopyruvate worked by generating ATP from ADP, we found that these compounds were better at restoring a normal sodium efflux than ATP itself. This was best shown by testing the potassium sensitivity of the sodium efflux, as has been done in Figure 34.5. Here it can be seen that, both in the normal axon and after the injection of

345

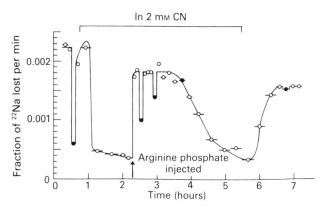

Figure 34.5. Effect of injecting arginine phosphate in restoring a potassium-sensitive sodium efflux to an axon poisoned with cyanide; filled circle, K-free external solution; open circle, external solution containing 10 mM K. The injection raised the concentration of arginine phosphate in the axon by 33 mM. (From Caldwell *et al*. 1960.)

arginine phosphate, the sodium outflow was cut down when potassium ions were removed from the external solution. However, the potassium sensitivity was lost fairly rapidly and had disappeared by the end of the plateau following the injection.

In a similar experiment with ATP there was no restoration of potassium sensitivity. It is thought that under these conditions, and others where one sees a large uncoupled sodium efflux, that internal and external sodium ions are exchanging across the membrane.

As a corollary to these experiments it was found that when phosphoarginine was injected into axons poisoned with cyanide it increased the inflow of potassium ions (which had been reduced by cyanide) to about the same level as in an unpoisoned fibre. On the other hand injection of ATP in the same type of experiment produced little or no potassium uptake.

The explanation of the greater effectiveness of ATP generators like phosphoarginine or phosphopyruvate over ATP itself, is that the pump requires not only that ATP should be present but also that the concentration of ADP should be low. An alternative statement is that the phosphorylation potential, which is proportional to $\log \frac{[ATP]}{[ADP]}$ should be large enough to permit formation of a bond with higher energy than the terminal phosphate bond in ATP. A high $\frac{ATP}{ADP}$ ratio helps to drive the reaction ATP + Z = ADP + ZP in the direction of ZP.

On this view phosphoarginine or phosphopyruvate act by regenerating ATP, but are more effective because they keep ADP at a low level by the reactions:

Arginine P + ADP = Arginine + ATP
Pyruvate P + ADP = Pyruvate + ATP.

Our work at Plymouth in the late 1950s clearly fitted well with the experiments of Skou (1957) in Denmark who showed that an essential component of the sodium-pumping mechanism was a membrane protein which catalysed the hydrolysis of ATP. This enzyme was activated by sodium and potassium and blocked by the cardiac glycoside ouabain, one of the standard inhibitors of the sodium pump. It is widely distributed in animal cells and is known as an Na,K-ATPase. The experiments done in Cambridge by Shaw, Whittam and Glynn also helped to emphasize the general applicability of Skou's result.

Internal perfusion of nerve fibres; experiments with Baker and Shaw, 1960–2

In the summer of 1960 Professor F. G. Young, the Head of the Department of Biochemistry, asked me if I would like to take on Peter Baker, who had just got a first in Part II Biochemistry, and wanted to do research for a PhD on a biophysical topic of some kind. I was delighted to do this, partly because Baker had an excellent academic record, but especially when I found that he was very keen on research and had already done some quite respectable entomological work on his own, partly at the Rothamsted Field Station and partly at home.

Baker welcomed the suggestion that he should divide his time between Cambridge and the Laboratory at Plymouth, where there were many biochemical experiments to be done on giant nerve fibres. I was planning to work at Plymouth later in the autumn, but arranged that in the meantime Trevor Shaw, who was on the Plymouth staff, would look after him. They started work together in September 1960 and almost immediately struck gold by showing that after the protoplasm had been squeezed out of a giant nerve fibre, conduction of impulses could be restored by perfusing the remaining membrane and sheath with an appropriate solution. This was done by the remarkable method illustrated in Figure 34.6.

347

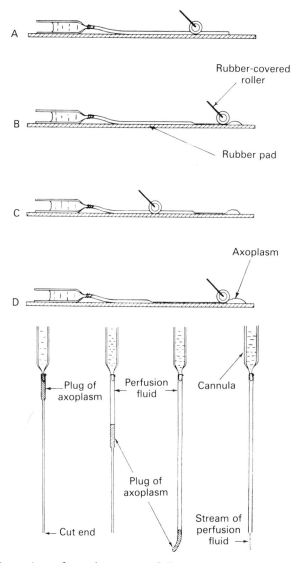

Figure 34.6. Extrusion of axoplasm (top) followed by reinflation of 'sheath' with perfusion fluid (bottom). (From Baker, Hodgkin and Shaw 1962a.)

One of the tricks that can be played with giant nerve fibres is to extrude axoplasm from the cut end. This is done by stroking with a glass rod or by running a device like a miniature garden roller over the fibre in a series of sweeps. In order to remove the axoplasm it is necessary to press on the axon quite firmly and until 1960 everyone assumed that the surface membrane must be severely damaged during extrusion. Baker and Shaw showed that we had been too

348

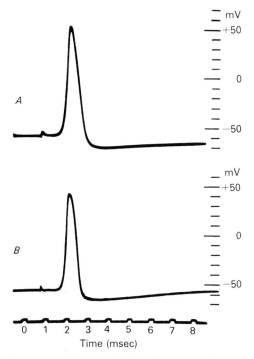

Figure 34.7. *A*, action potential recorded with internal electrode from extruded axon refilled with potassium sulphate solution (16°C); *B*, action potential of an intact axon, with same amplification and time scale (18°C). (From Baker, Hodgkin and Shaw 1962a.)

pessimistic, and that in a fair proportion of cases an action potential of normal size and shape could be obtained after moving on axoplasm or reinflating the membrane with an appropriate solution (Figure 34.7).

After making this discovery, Baker and Shaw very kindly asked me to join them. We spent some time working out the best method of changing internal solutions while recording action potentials with an internal electrode. It turned out that it did not much matter what solution was inside the nerve fibre as long as it contained potassium and little sodium. Provided this condition is satisfied a perfused nerve fibre is able to conduct nearly a million impulses without the intervention of any biochemical process. ATP is needed to pump out sodium and reabsorb potassium but not for the actual conduction of impulses.

The development of methods for perfusing giant nerve fibres at

349

Plymouth and in America allowed the ionic theory to be tested in several new ways.

Resting potential in perfused nerve fibres. Under many conditions the potassium gradient provided the main e.m.f. for generating the resting potential. With isotonic potassium chloride as the internal perfusion fluid, the resting potential was about $-55\,\text{mV}$ although the chloride equilibrium potential was close to zero. When the internal solution was replaced by isotonic sodium chloride the membrane potential fell to near zero. If now potassium chloride was gradually restored, the membrane at first behaved as a potassium electrode and then saturated as the potassium permeability declined. No resting or action potential was ever developed when the ionic composition was made equal on the two sides of the membrane. When the internal solution was isotonic sodium chloride and the external solution was isotonic potassium chloride, the interior became 40–60 mV positive to the external solution (Baker *et al.* 1962b).

Effect of internal sodium on the action potential. As expected on the sodium theory, the overshoot of the action potential was decreased by raising the internal sodium concentration and never exceeded the sodium equilibrium potential, although approaching it closely at high internal sodium concentrations (Baker *et al.* 1962b): all of which fitted well with the ionic theory.

A striking and unexpected finding was that reducing the internal ionic strength caused a dramatic shift in the operating voltage characteristics of the membrane (Baker *et al.* 1964). This effect, which finds a straightforward explanation in terms of the potential gradients generated by charged groups on the inside of the membrane, helped to explain several unexpected results that were sometimes thought to be inconsistent with the ionic theory of nerve conduction.

Today the Baker and Shaw method of perfusing a giant nerve fibre after rolling out its contents is not much used and has been largely replaced by other techniques, such as the internal dialysis method of Brinley and Mullins (1967). However, the method was important in its day, both because it led to new experiments and because it helped to remove a deep prejudice against the idea that the protoplasm inside a nerve fibre was of no immediate relevance to the transmission of impulses, apart from providing a conducting pathway. In 1963, Ragnar Granit told me that the internal perfusion experiments done with the Baker and Shaw method was one of the bits of evidence that finally convinced the Nobel Committee of the essential soundness of the ionic theory of nervous conduction.

Chapter 35

Moving house, holidays, travel, conferences, 1953–62

Moving house, 1953–4

Our fourth child, Rachel, was born in June 1951 and, although we just fitted into our house in Bentley Road, by 1953 we were starting to look for a larger one. After some disappointments we found one that suited us perfectly, although we didn't realize it at the time, in Newton Road only 200 yards away.

At first we thought it was too large and somewhat cheerless, with metered gasfires in every room and hideous decorations. But we soon realized that we must forget about the irrelevant effect of a wartime conversion to lodgings, and after seeing the house again on a fine autumn day we began to appreciate its tremendous advantages and to see that it was well worth the extremely modest sum that the house-agents were asking – partly as a result of large Edwardian houses going completely out of fashion. And so indeed it turned out. In the twenty-five years that we lived there, we never for a moment wanted to move, and when the time came when we had to sell it, our children could hardly forgive us, although they no longer lived there and the site had lost some of its original charm.

In 1954 when we moved in, 25 Newton Road was entirely surrounded by large gardens filled with a splendid variety of trees. And in the attic floor, which Sarah and Deborah rapidly took over as their own, you seemed to be living amongst the tree-tops. Of course the house needed to be painted from top to bottom and was terribly cold – but all that could be rectified in time and indeed was one of our main preoccupations for the next five years.

The proximity of our old and new homes was a great convenience for we found that we could move practically everything in a gigantic pram that we had inherited from the Rothschilds. When we finally got into the house at the beginning of February 1954 it was arcticly cold but we were too pleased to mind something that we knew we could put right.

Everyday life, travel, 1954–62

At that time the pattern of our lives was changing in many ways. All the children would soon be at school in Cambridge and in 1960 Marni would be able to return to her profession of editing and producing children's books, first at the firm of Hart-Davis and later at Macmillan; this took her to London, initially for two days a week and, later on for four or five. For the first ten years after the war I went very little to meetings in London but then Patrick Blackett, a sort of hero of mine, took me in hand and said in the nicest possible way that it was time I came out of my ivory tower and helped him in his drive to get more money for science. This involved service on a committee of the Department of Scientific and Industrial Research and later, membership of the Medical Research Council and Royal Society Council. This last is not normally hard work, except at the time of elections, but the second of my two years coincided with the Tercentenary of the Royal Society and occupied a week in London in July 1960 as well as a good deal of preparatory work. I had to give a lecture and a demonstration, for which I had rashly said that Caldwell and I would show action potentials recorded from squid giant nerve fibres. To be sure of getting something, the squid had to be caught in the morning at Plymouth, expressed to London, dissected and everything got going in the boiling heat of a summer evening, in a gallery where there was no gas, no water and no way to get rid of cooling fluid except for one of us to empty a bucket down a drain on the flat roof-top outside. Our nerve was just working but it was touch-and-go whether it still would be by the time that the Duke of Edinburgh came round. However, we just made it and how was he to know that the amplifier gain was five times higher than it should be!

The opening meeting was in the Albert Hall with scientists and academicians from every country in a magnificent display of doctoral gowns and hats. As a member of Council I was on the platform and felt that I let down the side rather because the best I could do was a plain silver Cambridge MA hood – I had meant to rectify this by taking an ScD but had left my application too late.

353

However, this didn't matter at all because my plain silk was so unusual amidst all the gorgeous colours that everyone thought it must be of immense distinction.

Sir Cyril Hinshelwood the President gave an excellent address of an historical kind. But he was a quiet scholarly man, not used to addressing vast gatherings in the Albert Hall in a glare of publicity, and those who knew him well could see that he was under a great strain. However, everything went off without mishap and the worries of those who feared he might crack up could be put at rest.

One nice thing for us was that, as a Foreign Member of the Royal Society, Marni's father Peyton Rous and her mother were invited to the celebrations. We spent a good deal of time with them, hunting in print shops or catching up on the minor sightseeing that you never do in your own capital city, for example visiting places like the Soane Museum or the Chelsea Physic Garden, or taking a river steamer to Greenwich.

With our children growing up, family holidays became increasingly important and occupied a part of the year which had formerly been devoted to 'squidding' at Plymouth. Happily, a potentially awkward clash was avoided by the discovery that at Plymouth, where the squid are a different species from those at Woods Hole, far and away the best time for catching large specimens is the late autumn, September to December, rather than the summer months as we had formerly thought.

Although my work often took me abroad I never became such an inveterate traveller as some of my colleagues, mainly because I preferred life at home to rushing from one airport to the next. The East Anglian countryside is perhaps not romantically beautiful, although Constable found it so, but it is of considerable interest to the naturalist and Sundays were regularly devoted to expeditions to places like Wicken Fen, the Devil's Dyke or the Brecklands in which the children could pursue entomological or botanical interests and I could prevent my knowledge of ornithology from growing too rusty.

In the summer of 1954, Bernhard Frankenhaeuser, from Granit's laboratory in Stockholm, worked with me in Cambridge and then Plymouth, on the action of calcium on giant nerve fibres. As usual we spent a long time analysing the data and found it necessary to meet occasionally to discuss the implications of what we had done. This led to my visiting Stockholm in the summers of 1955 and 1956 and incidentally to some most enjoyable sailing in the Swedish archipelago.

One of my most interesting bits of travel came rather by accident in the summer of 1959 when Richard Keynes and I found that we had a Gordon conference in New Hampshire followed a fortnight later by an International Congress in Buenos Aires. It seemed silly to return to this country in between, so we spent most of the intervening period in Peru, first visiting Cuzco and Machu Picchu and then taking a slow train down the Andes, followed by a steamer trip across Lake Titicaca at an altitude of 11,000 feet to Bolivia, where in La Paz we caught a plane to Buenos Aires.

Visits to Hallington, 1954 and 1958; George Trevelyan in old age

Two visits to Hallington, George Trevelyan's Northumbrian house, where he spent the summer, form a striking contrast to the rest of our lives. We had gradually come to be on very easy terms with the Master and Mrs Trevelyan and after his retirement used to visit them in their Cambridge house in West Road as often as we could – which I am afraid was not all that often. In the spring of 1954 they invited us to Hallington.

Anyone who wishes to know what George Trevelyan was like can quickly form a clear impression by reading two books that he published at the end of his life: *An Autobiography and Other Essays* (1949) and *A Layman's Love of Letters* (1953). Here one is reminded most forcibly of the three things in which he had an absorbing and passionate interest: history, the natural beauty of the countryside and poetry, particularly though not exclusively romantic poetry. In these essays you will learn that Trevelyan read Shakespeare's greatest plays some seventy times[1] and will guess if you did not know already that he learnt vast quantities of poetry by heart. I remember Noel Annan telling me that there really were tears in Trevelyan's eyes when he quoted the line from the Purgatorio, *Dolce color d'oriental zaffiro*, and continued 'Which means, being translated, "Sweet hue of orient sapphire." Yes, that is what it means. But those English words could never move me to tears, tears of pleasure in the sound of words, tears of gratitude for the beauty of the world in which we live, heaven above us, as do those four solemn, sweet Italian words:

355

Dolce color d'oriental zaffiro

They are of the same supreme order of poetry as Milton's moonlit waves of the sea:

> The sounds and seas, with all their finny drove
> Now to the moon in wavering morrice move.

Who could translate that in any language on earth without spoiling its essence?'[2]

To my mind, it was the intermingling of interests which made Trevelyan such a fascinating character. When he took you to a ruined Peel Tower you felt that his thoughts were on the cattle raiders of Northumberland and the harsh life which made the border ballads so much more savage than their southern counterpart.

I remember his indignation with his old friend, E. M. Forster, for calling Scott sentimental and for referring to *The Heart of Midlothian* as provincial and the anger with which he said, 'Personally I prefer the provincial to the suburban.' It is easy to see why Trevelyan should have been so upset at the attack, because Scott, as he points out in one of his last essays, was the first author to combine the poetic with the strictly historical approach to the events of the past.

As an undergraduate, Trevelyan was a member of the Cambridge Apostles where he must have met G. E. Moore, Bertrand Russell and Desmond MacCarthy. But he told me that in the early 1900s he broke with the Apostles because he disliked Lytton Strachey and the Bloomsbury set. However, in the 1930s Trevelyan had mellowed towards Bloomsbury and even occasionally attended meetings of the Apostles.

It must have been a great satisfaction to Trevelyan that his passion for the British countryside found a practical outlet in his work for the National Trust. Today one feels that this was started only just in time and that many parts of Britain, far more than Trevelyan ever guessed, were saved by his enthusiasm and force of personality.

When I knew Trevelyan he no longer tramped the great distances that he had walked as a young man. Some of his marches were astonishing. I remember him telling me that he had once walked from Urbino to Gubbio on a single hot sunny day. That must have involved forty miles of rough and hilly Apennine road and a climb of several thousand feet.

356

We had told the Master that we would arrive at about seven and considering the antiquity of our car thought we had done very well to reach Hallington at 7.20 p.m. However, this was not the Master's view, for we found him anxiously scanning the horizon fearing that we might be late for dinner. (I should explain that after he retired Trevelyan liked to be called Master by people of my generation. In answer to my suggestion that we continue to call him Master he replied, 'Quite right, once a Bishop always a Bishop, and similarly for a Master.') So we scurried upstairs and were ready for dinner with a minute or two to spare. We found that we were the only guests, apart from Janet Trevelyan's sister, Miss Dorothy Ward, who was on a long-term visit. She proved to be a great friend and ally, and was full of helpful advice about what would best suit the Master and Janet, who, it turned out, was growing increasingly frail and losing her memory, though still expert at Word-making and Word-taking. She thought the Master enjoyed showing his guests some of the many historical sites and that we should encourage him to do so, which was exactly what we wanted to know.

Dorothy Ward had inherited all Matthew Arnold's papers and there was a good deal of chat about 'Uncle Matt' which we felt lent tone to the conversation. In a letter to her parents, Marni recorded the following dialogue between the two sisters: Mrs T., 'What was in that parcel that came for you yesterday, Dorothy?' Miss Ward, gloomily, 'Another book of criticism.' Mrs T., 'Not Uncle Matt again.' Miss Ward, 'Well, no, for a wonder, but it's because of some help I gave him on another book that *was* Uncle Matt.'

As Miss Ward had predicted, the Master was wonderfully conscientious in showing us the sights. On our first day he took us to a Peel Tower at Cock Law, which he explained was the normal sort of dwelling-house for a landowner, not grand enough for a castle, who, in the South, would have lived in a Manor House but in Northumberland needed protection against the depredations of the border brigands known as Moss Troopers. Cock Law was an especially good example, being wonderfully preserved, he continued, as we picked our way through the nettles and old iron of a dilapidated farmyard which surrounded the ruin, and so it proved to be. The tower appeared to have been used as a pigeon loft for two or three centuries, to judge from the vast quantity of guano and the thunder of wings which greeted our arrival. The cattle were evidently stored in the bottom of the tower and the living quarters began on the floor above, connected by a ruined staircase up which we presently climbed. 'The Master went first, groaning dreadfully, and we behind him in a shower of guano, for several centuries of pigeon dung had rendered the stairs invisible and perilously slippery. And if going up was bad, coming down was even worse. At any moment we thought the Master was going to take a header down the shaft. However, he didn't, and of course it was worth it.'

On the following day the weather looked ominous, at least the Master assured us that it did and we elected for an indoor pursuit: to visit Wallington, the Trevelyan seat where Sir Charles, his eldest brother, and Lady Trevelyan lived. It was great fun to see this beautiful Stately Home being lived in by three generations of that branch of the Trevelyan family, but of all the interesting objects that it contained, the only one that sticks in my mind was a gigantic, contemporary tapestry worked by Lady Trevelyan to whom we were shortly introduced. It took her from 1910 to 1933 to complete this colossal project, which depicts the legend of the first Trevelyan who swam his horse from St Michael's Mount to the mainland of Cornwall.

Afterwards we related the meeting with Lady Trevelyan, whose christian name was Molly, to Dorothy Ward who explained that 'we call her Molly the Great to distinguish her from Molly the Less [George T.'s daughter in law], though the latter is certainly not less in esteem, you understand.'

Our last expedition was to the Roman Wall which I knew well but Marni had never seen. The Master suggested thoughtfully that we take our own car as well so that we could walk on the Wall for a bit after he had shown us Housesteads. We took his advice and, feeling in the need of exercise, fairly sprinted along the Wall, earning a good mark when we described where we had got to, particularly as we managed to get home exactly on the dot of 12.57, just three minutes before lunch. The Master was out watching for us and preparing to fuss, but we didn't let him down.

On our way to the Wall the Master, whose car we were following, stopped so that he could show us the house where my grandfather started to write *Italy and her Invaders*, which I am ashamed to say I had never seen before.

Marni writes: 'The general conversational tone of the party was splendidly intellectual. As we sat with the Master in good Cambridge silence (not uncompanionably) he rose and, giving the fire a moody kick, said "I've been wondering, lately, what our civilization would have been like without Christianity." How's that for a conversational opening? Some other almost equally daunting openings were "I read history blacker than I used to do" or "all novelists since Conrad are cads".'

Second visit to Hallington, June 1958. Our second visit to Hallington was after Janet's death. Trevelyan's eyesight was now so poor that he could read very little, but he still derived consolation from the vast quantity of poetry which he knew by heart. I remember that he would sometimes stop in the dripping woods and recite poetry for ten minutes at a time: *Pippa passes* and *Marmion* are two that I remember.

On this occasion we were not the only guests, the others being Trevelyan's daughter Mary and her husband Canon Moorman, who reminded me strongly of Mr Arabin in *Barchester Towers*. Mary had been described to us as formidable, but we liked her very much. They were both writing books: she was into the second volume of a life of Wordsworth, having just received a prize for the first volume, and he was in the middle of a History of the Franciscan Order, which looked like being a very Magnum Opus indeed.

Marni wrote :

> They [the Moormans] are of course True Believers in the strongest Anglican sense, but of the sort that remains on terms of humorous and often witty magnanimity with us non-believers. And vice versa. This morning, at breakfast, Mary was describing, as a horror story, a clergyman of their acquaintance who collected butterflies and was heard to say, 'Oh, you won't find that in North Wales; it used to be there, but I bagged the lot.' I said that I thought he was a 'disgrace to the cloth', at which the Master who seemed not to have been listening, suddenly barked 'The fellow was too much in love with the works of God!', which made us all laugh, believers and non.
>
> The Master is in fairly cheerful form but unfortunately his eyes have gone back on him so he can only read a very little of the day, and cannot see birds, etc. and this is hard to bear. He reads, he told us, ten minutes every night in Gibbon's *Autobiography*, and in the brief time he can in the day he is reading a book called *On Sherman's Track* by an Englishman called Kindersley who went from Atlanta to the sea two years after the war. I told him about our family photographs and Fanna's [Marni's grandmother] story about 'Mr Lincoln has been shot; the South has lost its only friend.' At which – sitting at tea with A. and

me for the Moormans were out – he suddenly gave a loud strangled coughing sob, and wiping tears from his eyes said, 'You have greatly moved me. I knew that it was so, but to hear it told like this moves me deeply'. He is fine nineteenth century stuff.

A false alarm, October, 1961

Life was proceeding on its usual course when our peace was disturbed by a telephone call from a Dr Bolander of Sweden who spoke to Marni from London. Could he leave a message? No, he had really better speak to me personally, but Marni could tell me that Dr Bolander had a piece of *very good news from Stockholm*. This told me what Dr Bolander was on about and also put me on my guard, for I knew that in a previous year my cousin Dorothy Crowfoot Hodgkin had been photographed with her family on the basis of a false alarm – she didn't get the Nobel prize until 1964 – and that in 1960 the Australian radio had broadcast an equally incorrect rumour about Jack Eccles. After a while another Swedish journalist, a Mrs MacFarland, joined in the fray and it eventually transpired that on Thursday it would be announced that Eccles, Andrew Huxley and I had been awarded the Nobel Prize for Physiology or Medicine and would I please come for an interview and photograph in London.

As I had to be in London anyhow for a PhD viva at University College in the afternoon, I went in the morning to *The Times* building where the Swedish Journalists had an office. Here I was photographed and asked every conceivable question about my life, family, what I would do with the money and would I bring all the children to Stockholm. After lunch with Andrew Huxley, who was now Jodrell Professor of Physiology at University College, I started to examine my candidate on his thesis. In the middle of the interview I was summoned to Andrew's office, outside of which several secretaries were twittering, evidently aware that something was up and there were Bolander and MacFarland with faces as long as your arm, who began glumly: 'Oh Professor Hodgkin, we have bad news for you; it has gone to an American called Bekesy for work on the ear.'

So that was that. Mrs MacFarland said she hoped I wouldn't take it out on my PhD candidate – I didn't – and Andrew and I made suitable remarks about never having counted on it in the first place and went back to our ordinary lives.

Chapter 36

Stockholm, 1963

TWO YEARS after the false alarm about the Nobel Prize, I had just joined Knox Chandler and Hans Meves at Plymouth, when I was once more rung up by the same Dr Bolander who had alerted me before. Of course I had rehearsed various answers that I should give in this situation, but in the event I said mildly, 'Well how do you know so much more this time than last?' – this was after Dr Bolander, who had the grace to be slightly embarrassed, had assured me that this time it really, really was true. Then a Swedish newspaper called, and then Bryan Matthews from Physiology in Cambridge, to say that a cable had come in at the lab from the Karolinska Institute. This was more like it, but not quite enough, as I had heard that what had happened last time was that the Karolinska had voted for Andrew and me, but not the Nobel committee, which had the final say.

The reason the situation was confused was that Marni and I had driven down to Plymouth with some equipment where she spent the night and she was taking the car back to Cambridge when the news came through. This meant that the only people at home to field the Press, which they did with great efficiency, were our two elder daughters, Sarah and Deborah. They remembered the previous false alarm of 1961 and were initially cautious. But they prudently rang me at Plymouth to report that the Press including Reuter's had called with the good news from Stockholm. Apparently I replied, 'Reuter's isn't good enough. Ring me again if a positive cable comes from the Nobel Committee.' This cable presently arrived, my daughters rang me again and at that point I decided to return to Cambridge. By this time the Laboratory at Plymouth was starting to hum and presently the BBC was arranging to send round a camera crew. Meanwhile Andrew Huxley had given a masterly interview in London explaining what we had done in layman's language.

Eventually, I caught the late afternoon express from Plymouth and arrived home near midnight to be met by Marni, Sarah and Deborah on Cambridge station, all of whom had been very busy answering the 'phone and taking down the telegrams of congratulations that were pouring in. We couldn't reach the younger children, who were away at school, but knew they would hear by the following morning at the latest.

Naturally I was terrifically pleased and almost my first words on Cambridge station were that we must have a party – which we did next day on 18 October. Later, Marni wrote:

> We had the most tremendous party you can imagine, for both A & I thought that it was then or never, so we ordered four dozen bottles of champagne and circularized the lab and A. had lunch in Trinity and asked everyone he saw, and I spent the afternoon, more or less, on the telephone – the family and Mrs Hume worked all the morning getting the place ready, while A. was being interviewed yet again in his study. The result of all this was a party which began at six and went on until midnight so that cars in the road had to be seen to be believed – so much so that Anglia television announced that Professor Hodgkin was giving an enormous party at his residence in Newton Road.

The next three weeks seemed to go by in no time with several Societies wanting to celebrate another Nobel Prize and masses of congratulatory messages to answer. Everyone agreed that children were warmly welcomed by the Swedes and that we should bring them to the Nobel celebrations if at all possible, at any rate if they were over ten, which all of ours were. So one of the first things we had to do was to get permission for each of the three younger children to miss a week of school – Sarah had already left school and was in the middle of interviews for a university place. Then there was the Nobel Lecture to write, which was tricky as Huxley and I had to divide the field, and each lecture needed to be reasonably comprehensive without becoming too heavy.

Still we got through somehow and, after a dishevelled start occasioned by the failure of our hotel to order the taxis we asked for, we got to the airport just in time where we found the Huxleys, with their three elder children, anxiously waiting for us.

We had told the children that there would be reporters and

Figure 36.1. Arriving in Stockholm, 6 December 1963.

photographers to meet us, but hadn't realized ourselves how dramatic would be the contrast between leaving London as modest private citizens and arriving in Stockholm as full-blown celebrities (Figure 36.1). As soon as we climbed out of the plane there were my colleagues, Frankenhaeuser and Granit, to welcome us, together with Dr Stöhle, Head of the Nobel Foundation, the Scientific Attaché at the British Embassy, the Australian Chargé d'Affaires and a host of photographers, all snapping and flashing. We were lined up and grouped and re-grouped; and then Jack Eccles went up the steps of the plane to snap us all, which the reporters adored and they snapped him snapping us, and altogether, quite suddenly, we were VIPs.

The children tried for a short time to be modest but then confessed they were simply loving it. We were finally urged indoors while our passports were stamped, the principals were interviewed and the rest milled around being introduced to various people, in particular to their own personal attachés. My family had an immensely tall, solemn young man called Rasmus Rasmussen, with a very pretty young wife, Ulla, who could come only to the grander receptions as she had to look after a fourteen month old son. 'Ras' was immensely helpful and we should never have got through without him. We also managed to end up with our own driver, who confided that he was not really a driver, though most impressive in his livery, but a Forestry student earning some pin money.

Eventually we were ushered out of the airport and into a fleet of those Corps Diplomatique cars, our children in one and Ras, Marni and I in another in which we drove slowly through a landscape of bare birch and fir trees, to the Grand Hotel in Stockholm where we were to stay for a week.

After we had once more been thoroughly photographed, we were ushered up to our rooms on the third floor. Marni and I had a suite, with banks of windows looking out across water and traffic to the Royal Palaces and the Old City. The children all wanted to be separate, so their rooms faced inwards, but they had the most exquisite miniature suites, each with its own little hall and bathroom. They usually had breakfasts and sat about, in such time as there was, in our beautiful airy drawing room.

In the afternoon I was scheduled for a Press conference with Jack and Andrew, but after we had held forth for a while, the rest of the families were summoned, interviewed and again photographed.

After the Press Conference, we managed to finish our unpacking, and thus discover what we had left behind, before being driven to the Frankenhaeusers about twenty miles from Stockholm. Their drive was lit with flares, which were characteristic of Sweden in the winter. There were flares all round the Opera House and banks of them burning away in the parks. The Frankenhaeusers were giving a 'small', informal dinner party for the physiologists – no mean number with two Eccleses, five Huxleys and six Hodgkins – with their own daughter Carola and two guests. So we ended with ten grown-ups in one room, and eight children in what was actually Carola's bedroom but converted into a dining room for the occasion. We were slightly nervous about this arrangement as the Huxley and Hodgkin children hadn't met all that often, let alone with a stranger like Carola thrown in to make it harder. However, they found plenty of common ground and you could see the grown-ups relax visibly when we heard an absolute squeal of united mirth coming out of the children's room.

We first met the other laureates at a reception at the Swedish Academy where I had a long talk with Maria Goeppert Mayer who shared the Physics prize with Eugene Wigner and Hans Jensen. I liked her, but soon found that her speech was slow and that there was a curious, unfocused, wandering quality about her. Another of the Prizewinners, Giulio Natta, the Italian (Chemistry with Karl Ziegler) had Parkinson's disease and could only move with the utmost difficulty, though a personality of considerable charm came sadly through the physical wreck. I had a painful conversation with Signora Natta, who knew that I was receiving the prize for work on nerve conduction, in which she asked sadly, but without bitterness, why nothing could be done for her husband. Both Marni and I liked her tremendously and we parted with fond promises to let her know if ever we came to Milan.

Surveying the scene, Jack Eccles trumpeted with his splendid mow-em-all-down tactlessness, 'Looks like the Physiological contingent has the best health round here!'

Tuesday was the great day, with a rehearsal in the morning and the prizegiving in the afternoon. The rehearsal wasn't much use as men and women of distinction are notoriously unreliable in matters of protocol.

Time has blurred my memory of the actual ceremony but it is described in vivid detail in one of Marni's letters to her parents:

> Eventually we are all ready, and are driven pompously away to the Concert Hall where the doings take place. Here we are ushered into a kind of robing room where dignitaries are getting into gowns of various kinds. All the Swedes are wearing those wonderful pleated top hats with gold buckles at the front such as one saw in the academic procession in *Wild Strawberries* and there is a dazzling profusion of medals, stars, cockades, ribbons, etc. The hall is filled with a capacity crowd, all in full evening dress, (it is now about 4.15) and all the photographers are in white tie too. Lady Eccles is behind us, with a camera and permission to snap – still more transparencies! The stage gradually fills with dignitaries and eventually the laureates shamble in and take their places. Fanfare of trumpets. Enter the Royal Family. We all stand for the National Anthem, in which the Swedes, including Princesses, actually sing the words. We're off! Procedure: piece of very light music played by orchestra in gallery at back. Lengthy citation is read in Swedish (we have translation), ending with short citation in laureate's native tongue. Laureate is bidden to descend from platform to receive medal from King. Laureate advances, bows, descends staircase, shakes hands with King, bows to ladies of Royal Household, receives medal and enormous portfolio, advances to other side of staircase, *supposedly* gives up medal and portfolio which will be on display in City Hall the same evening, ascends other side of staircase, bows to audience

and returns. Well, this may sound fairly straightforward, and perhaps is with a single recipient, but with these divided prizes, three people are trundling down the stairs etc. and is the first to wait, or go on, and does he bow with the trumpets? After the trumpets? Wait for the trumpets? There is nothing like an ear-splitting fanfare to tingle the blood, but also to upset the routine. Well, anyhow, nobody got it quite right. Even the Ambassadorial poet, Seferis, who spent six years at the court of St James, forgot to bow.

To do the Swedes justice, they are very fond of formalities, but they appear not to mind if things don't go quite according to plan. Also there were several things put in to make it harder. The Physics people came first, and Mrs Goeppert Mayer, whose shoulder straps came right off as she sat, so that she began to twitch them up in a vague way when it was her turn to rise, forgot all about returning the medal and portfolio and drifted back to her seat carrying them in a slightly absent-minded way.

Then there was Professor Natta. He could get as far as the top of the stairs but couldn't possibly descend, but it had been arranged that his son should go down the steps to receive the actual medal. Well, when this actual moment came, the King, who is 81 and well over six feet tall, gave one look and like a veritable moose he *bounded* – there is no other word for it – up the stairs and no nonsense about the boy standing deputy for the father. I found this very moving: Natta being so pathetic in his hour of glory and the King so nice. Mr Erikson, our chauffeur, confided to me that everyone, even the extreme republicans, love the King, and indeed just looking at him you can see that he is a truly lovable man.

Anyhow, after this, the routine of the procedure was rather shaken, but never mind, apart from not knowing whether to give up the portfolio and medal, our own Hero acquitted himself splendidly, did remember to bow and looked younger than anyone else, so that even his critical children felt that he had covered them with reflected glory. So more music, more citations, National Anthem again and finally it is all over.

Between the Prizegiving and the Dinner, the laureates and their wives and children get to be introduced to the Royal Family. In our case this was done by the British Ambassador, HE Sir Ponsonby Moore Crosthwaite, whom we had met earlier in the week. He is known as Sir Moore, which sounds wrong but probably isn't – in contrast to Sir Eccles, which was also used and definitely is wrong. Sir Moore had just arrived in Stockholm and confided to Marni and me that, as he was nervous about the introductions, he had thought

Figure 36.2. With Princess Margaretha at the Nobel Dinner.

up mnemonics for the various sets of children, ours being 'Some Don't Refuse Jam' for Sarah, Deborah, Rachel and Jonathan. He cast us a look of triumph as he negotiated this task successfully.

At the dinner I was lucky enough to sit next to Princess Margaretha, who was both attractive and good company, which took my mind off the speech that I had to give later (Figure 36.2). The speech was not just a matter of standing up in my place, but involved a walk of some 200 feet around a long table to a floodlit lectern. To complicate things, Princess Margaretha mischievously insisted that I walk the longer of the two ways round the table – 'To make a change,' she said. However, the speech went off all right and the audience were so surprised to hear a joke that they actually laughed.

On the following day you might think we could take it easy, but I had my Nobel Lecture to give and wanted to make certain that the projection room could handle my bit of film and slides. There were only two minutes of film and the slides fitted all right, but even so when it came to the actual lecture, the slides were out of focus and at one point the projector broke down, plunging us all into total darkness.

It turned out that the lecture was nothing like as formal or as large as I expected. Jack Eccles was rather put out. 'I spoke to audiences three or four times this size in Minnesota,' he said crossly.

I was the first to speak and as I got up to the podium the audience began to laugh. I smiled back but the laughter recurred whenever I turned to point to the screen. What had happened was this. The three lecturers had been ushered to reserved seats and by some freak of static electricity a large RESERVED sign stuck to the seat of my trousers and went up to the platform with me, until eventually plucked off by Professor von Euler. As it happened, this mishap paid off handsomely, for not only did it put the audience in a good humour, but the story was a tremendous hit with the Royal Family that evening. Each Royal lady asked me to tell the next who came round, and urged me to tell the Queen, each exclaiming, 'Not true!' in incredulous delight, before bursting out laughing. As one of my friends said later, 'Nothing is more sure-fire than the bishop on the banana peel.'

Our last major event was the 'intimate' dinner for 120 people at the Palace that evening. In a strange way it really was rather intimate. This dinner was given by the King and there were no speeches and no photographers: 'It is a family affair,' everyone said. The food was magnificent, and all the people we met towards the end of our stay said, 'The best meal was with the King, no?' for they take great pride in the traditional excellence of the Palace food – even people who have never been there. We were waited on by servitors in late eighteenth century dress, blue liveries with white facings and tricorne hats turned up in front; and the Queen was waited on by a special servitor who had a plume about three feet high going up from this same hat – something you see in prints but never expect in real life.

The last two days were not quite so full and at last there was time to see the lovely Chardins and Rembrandts in the National Gallery and to do some shopping. But the engagements I had were not uninteresting or unimportant: collecting the medal and cheque for one thing: £6,000, which doesn't sound so much now, but would easily put central heating into our house which was our priority number one. Then there was a joint broadcast by all the Nobel Prizewinners of the *Whither Science?* variety.

We ended with the Medical Students' Dinner and Ball on St Lucia's day; another excellent dinner concluding with speech by a student and witty reply by Andrew; after which we all danced. Suddenly it was 1.15 a.m, but to quote from M's letter, 'The Cinderellas stagger from the Ball, but the coach hasn't yet turned back into the pumpkin. That's on the morrow. Plane at 1.30. Bumpy flight. Home at last. Icy weather. Icier house. Real life once more.'

Chapter 37

Postscript

IF THIS book were a biographical novel rather than an autobio-graphy, I should not have pursued it beyond the point at which the central character has recovered from the war and completed the scientific project which had been interrupted for six years. This is partly because the beginning of people's lives is generally more interesting than the end, and partly because I wanted to give a detailed picture of the pursuit of science in the mid-twentieth century. But in an autobiography, as opposed to fiction, there is the possibility of saying what actually happened to the author in the latter part of his life. The final chapter contains such an account, but will of necessity have to omit many interesting people and events.

In congratulating me on the Nobel Prize, John Kendrew who had received the Prize in 1962 said that it might be a year before I got back to any scientific research. It wasn't quite as long as that, but there were many unexpected things to do and letters to write. Perhaps the most difficult were the letters from people with neurological diseases who hoped that our work might help them, which in most cases it could not do, even though, in the long run, understanding nerve conduction and the part played by sodium, potassium and calcium has been of considerable benefit to medicine. But if you are ill, it is not cheering to be told that future advances may cure your condition, so I relied on advice from my medical friends at Queen Square or the Maudsley as to the best way to answer the letters. I can't pretend that this did much good but it would have been worse to do nothing, and in one case a suggested consultation did lead to a cure.

A very nice thing that happened was that in 1966 my father-in-law, Peyton Rous, who was then 87, was awarded the Nobel Prize (together with Charles Huggins) for his work on cancer-producing viruses. As Peyton's most important work was carried out in 1910, this was stretching the Nobel Foundation's rule about discoveries recognized by the award being within the last ten years! But nobody minded as Peyton had continued to work in the same general field for the next fifty years. So, although we were no longer VIPs, Marni and I joined her mother and father in Stockholm and enjoyed it quite as much as in 1963, though in a different way. Our only small disappointment was that the flowers in the Concert Hall, although lavish and beautiful, were not in the extraordinary profusion that we remembered. We found out presently that in 1963 the florists in North Italy where Professor Natta lived were so delighted at his success, and saddened by his illness, that they sent an almost overwhelming quantity for the Hall.

Research on muscle and nerve: Cambridge and Plymouth, 1956–70

In Cambridge I worked mainly on single muscle fibres with Paul Horowicz, who was extremely good at dissecting them. I never got on top of the dissection, which was more difficult than anything I had tackled before. So I earned my keep in the laboratory by making electrodes and preparing the large number of solutions that we needed. These experiments, which started in 1956, led on to the work with Richard Adrian and Knox Chandler on voltage-clamping muscle, and to the introduction of a technique which though not ideal has been fairly widely used. The electrical properties of muscle are sufficiently different from those of nerve to make it both challenging and interesting to explore them with the same general strategy that we had employed for nerve. We were particularly interested in the electrical properties of the system of membrane-lined tubules which conveys the influence of the surface action potential to the interior of the muscle and helps to activate the contractile machinery.

My last experiments on muscle were done in collaboration with K. Nakajima in 1969–70. I thought Nakajima should publish on his own, but he was very keen to involve me and used to collect me at 7.30 p.m. to work the camera for an hour, which he could have perfectly well have done himself. This got me involved in his problem and resulted in my doing a reasonable amount of work on the theory and analysis of his experiments.

I worked at Plymouth on squid nerve nearly every autumn between 1958 and 1970, sometimes as late in the year as December, on one occasion returning after Christmas for a fortnight in January. I found, as other people have done, that it is easier to keep going with experiments when you are away from home and the laboratory has absolute priority. My scientific partners during this period included P. F. Baker, T. I. Shaw, H. Meves, W. K. Chandler, M. Blaustein and E. B. Ridgway. To begin with, we worked mainly on perfused nerve fibres, but later studied calcium movements in intact axons, using either radioactive calcium or the calcium-sensitive protein aequorin, extracted from certain jellyfish, which emits light in the presence of calcium ions.

Move to visual research

The early autumn of 1970 was the last time that I did any experiments at Plymouth. After that I switched my interest very largely to visual research, which I could do in Cambridge with the help of visiting post-doctoral scientists. The reason for this drastic change was that on 30 November I was due to become President of the Royal Society, with a tenure of five years, and was already much involved in the Society's business. I thought that with the right colleague I could keep some experimental work going in Cambridge, and combine a London life with a Cambridge one, but saw no way in which I could also devote long periods to experiments at Plymouth. I felt that if I worked on the electrical and ionic properties of nerve or muscle I should all the time be hankering after the squid giant axon, and that I had better make a complete change.

As a student at Cambridge in the 1930s I had been influenced by the work of Adrian and Rachel Matthews on the retina and by Hartline's experiments on the eyes of *Limulus*, the horseshoe crab. Later, I saw Hartline's experiments at first hand when I visited his laboratory in 1938 and 1948. Finally, I had been impressed by the work of Hartline, McNichol and Wagner on generator potentials, which I first heard about at the 1952 Cold Spring Harbor conference.

In making the move I was helped by an Italian physiologist, M. G. Fuortes, whom I first met in Cambridge in the late 1940s. Fuortes, who died in 1977, spent the latter part of his life in America, but remained in touch with his friends in Italy and

England. In 1961 we started to correspond about work that he was doing on the eye of *Limulus*. For this he used the microelectrode technique that I had helped to develop. I was due to lecture at Woods Hole in 1962, and Fuortes asked me to join him in electrical measurements on the visual cells of *Limulus*. We became interested in the long delay between a light flash and the appearance of an electrical response, which we thought might arise from the time taken for a signal to pass through a cascade of intermediate chemical reactions, possibly stages of chemical amplification. We were also anxious to learn how the delay might alter as the eye became light-adapted. It turned out that in the *Limulus* eye, as in most eyes, there is a trade-off between time resolution and sensitivity: the eye loses sensitivity but gains time resolution as it adapts to light.

Denis Baylor had been a post-doctoral-student of Fuortes at the National Institutes of Health and had done distinguished work with him on the rods and cones of terrapins. He joined me in Plymouth in October 1970, but after a few weeks there we moved back to Cambridge, where after a long period assembling optical equipment, in which we were greatly helped by Andrew Huxley, we eventually started impaling the rods and cones of terrapins.

This work with Baylor was the first of a series of experiments on the vertebrate retina which lasted for seventeen years, and might have continued longer, but for the untimely death of my last collaborator, Brian Nunn, in 1987 soon after his return to America. During the intervening period I greatly enjoyed working with T. D. Lamb, P. A. McNaughton, P. M. O'Bryan, P. B. Detwiler and K.-W. Yau. Summaries of this research can be found in lectures given in 1971, 1974, 1981 and 1988 (Figure 37.1).

I realize now that the move to visual research was a fortunate decision and that without the help of people like Denis Baylor who made it possible I should never have been able to swim against the tide of administration that was to flow strongly in my life between 1970 and 1984.

The Royal Society, 1970–5

In the 1960s there were substantial changes in the composition and position of my research group in Cambridge. In 1960 we lost both Andrew Huxley and Richard Keynes: Andrew to University College London and Richard to the Institute of Animal Physiology at

Figure 37.1. At the Physiological laboratory, Cambridge, 1976.

Babraham, although as both continued to live in Cambridge I did not feel the loss as acutely as I otherwise would have done. Their place in my group was filled naturally, without any special action on my part, by R. H. Adrian, I. M. Glynn and later P. F. Baker, all of whom had University teaching posts. In 1970 we formed an alliance with a group led by D. A. Haydon who were studying the properties of bimolecular films of lipid between two watery solutions. This work, which had little contact with biology, was done across the road in Colloid Science, a research department which the University Authorities were running down and, rightly or wrongly, were anxious to close altogether. I was very keen on Haydon's work as I felt he had gone a long way towards making an artificial cell membrane, and thought that the right course would be for him to move into some vacant space next door to Physiology. Oddly enough, Colloid Science was well endowed, with finance made available to the department by the Oppenheimer Foundation, in recognition of the value to the mining industry of advice from Jack Shulman, a former Cambridge colloid scientist, who died in 1967.

In 1969 Cambridge University offered me the Plummer Chair which the appointing committee wanted filled by a biophysicist. Privately the Vice-Chancellor, Ashby, advised me against taking the chair as I was very well placed with my Royal Society research chair and he could see no point in the move. Actually there was a point, as I learnt from Swinnerton-Dyer, then the eminence grise of Cambridge politics. If I took the chair the University would transfer Haydon and his technical staff to Physiology and make the necessary financial provision for supporting them. They would also provide a permanent position for my instrument maker and part-time secretary, about whose future I had been growing increasingly concerned. All this was agreeable to Bryan Matthews, the Head of Physiology, provided it did not involve the department in extra expense and I consulted him about major administrative decisions. These arrangements were made in the first part of 1969 and I agreed to take up the Plummer Chair at the beginning of 1970.

I mention these details because, as things turned out, the Royal Society would not then have wanted its President to be the holder of one of its Research Chairs, and from that point of view it was convenient that I had become a Cambridge Professor in 1970. However, this consideration did not enter my mind when I accepted the Plummer Chair at Cambridge a year earlier. At that time, I was firmly convinced that the obvious next President of the Royal Society was Peter Medawar and it was not till after his tragic disabling stroke in the late summer of 1969 that I realized I might be considered for that position. Even so, I was surprised when Patrick Blackett summoned me to London in February 1970 and asked if I would succeed him as President.

It didn't take me long to make up my mind, and after talking things over with Marni, who was enthusiastic about the prospect, I let Blackett know that I would be glad to accept. He said that it was customary for the incoming President to attend Council and Officers Meetings which I was happy to do, as I felt that I badly needed education in Royal Society affairs. As anyone who knew him would expect, David Martin, the Executive Secretary, was enormously helpful, and continued so, in spite of minor disagreements, during my five years as President.

Three of the officers, Harold Thompson, Foreign Secretary, Fred Bawden, Treasurer, and Harrie Massey, Physical Secretary, together with David Martin, had accepted an invitation from the Japan Academy of Science to visit Tokyo in March, to discuss the question of extending the Royal Society's exchange scheme to that country. Blackett, who was a sick man, did not want to go, and Thompson who led the group was keen that I should join it, which I agreed to do. My presence puzzled the Japanese. Was I the incoming President or was I not? And why was I not a Sir like all the others? However, I had been to Japan before, had many Japanese colleagues and friends and so got on all right. The visit was a rather grand affair and included an hour in the Imperial Palace with the Emperor. We were strongly advised to take visiting cards in Japanese and morning dress – another visit to Moss Bros – and I was given the special task of finding a present for Hirohito, for which I eventually chose F. S. Russell's beautifully illustrated book on jellyfish, which I hoped would be suitable, in view of the Emperor's interest in marine biology.

My letter to Marni gives the flavour of this visit to Japan:

> We leave for another reception in about twenty minutes but perhaps there is time to get a letter written. The high point so far, at any rate for interest, if not for enjoyment, has certainly been the visit to the Emperor but I must lead up to this gradually. We started with a reception and lunch with the Academy which was a pretty formal affair. (I'd already spent a useful hour with Sato and Kikuchi.) The President of the Academy, Shibata, is 88 and the average age must be well up in the seventies. Lunch was followed with a joint meeting with the Academy and Science Council, run by the Academy, who are probably not the right people to deal with anyway. After a slight break we went to a highly successful cocktail party at the British Embassy. Everyone became reasonably gay, which is something of an

achievement with a group of Japanese and English scientists all
subjected to considerable language difficulty. The Ambassador, Sir
John Pilcher, knows Japan well and is an exceedingly amusing and
interesting person. He explained to us about the Imperial Audience and
sent us the enclosed seating plan sent by the Palace. Note the loving
care with which the chairs, type carefully graded according to rank, are
arranged, and the attention to details which show even the type of chair
on the plan. [There were three grades: as non-knights David Martin
and I occupied the lowest grade.]

The Ambassador explained that our remarks would be put into low
gear by the interpreter and that we needn't bother to insert 'this
Honourable Personage', etc. In this connection he told us that when
George Brown [deputy Prime Minister] was here with his wife and said
something like 'I am so glad to be here and to be accompanied by my
wife', it was translated by '. . . accompanied by my dilapidated
spouse.' I am afraid that I shall never know exactly how our remarks
got translated. [I once had a letter from Japan ending, 'Longing for a
kind guidance. Yours with wormful fraternity etc.' – presumably the
equivalent of 'Your humble servant'.]

Anyway after getting into our morning coats we were driven into the
palace group – a vast walled and moated area unapproachable by the
normal person. We were received in the new palace, which is a very
successful modern version of a classical palace. It is surrounded by a
Japanese Garden with waterfall and lakes – all cunningly contrived so
that the scenery seems to be a continuation of the rooms inside the
Palace. The rooms are on a gigantic scale but furnished in such a way
that they seem of normal size and you feel small. In one corner of the
room where we were received there was an urn – the sort of thing that
is usually three or at most four feet high. But here it was in scale with
the room and was about twelve feet high. In two corners of this room,
about 70 feet apart, there were two flower arrangements of really
stunning beauty and simplicity. I remember the thing that *made* one
arrangement was a dead branch of a particularly beautiful shape. Apart
from these three objects I remember nothing except a very beautiful
carpet and eleven chairs arranged as shown.

Presently Hirohito came in and asked us pleasant and intelligent
questions about our work. We had been told that we must give our
present (F. S. Russell's book suitably bound) rather casually, in the
course of conversation as it were. 'By the way, we have here something
which your Imperial Majesty might conceivably wish to accept' – that
kind of thing. We'd decided to do this via nerve conduction, squid,
marine stations etc. so my bit of talking was directed to that end, and
the book was eventually given to the Emperor. He seemed an

intelligent, sweet-natured man, but extraordinarily remote – rather what one would expect of the 125th descendant of a deity. Apparently, he virtually never appears and there was a great argument as to whether he should open Expo. 70.

[Continued later in Kyoto] . . . After the Palace I was whisked off to a reception at the Ministry of Education where I got picked up (more or less) by an extraordinarily ebullient Japanese millionaire who is a central figure in medical research. He insisted on taking me to a very smart Tokyo bar where he was given a tremendous reception. The result of all this was that I never got any dinner and had more to drink than I wanted in view of an early start on the following day . . . I got packed early and we caught the 150-mile-an-hour express to Kyoto . . . Wonderful views of Mount Fuji on the way and pleasant sights of rural Japan – some of it deep in snow. We spent yesterday afternoon doing the Kyoto temples (10th–12th century) and the whole of today (Sunday) at Nara (7th–8th century). The high point so far as I was concerned was a drive up into the wooded hills to see two enchanting temples in exquisite surroundings. The weather remains very cold and it was snowing by the time we got to the second temple. Beautiful but cold.

It took us some days in Tokyo to achieve our objective, which was to start a scientific exchange system of the type which the Royal Society had with many countries. We were formally correct in starting with the Japan Academy, but neither they nor the Japanese Science Research Council, which was represented at our meetings, were really the right people to talk to. It turned out that the organization which would provide funds and with whom we should make detailed arrangements was a small group attached to the Prime Minister's Office called the Japan Society for the Promotion of Science. When we had met them it did not take long to sketch out the framework of a mutually satisfactory scheme, though this could not be put into action until a high-level delegation from Japan, including representatives of the Academy, had visited Britain, been received by The Queen and finally signed the appropriate agreement at the Royal Society. This happened in March 1971.

Apart from its intrinsic interest this visit to Japan was important to me because it taught me about the way that the Royal Society conducted its overseas affairs, as well as helping me to get to know the people who ran it. The Biological Secretary of the Royal Society, Bernard Katz, was an old friend, and after the Japan trip I knew that I should have no difficulty in getting on with all the Officers.

In 1970 the Royal Society had been in Carlton House Terrace for three years and the Blacketts had furnished and lived in the President's flat on the third floor. This consisted of one large room with a splendid view looking out across St James' Park to Westminster; attached to this were a small bedroom, bathroom and kitchen. David Martin thought that I would probably need to spend two or three nights in London – an estimate which proved to be about right. At that time Marni was running children's books at Macmillan and as a rule commuted four days a week from Cambridge. She welcomed the idea that the Royal Society should be our London home and we lived there happily in the mid-week for the next five years. With David Martin's help we arranged that small Presidential lunches could be given in the large room, so justifying rather grander pictures and furniture than we would otherwise have asked for.

I was very keen to keep my experimental work going in Cambridge, both because it was going well and because, unless I have some research to think about, I become too obsessively involved with administration – and too upset when things go wrong, as they invariably do. With the help of Denis Baylor and other visiting scientists I managed to do this reasonably successfully, though it often meant working for much of the weekend.

The task that I was most nervous about was the speech at the Anniversary Dinner on 30 November when I became President. I was used to lecturing to quite large audiences, but this was a different matter. To make life harder I found that the Society had invited the speakers some six months earlier when a Labour Government was in power, without foreseeing that we should have an election and that for the first time we would have a scientist, Margaret Thatcher, in charge of Education and Science. By 30 November it was clear to everyone that she, rather than Shirley Williams, should have been one of the speakers at the dinner. However, we asked Mrs Thatcher to give the principal speech in 1972 and she laughed when, in welcoming her in 1970, I said that she had learnt her chemistry but not her politics from my cousin Dorothy Hodgkin.

When I became President, David Martin asked me rather nervously whether I had a policy. I said I hadn't but thought that my predecessors, Florey and Blackett, had formulated objectives which would keep us busy for five years. In general these were that the Society should take a greater part in promoting research, particularly in its international aspects or in connection with individual appointments of outstanding distinction, such as the Royal Society Research Professorships: also that it should attempt to make its meetings and discussions more interesting and accessible to all concerned with pure and applied science. This had been difficult in the confined space at Burlington House, but would be easier in our new premises with their large lecture hall. Lord Florey had stated his objectives more concisely in conversation by saying that he would take the Royal Society into the twentieth century, even if he had to drag it there kicking and screaming.

If I were asked what I had enjoyed most during my five years as President, the most honest answer would be 'entertaining friends and colleagues in this beautiful building'. Next to that, and on a more serious plane, I would put the sense of historical continuity, and the satisfaction of taking part in first-class scientific discussions in a Society which counted Boyle and Newton among its earliest members.

The Royal Society: foreign affairs. International relations continued to occupy a prominent position in the Society's activities and I found myself bombarded with invitations from different countries. During the next six years I visited India, USA, Canada, Australia, China, Kenya and Iran (the last two after my presidency but on Royal Society business) as well as several European countries. A delegation was also invited to Moscow in 1975 but was postponed at the last minute to a date that I couldn't manage. However, I had spent May 1967 in the USSR and Georgia and didn't particularly mind missing this trip.

The Officers attached high priority to restoring the links with China which had flourished before the cultural revolution but disappeared completely after it. One or two Fellows did manage to go to China and we helped them to get visas, though the Chinese Chargé d'Affaires hated to put anything on paper and preferred to make a solemn declaration in our premises that it was perfectly in order for Dr So-and-so to visit China.

Eventually the Chinese Academy of Science invited a small delegation to visit China and discuss scientific exchanges in May

1972. Kingsley Dunham (the new Foreign Secretary), Martin and I accepted at once and booked tickets on the overland air route through Siberia. However, at the very last minute we were told by the Chargé d'Affaires that permission was withdrawn and we must cancel our visit. This we refused to do, cabled that we were coming and went ahead on the flight through Moscow, Omsk and Irkutsk to Beijing. We were greeted in a friendly way, put up in a comfortable hotel and, for several days, taken sightseeing, or shown various university departments, which were in a state of disarray with students learning through work, for example, electrical engineers mass-producing cathode ray oscilloscopes on a crude assembly line. From these visits we concluded that 'the universities were more disorganized by the Cultural Revolution than most other institutions. This is not surprising since one of the aims of the Cultural Revolution was to prevent the re-emergence of an intellectual élite'.

The sightseeing was interesting but not what we came for. After several days it became obvious that the Cultural Revolution was still very much in force and that members of the Chinese Academy were frightened of arranging any sort of meeting. After consulting the British Ambassador, we sent a letter asking for a meeting to Chin-Li-Sheng, the Deputy Secretary General of the Academia Sinica, which was written in the grandest handwriting and phrased in the politest language we could manage. This did the trick. An evening meeting was arranged for our penultimate day and an exchange agreement was discussed and supported, on the understanding that it would be developed later by a Chinese delegation to the Royal Society – an event which took place in October of the same year, and formed the initial basis of the numerous visits which have been made since then by both British and Chinese participants.

On our last evening in Beijing we went to a formal banquet in the Hall of the Peoples, where the delegation was received by the President of the Academy, Ko-Mo-Jo, with whom gifts of books were exchanged and where he stated that the Academy looked to the United Kingdom scientists for help in developing the study of fundamental sciences in China.

Before returning home we spent the last few days of May in and around Shanghai seeing scientific institutions and university departments, and confirming the same general conclusions that we had reached in Beijing. We had planned one day's sightseeing in

Hangchow, but this was cancelled by the authorities for no obvious reason and we went to Soochow instead; at first we were disappointed, but it was fun to see a much smaller city than Beijing or Shanghai with perhaps something of old China about it.

At the end of our second day in Beijing, the Ambassador, J. M. Addis, a very pleasant man with a passion for Ming and Sung pottery and an obvious affection for China, but with no illusions about the dark side of the Cultural Revolution, asked me for my first impression of China. My diary records that I replied, 'It's rather like falling in love with a beautiful and attractive girl who is also a revivalist of the most dogmatic and proselytising kind.'

After we got back to London, we wrote a report for the Royal Society Council from which the following quotations are taken:

From the Introduction (pp. 1–2):

In its relations with foreign academies, the Royal Society is not normally concerned with politics. However, there have been great changes in China since the visit of the Society's previous delegation in 1962 as a result of the cultural revolution which has intervened. The relation of this to scientific teaching and research is so little understood in this country that a factual statement, based on what our Chinese colleagues have told us, is necessary for the purpose of this report. It is not our intention here either to commend or criticise what has happened.

One of our Chinese friends remarked that, although there had been less bloodshed, the effects of the revolution penetrated far deeper than those of the Maoist Revolution of 1949. That revolution led to a large expansion of scientific endeavour and most of the institutes of the Academy date from about that time. The small corps of experienced scientists then available was encouraged to build up substantial research organizations.

The Great Cultural Revolution has been deliberately aimed at establishing the leadership of the workers, peasants and soldiers by encouraging them to criticise all 'bourgeois' or 'revisionist' tendencies found among those engaged in teaching or research and those who have had the benefit of a university or professional education. Such tendencies include among others: (i) the idea that workers are incapable of directing scientific research; (ii) the cult of the superiority of all things foreign; (iii) the conduct of scientific research for the personal satisfaction of the researcher; (iv) theory without practice. The struggle has been actuated by certain practical measures. Established teachers and researchers have been caused to go to May 7th camps for periods of

382

six months, to contemplate their work while engaging in manual labour for periods of 6 months, or to undertake factory or farm work for periods of 1 to 3 years. At the head of every organization whether city, factory, peoples' agricultural commune, university or research institute, a revolutionary committee has been installed as governing body, the members being drawn from factory workers, peasants, members of the Peoples' Liberation Army together with junior and senior representatives from the organization concerned. We assume that these committees are appointed from members of the Communist party but did not establish this point with certainty. The effect upon scientific research may be summarized as follows: (a) as a short-term effect, severe disruption has been caused by the prolonged absence of staff members; some have presumably not survived the process of political reorientation; (b) scientific publication ceased completely from 1966, and is only now being resumed in a small way; we were told however that there will be no objection to papers carrying their authors' names; (c) the demand is now much more for work of immediate practical relevance to the community; (d) laboratories are being encouraged to go into industrial production if they can, even on a small scale, and otherwise to become closely integrated with industry; (e) of the 120 former institutes of the Academia Sinica, only about 60 remain under its control.

It must be added that although there may have been a short-term loss of scientific efficiency, this is regarded by workers and scientists alike as a small price to pay for the unity they now experience in the service of the state.

The Report concludes (p. 15):

Apart from seeking acquaintance with the Chinese scene the delegation's purpose was to assess the prospects of encouraging exchanges between United Kingdom and Chinese scientists. At the Academy Institutes we visited we found scientists who could speak English and visits to such institutes could be useful and can be encouraged. As however some of them are still in an unsettled state it may be some time before this can happen to a significant extent.

There was nothing in what we saw at Peking National University to lead us to think that exchanges with scientists in China could profitably be contemplated in the near future.

The Royal Society and the Rothschild Report. Soon after I became President it became clear that there was going to be trouble with the financing of the Research Councils, especially the Agricultural

Research Council, in which the Ministry of Agriculture had a particular interest. Early in 1971, at a time when he was building up the Central Policy Review Staff (the Think Tank), Victor Rothschild asked David Martin and me to a very pleasant dinner in a private room at the Ecu de France. Other people there were the Chief Scientific Adviser, Sir Solly Zuckerman, the Secretary to the Cabinet, Sir Burke Trend, and the Head of the Civil Service, Sir William Armstrong. Victor was always good at putting people at their ease when he wanted to, which he did on this occasion, and we had a long agreeable conversation starting with Agricultural Research and then ranging over the funding of Science in general. Reading between the lines, I felt that William Armstrong considered that Science had grown too big for its boots and needed shaking up, but that Burke Trend was sympathetic to the case for preserving the independent Research Councils; this was borne out by a brief letter written by him a day or two later which ended, 'Good luck in the battles that lie ahead.'

The next major event on the Research Council front was the publication at the end of November 1971 of the Rothschild report, *The Organisation and Management of Government R. and D.* This appeared as part of a Government Green Paper called *A Framework for Government Research and Development.* The Royal Society was asked for written comments to be sent by February to Mrs Thatcher, then Secretary of State for Education and Science. The Rothschild report was primarily concerned with the Medical, Agricultural and Natural Environment Research councils, but it indicated that both the Science and the Social Science Research Councils might be studied later. The report was based on the principle that applied research and development must be done on a customer–contractor basis. 'The customer says what he wants; the contractor does it (if he can); and the customer pays.' It advocated building up the scientific strength of the appropriate Ministries and then transferring to them a substantial fraction of the funds that would previously have gone to the Research Councils. The Ministries could then use the transferred funds to commission research by the appropriate Research Council. For the three Research Councils considered in the Green Paper the annual transfers should approach the following limits over a five year period: Medical Research Council, about one quarter to be transferred to Health Departments; Agricultural Research Council about three-quarters

384

to Ministry of Agriculture, Fisheries and Food; Natural Environment Research Council, about half, mainly to Departments of Environment and Trade and Industry. At the Anniversary meeting on 30 November 1971 we had only just received the Green Paper, and all I could do then was to encourage Fellows to let the Royal Society know their views. More than a hundred sent in written comments, and 170 Fellows attended a discussion meeting on 11 January 1972, from which it appeared that the Rothschild proposals was supported by some physical scientists but that the majority of the Fellowship were against them. A committee appointed by Council met four times between mid-December and mid-January and its draft memorandum was discussed at a special meeting of Council. Our report was sent to the Secretary of State early in February and published soon afterwards. Advance copies were also sent to the Select Committee on Science and Technology, before whom Sir Harrie Massey and I appeared on 9 February and were questioned for several hours in a friendly but thorough manner.

The line taken by Council's report, which was unanimous, was to oppose extensive use of the customer–contractor principle, which they felt unsuited to many medical and environmental problems; they were also against the large transfer of funds to user-ministries. However, they welcomed the proposals for improving the dialogue between customers and contractors, and for strengthening the scientific personnel at user-ministries.

At the time, many of us felt that the period of consultation was too short and that a major reorganization of science was being rushed through without adequate consultation. In retrospect, I came to feel that the rush was not altogether bad, as a prolonged debate would have inhibited many normal scientific activities at the Royal Society and elsewhere. It also appeared that the Government had largely made up its mind and that the most we could hope for was some amelioration of the more stringent recommendations of Lord Rothschild's report. This we probably did achieve, for the White Paper published in July 1972 recommended smaller transfers than those suggested in the original Rothschild report.

A development that was widely welcomed took place a few years later when the Department of Health and Social Security decided to forgo the transferred funds and have priorities for clinical and applied work settled by the Medical Research Council, of which its Chief Scientist was a member.

385

The Royal Society: finances. Before I became President, Patrick Blackett assured me that I had no need to worry about finance. The money for converting our new premises had been collected and the day-to-day finances were looked after by the Treasurer, Fred Bawden, and a highly efficient office staff. It was not until after Fred's death on 8 February 1972 that the new Treasurer, Jim Menter, and I became seriously worried about the Society's financial position. The new premises had enabled the Society to expand its activities in the directions that Lord Blackett hoped it would. But this expansion was expensive and inevitably involved the Society in costs that were only partially met by the Parliamentary Grant-in-aid. The Society had several substantial Trust Funds devoted to the support of Science, but its General Purposes Capital amounted to only about £0.8 million, on which the interest was quite inadequate to maintain the wide range of the Society's activities. There had been a financial appeal in 1967 and over the next five years the Society avoided depleting capital only by using the covenants given in reponse to the appeal. In 1972 the deficit without drawing on appeal money, which would shortly come to an end, was £100,000 and our expenses were undergoing a sharp inflationary rise. The Treasurer and the Office Staff therefore took a number of steps to reduce the deficit. Briefly, these involved economies in administration, raising journal prices and reducing the free issue of journals. In addition, negotiations with the Department of Education and Science and the Advisory Board for the Research Councils led to a better way of financing activities like the National Committees without any loss of financial independence. By these means the Office Staff (particularly the Deputy Executive Secretary, Ronald Keay) reduced the immediate financial danger that alarmed us so acutely in 1972.

However, these measures did not solve the problem that there were many desirable activities that could not be charged to Government funds and, more basically, that our General Purposes capital of £0.8m was much too small. In the summer of 1973 we decided to launch a development appeal aimed at raising £1m. After taking this decision the financial climate deteriorated sharply and the Officers had to decide whether or not to postpone the appeal. We decided to push on, in spite of the worsening position, and the appeal was launched in July 1974. The arguments which carried the day were first, that there was no immediate prospect of recovery of the

Figure 37.2. In 25 Newton Road, spring 1978.

financial position of Britain and secondly that we thought it no bad thing for the Royal Society to show some degree of optimism in a period of national gloom. In much the same spirit a former President, Sir Henry Dale, set up a committee to consider post-war science in 1943. Finally we had confidence in the future of science-based industry and knew that many firms as well as some of the nationalized industries appreciated our efforts to strengthen the links between science and technology on the one hand and industry and agriculture on the other. In the event our confidence was justified and the appeal brought in over £800,000 by January 1979.

Trinity College, Cambridge, 1978–84

As getting a Trinity Scholarship in 1931 was the event which opened up the prospect of a career in science for me, there was something rather appropriate about ending my academic life as Master of Trinity, where so many distinguished scholars had found interest and happiness. So I had no hesitation in accepting the offer of the Mastership, although it meant a great change in our way of life. My wife gave up her publishing job with Macmillan and we had to sell our over-large but much-loved house in Newton Road. This saddened our children and grandchildren, who were deeply attached to our old home although they no longer lived there, but they soon came round to the view expressed by one of my American friends that the Lodge was 'not a bad pad'. Marni and I soon overcame the feeling that the Lodge was a palace rather than a home, and acquired a strong affection for it. Two practical things helped. One was that the Bursars had the bright idea of making a housekeeper's flat in a derelict part of the Lodge. The other was that we acquired a splendid housekeeper who positively enjoyed looking after the rooms and furniture in the Lodge, as well as us.

Even if you do not love grandeur you would have to be very dull of soul not to succumb to the charm of living in the splendid house, described by Trevelyan as 'built by Nevile's love and Bentley's pride'. It is true that in summer time the courts were full of tourists and one wished that more visitors would accept Baedeker's advice that 'Cambridge is less attractive than Oxford and may be omitted altogether if the visitor is short of time.' But even at the height of the tourist season, peace returned in the evening; and in the very early morning a kingfisher or a heron could occasionally be seen sitting on a wall at the bottom of the Master's garden.

Transcending these details was the feeling that the Master's Lodge was part of the College and belonged to its history, or even its prehistory, as in the Comedy Room wall, where, to quote Trevelyan once more, 'the bees have made their hives in blocked-up windows that once looked out on the Wars of the Roses.'

Most country houses or palaces are lived in for only a few months in the year and are frequently empty for long periods of time. But Trinity Lodge, so far as I can make out, has been lived in more or less continuously for nearly four centuries and must have seen some fifty thousand undergraduates come and go in Great Court. Partly

Figure 37.3. With grandson Felix Hayes in a Trinity punt, 1979.

for that reason we adopted the practice of keeping the picture-lights on in the Lodge, so that on winter evenings the undergraduates crossing Great Court could catch glimpses of the great portraits and be reminded of Elizabeth and Essex or Marvell and Newton.

One change that I remember with satisfaction was the coincidence between the beginning of my Mastership and the entry of women undergraduates to Trinity in 1978. I believe that this change, about which many people were nervous, has been a resounding success and that it will be of enduring benefit to the College.

Another satisfying development was the continued growth of the Trinity Science Park on land given to King's Hall (the precursor of Trinity) in 1443. Both the creation and the development of this major enterprise were the work of the Senior Bursar of Trinity, John Bradfield; I am glad that I may have been able to help him a little with a project that I believe is of major importance in bridging the gap between science and industry – not only in Cambridge but in the country as a whole.

389

People say that on leaving a Master's Lodge, you must either move entirely into the country or stay as near the centre of the city as possible. We chose the latter course and found an oldish house between the Fitzwilliam Museum and the Botanical Gardens. Although quite unlike our previous homes, it suits us down to the ground.

NOTES

PART I

Chapter 1
Apart from the long letter from my mother to her parents, this chapter is largely based on *The Life and Letters of G. L. Hodgkin* by his sister L. V. Hodgkin (1921).

Chapter 2
The quotations about my paternal grandmother's family are from *A far serener clime: Memories by Ellen Bosanquet* (1957). Information about Edward and Catharine Hodgkin comes mainly from the same source or from *Catharine Hodgkin*, ed. Mary Smith (1948).

Chapters 3 and 4
These are based on memory helped by letters and note books. Information about Reader Bullard comes partly from his autobiography *The camels must go*. The quotations about my stepfather, A. L. F. Smith, in Baghdad are from the book about him by E. C. Hodgkin and others (1979).

Chapter 5
p. 62. After a career in the Colonial service in East Africa, Paul Cotton joined the RAF soon after the outbreak of war; he was awarded the DFC in 1941 but sadly, was killed in action in 1944. Kenneth Chapman became an academic zoologist, working partly in South Africa and the Sudan. John Pringle FRS, who worked at TRE during the war, was a distinguished entomologist and became Linacre Professor of Zoology

at Oxford in 1961. He helped to build up the important International Centre of Insect Physiology and Ecology (ICIPE) in Nairobi. He died in November 1982.

Chapter 6
p. 74. Interaction between active and inactive nerve fibres of the kind expected by Erlanger is conspicuous in certain situations, for example in the case of two crab nerves immersed in oil where the excitability changes induced in the inactive fibre conform quantitatively to the expectations of the local circuit theory (Katz and Schmitt 1940).

Chapter 7
p. 84. 'Curiously memorable though undistinguished lines', *Oxford Companion to English Literature*, ed. Margaret Drabble (1985). The account of John Cornford and quotations from him are from Sloan (1938).

Chapter 11
p. 129. Aunt Helen was not a real aunt but Helen Sutherland, one of the very few patrons of the arts in the interwar period; particular protegés were Kit Wood, David Jones, Ben and Winifred Nicholson, Barbara Hepworth and Vera Moore. She lived in Lowndes Square (London), Rock Hall (Northumberland), and finally in a large house in a remote part of the Lake District; she was keen on my stepfather, Lionel, and my younger brothers, but didn't care for me or my mother. Helen's female unmarried protegées tended to

become pregnant, to the consternation of the quiet rural communities in which she lived.

PART II

Chapter 12
1, 2 Poston, Hay and Scott (1964).

Chapter 13
1 Cockcroft report of 3 November 1940 PRO/AVIA/10.

Chapter 14
1 Hodgkin, Radar Notebook I, p. 121.
2 'We' is George Edwards, Downing and myself.
3 An exception was the high-altitude reconnaissance version of the Mosquito.
4 Other names used before this were Bawdsey research Station (BRS). Air Ministry Research Establishment (AMRE) and Ministry of Aircraft Production Research Establishment (MAPRE). For simplicity I use TRE throughout.
5 After we had moved to Malvern, Kendrew was succeeded by MacPherson who later became Registrary at Cambridge, a position which again required much tact and administrative ability.

Chapter 16
1 A. P. Rowe (1948), *One Story of Radar*.
2 Curran. Obituary of Dee, *Biographical Memoirs of Fellows of the Royal Society*, (1984), Vol. 30, p. 139.

Chapter 17
1 Marris was the administrative head and Espley the technical leader of the GEC television team.
2 Jesty at GEC was the important contact for development of cathode ray tubes.
3 Rowe, op. cit., p. 122.

Chapter 19
1 E. G. Bowen (1987), *Radar Days*, p. 187.
2 IFF stands for Identification: Friend or Foe.

Chapter 20
1 This account of Jackson is based on the memoir by Kuhn and Harley in *Biographical Memoirs of Fellows of the Royal Society*, (1983), Vol. 29, p. 269.
2 The principle of Window was possibly discovered as an accidental result of dropping metal-coated leaflets over Germany in 1939–40. In 1942 Joan Curran at TRE showed that bundles of thin copper strips were highly effective against British radar stations. It was first used on a large scale in the raid on Hamburg on 24 July 1943, and in the invasions of Sicily and Normandy.
3 Jackson, quoted by Kuhn and Hartley, ibid.
4 Kuhn and Hartley, ibid.; a Cambridge psychologist, Squadron Leader MacPherson, played a major part in getting SCR 720 going in Britain.

Chapter 21
1 This account follows closely on that in the Air Ministry Air Historical Branch Monograph *Signals, Vol. V. Fighter Control and Interception* (1952), Chapter 10, Centimetric AI (from which all quotations are taken).
2 AHB *Signals V* (1952), p. 152, referring to FIU, ORB 7 December 1941.
3 AHB *Signals V* (1952), p. 153, FIU Report 105.
4 *The History of TRE Post Design Services 1941–1945*. AHB IIE/244. A. M. File CS 13884, Encls 1A and 12A.
5 FIU, ORB April 1942.
6–10 AHB *Signals V* (1952), p. 153.
11 Gunston, W.T. *Night Fighters* (1976).
12 AHB *Signals V*, p. 156. 3 February 1943 is the official record of the first enemy aircraft downed by AI Mark VIII but Burcham's diary records that on '21st January 1943 G. G. Roberts [TRE] in E2180 got a Dornier 217 last night.'
13 AHB *Signals V*, p. 156 and Burcham (1945).
14–16 AHB *Signals V*, pp. 157–62.
17 AHB *Signals V*, p. 162 and Gunston (1976).

18 I am much indebted to Miss Dudley for the information in this paragraph which she obtained from Vols V and VI of *The Mediterranean and Middle East*. London: HMSO.

19 AHB *Signals V*, p. 162, Burcham, (1945); see also Chapter 20.

20–24 AHB *Signals V*, pp. 163–9.

25, 26 Ibid. p. 228.

27–29 Ibid. p. 157.

30–31 Ibid. pp. 228–9.

32 Ibid. pp. 166, 167.

Chapter 22

1 A. P. Rowe (1948), *One Story of Radar*, pp. 128–9.

2 R. V. Jones (1978), *Most Secret War*, p. 247.

3 A possibility first suggested by B. V. Bowden.

4 Rowe, op. cit. p. 129.

5 Ibid. p. 85.

Chapter 23

1 *Biographical Memoirs of Fellows of the Royal Society* (1960), Vol. 6, p. 259. H. W. B. Skinner by H. Jones. p. 259.

2 Published in the Harwell Magazine, *Harlequin*. Quoted in *Biographical Memoir* above.

3 E. G. Bowen (1987), *Radar Days*.

4 A. P. Rowe (1948), *One Story of Radar*.

5 Lewis, Dee, Skinner and Lovell at TRE; Blackett at Coastal Command; E. G. Bowen and others at MIT Radiation Laboratory in America.

6 Burcham recorded in his diary that he flew with Lindemann from Christchurch on 25 April 1942; this may be the flight that I have referred to or there may have been an earlier flight with Hensby as well.

7 Churchill (1951), Vol. IV, pp. 250–7.

8 Sometimes said to stand for Home-Sweet-Home. R. V. Jones gives an alternative and to my mind, more plausible explanation of the origin of the acronym (Jones, op. cit. p. 319).

9 See Lovell's excellent biography of Lord Blackett (*Biographical Memoirs*, 1975), on which my account is largely based.

10 A powerful searchlight mounted in a retractable turret in the belly of a Wellington aircraft – the idea of Wing-Commander Leigh.

11 Lovell (1975), *Biographical Memoir* of Blackett.

12 Blackett, 1953, quoted by Lovell, op. cit. as ref. (65).

13 Churchill, Vol. IV, p. 255.

14 Rowe, op. cit. p. 160.

15 Bowen, op. cit., p. 116.

16 Doenitz, quoted by Bowen, op. cit. p. 116.

17, 18 Blackett (1961), *Science and Government*, Scientific American, April, quoted by Lovell op. cit. as ref. (89).

19 Lovell, op. cit. p. 66.

Chapter 24

1 In 1942 my colleagues and I failed to anticipate the extent to which German nightfighters would home in on radar devices like Monica and H$_2$S.

2 The dorsal turret with two machine guns was not much help.

3 A 30 mm shell is some sixty times heavier than one 7.7 mm in diameter.

4 *Biographical Memoir* of Wallis (1981).

5 The small team included, among others, W. R, Beakley, F/Lt R. Willmer, C. J. Carter, H. D. S. Faulder and Bamford.

6 EMI (Electrical and Musical Industries) and HMV (His Master's Voice or the Gramophone Company) were closely allied, with research done mainly in EMI and production in HMV.

7 Figures for casualities from flying bombs are from R. V. Jones, op. cit.; the percentages of bombs destroyed are from the same source and E. G. Bowen, op. cit.

PART III

Chapter 35

1, 2 See pp. 2 and 27 of *A Layman's Love of Letters* by G. M. Trevelyan (1954).

BIBLIOGRAPHY

Adrian, E. D. (1947). *The physical background of perception*. Oxford: Clarendon Press.

Adrian, E. D. (1928). *The basis of sensation*. London: Christophers.

Adrian, R. H., Chandler, W. K. & Hodgkin, A. L. (1970). Voltage clamp experiments in striated muscle fibres. *J. Physiol.* **208**, 607–644.

Aidley, D. J. (1989) *The physiology of excitable cells*, third edition, Cambridge: Cambridge University Press.

Air Historical Board (1952). Signals, Vol. V. *Fighter Control and Interception*. Air Ministry A.H.B. Monograph.

Atwater, I., Bezanilla, F. & Rojas, E. (1969) Sodium influxes in internally perfused squid giant axons during voltage clamp. *J. Physiol.* **201**, 657–664

Armstrong, C. M. & Bezanilla, F. M. (1974) Charge movement associated with the opening and closing of the activation gates of the Na channels. *J. Gen. Physiol.* **63**, 675–89.

Baker, P. F., Hodgkin, A. L. & Shaw, T. I. (1961). Replacement of the protoplasm of a giant nerve fibre with artificial solutions. *Nature, Lond.* **190**, 885–887.

Baker, P. F., Hodgkin, A. L. & Shaw, T. I. (1962a). Replacement of the axoplasm of giant nerve fibres with artificial solutions. *J. Physiol.* **164**, 330 –354.

Baker, P. F., Hodgkin, A. L. & Shaw, T. I. (1962b). The effects of changes in internal ionic concentrations on the electrical properties of perfused giant nerve fibres. *J. Physiol.* **164**, 355–374.

Baker, P. F., Hodgkin, A. L. & Meves, H. (1964). The effects of diluting the internal solution on the electrical properties of a perfused giant axon. *J. Physiol.* **170**, 541–560.

Baker, P. F. & Shaw, T. I. (1965). A comparison of the phosphorus metabolism of intact squid nerve with that of the isolated axoplasm and sheath. *J. Physiol.* **180**, 439–447.

Banwell, C. J. (1946). The use of a common aerial for radar transmission and reception on 200 Mc/s. *Proceedings of the Institution of Electrical Engineers.* **93**, **IIIA**, 545–551.

Barcroft, J. (1914). *The Respiratory Function of the Blood*. Cambridge: Cambridge University Press.

Barcroft, J. (1932). 'La fixité du milieu intérieur est la condition de la vie libre'. *Biol. Rev.* **7**, 24–87.

Baylor, D. A. & Hodgkin, A. L. (1973). Detection and resolution of visual stimuli by turtle photoreceptors. *J. Physiol.* **234**, 163–198.

Bear, R. S. & Schmitt, F. O. (1939). Electrolytes in the axoplasm of the giant nerve fibres of the squid. *J. cell. comp. Physiol.* **14**, 205–215.

Berg, H. C. (1975), Chemotaxis in bacteria. *Ann. Rev. Biophys. Bioeng.* **4**, 119–136.

Blackett, P. M. S. (1953). Recollections of problems studied 1940–45. *Brassey's annual* 1953 (Reprinted in *Studies of War*. Edinburgh: Oliver & Boyd, 1962).

Blackett, P. M. S. (1961). Science and

Government. *Scientific American*, April 1961.

Blinks, L. R. (1930). The direct current resistance of *Nitella. J. gen. Physiol.* **13**, 495–508. See also Blinks (1936).

Blinks, L. R. (1936). The effect of current flow on bioelectric potential. III. *Nitella. J. gen. Physiol.* **20**, 495–508.

Bosanquet, Ellen S. (1957), *A far serener clime: memories.* Privately printed.

Bowen, E. G. (1987). *Radar Days.* Bristol: Adam Hilger.

Brinley, F. J. & Mullins, L. J. (1967). Sodium extrusion by internally dialysed squid axons. *J. gen. Physiol.* **50**, 2303–2332.

Bullard R. (1961), *The camels must go.* London: Faber & Faber.

Burcham. W. E. (1945). The development of centimetre AI (*TRE Journal 1945*) (Reprinted in *Proc. Inst. Elect, Engineers,* 1985, **132** Part A, 385–393).

Caldwell, P. C. (1960), The phosphorus metabolism of squid axons and its relationship to the active transport of sodium. *J. Physiol.* **152**, 545–560.

Caldwell, P. C. & Keynes, R. D. (1957). The utilization of phosphate bond energy for sodium extrusion from giant axons. *J. Physiol.* **137**, 12P.

Caldwell, P. C., Hodgkin, A. L., Keynes, R. D. & Shaw, T. I. (1960). The effects of injecting of 'energy-rich' phosphate compounds on the active transport of ions in the giant axons of *Loligo, J. Physiol.* **152**, 561–590.

Callick, E. B. (1990). *Metres to microwaves.* London: Peter Peregrinus Ltd, for the Institution of Electrical Engineers.

Campbell. G. A. (1937). *The collected papers of George Ashley Campbell.* New York: American Telephone and Telegraph Company.

Campbell, G. A. & Foster, R. M. (1931). *Fourier Integrals for practical applications.* Bell Telephone system Technical publications monograph, B. **584**, 162.

Catterall, W. A. (1986). Voltage-dependent gating of sodium channels: correlating structure and function, *Trends Neurosci.* **9**, 7–10.

Chandler, W. K., Hodgkin, A. L. & Meves, H. (1965). The effect of changing the internal solution on sodium inactivation and related phenomena in giant axons. *J. Physiol.* **180**, 821–836.

Churchill, W. S. (1951). *The second world war.* vol. IV.

Clark, R. (1965). *Tizard.* London: Methuen.

Clay, J. R. & DeFelice, L. J. (1983), Relationship between membrane excitability and single channel open close kinetics. *Biophys. J.* **42**, 151–157.

Cole, K. S. (1941),. Rectification and inductance in the squid giant axon. *J. gen. Physiol.* **25**, 29–51.

Cole, K. S. (1949). Dynamic electrical characteristics of the squid axon membrane. *Arch. Sci. physiol.* **3**, 253–258.

Cole. K. S. (1972), *Membranes, ions and impulses.* Berkeley, Los Angeles: University of California Press.

Cole, K. S. & Curtis, R. J. (1939). Electric impedance of the squid giant axon during activity. *J. gen. Physiol.* **22**, 649–670.

Cole, K. S. & Hodgkin, A. L. (1939). Membrane and protoplasm resistance in the squid giant axon, *J. gen. Physiol.* **22**, 671–687.

Conway, E. J. (1946). Ionic permeability of skeletal muscle fibres *Nature, Lond.* **157**, 715.

Cornford, F. M. (1908). *Microcosmographica Academica.* Cambridge: Bowes & Bowes.

Creighton, Louise (1917): *The life and letters of Thomas Hodgkin.* London: Longman's Green & Co.

Curran, S, C. (1984). Philip Ivor Dee. *Biogr. Mem. Fellows R. Soc. Lond.*, **30**, 139–166.

Curtis, H. J. & Cole, K. S. (1940). Membrane action potentials from the squid giant axon. *J. cell. comp. Physiol.* **15**, 145–157.

Curtis, H. J. & Cole, K. S. (1942). Membrane resting and action potentials from the squid giant axon. *J. cell. comp. Physiol.* **19**, 135–144.

Draper, M. H. & Weidemann, S. (1951). Cardiac resting and action potentials recorded with an intracellular electrode. *J. Physiol.* **115**, 74–94.

Feng, T. P. & Liu, Y. M. (1949a). The connective tissue sheath of the nerve as effective diffusion barrier. *J. cell. comp. Physiol.* **34**, 1–16.

Feng, T. P. & Liu, Y. M. (1949b), The concentration-effect relationship in the depolarization of amphibian nerve by potassium and other agents. *J. cell. comp. Physiol.* **34**, 33–42.

Feng, T. P. & Liu, Y. M. (1950). Further observations on the nerve sheath as a diffusion barrier. *Chin. J. Physiol.* **17**, 207–218.

Frankenhaeuser, B. (1957). A method for recording resting and action potentials in the isolated myelinated nerve fibres of the frog. *J. Physiol.* **135**, 550–559.

Frankenhaeuser, B. & Hodgkin, A. L. (1957). The action of calcium on the electrical properties of squid axons. *J. Physiol.* **137**, 218–244.

Fuortes, M. G. F. & Hodgkin, A. L. (1964). Changes in time scale and sensitivity in the ommatidia of *Limulus*. *J. Physiol.* **172**, 239–263.

Geren, B. B. (1954), The formation from the Schwann cell surface of myelin in the peripheral nerves of chick embryos, *Exp. cell. Res.* **7**, 558–562.

Goldman, D. E. (1943). Potential, impedance and rectification in membranes. *J. gen. Physiol.* **27**, 37–60.

Gray, J. (1931). *A text-book of experimental cytology*. Cambridge: Cambridge University Press.

Gunston, W. T. (1976). *Night fighters*. Cambridge: Patrick Squires.

Hill. A. V. (1932). *Chemical wave transmission in nerve*. Cambridge: Cambridge University Press.

Hille, B. (1984). *Ionic channels of excitable membranes*. Sunderland, Massachusetts; Sinauer Associates.

Hodgkin, A. L. (1936). *The electrical basis of nervous conduction*. Fellowship dissertation. Library of Trinity College, Cambridge.

Hodgkin, A. L. (1937a). Evidence for electrical transmission in nerve. I. *J. Physiol.* **90**, 183–210.

Hodgkin, A. L. (1937b). Evidence for electrical transmission in nerve. II. *J. Physiol.* **90**, 211–232.

Hodgkin, A. L. (1937c). A local electric response in crustacean nerve. *J. Physiol.* **91**, 5–6P.

Hodgkin, A. L. (1938). The subthreshold potentials in a crustacean nerve fibre. *Proc. Roy. Soc. B.* **126**, 87–121.

Hodgkin, A. L. (1939). The relation between conduction velocity and the electrical resistance outside a nerve fibre. *J. Physiol.* **94**, 560–570.

Hodgkin, A. L. (1947). The effect of potassium on the surface membrane of an isolated axon. *J. Physiol.* **106**, 341–367.

Hodgkin, A. L. (1949). Ionic exchange and electrical activity in nerve and muscle. *Arch. Sci. Physiol.* **3**, 151–163.

Hodgkin, A. L. (1951). The ionic basis of electrical activity in nerve and muscle. *Biol. Rev.* **26**, 339–409.

Hodgkin, A. L. (1954). A note on conduction velocity. *J. Physiol.* **125**, 221–224.

Hodgkin, A. L. (1957), The Croonian Lecture. Ionic movements and electrical activity in giant nerve fibres, *Proc. Roy. Soc. B.*, **148**, 1–37.

Hodgkin, A. L. (1964a). *The Conduction of the Nervous Impulse*. Liverpool University Press.

Hodgkin, A. L. (1964b). *The ionic basis of nervous conduction*, Les Prix Nobel en 1963. pp. 224–241. Kungl. Boktr. Stockholm.

Hodgkin, A. L. (1971). Anniversary Address of the Royal Society (30 November 1971). *Proc. Roy. Soc. A.*, **326**, v–xx.

Hodgkin, A. L. (1974). Anniversary Address of the Royal Society (30 November 1973). *Proc. Roy. Soc. B.*, **185**, v–xx.

Hodgkin, A. L. (1975). The optimum density of sodium channels in an unmyelinated nerve. *Phil. Trans. R. Soc. B.*, **270**, 297–300.

Hodgkin, A. L. (1976). Chance and design in electrophysiology: an informal account of certain experiments on nerve

carried out between 1934 and 1952.
J. Physiol. **263**, 1–21.

Hodgkin, A. L. (1982) The physical basis of
vision. *Royal Institution Proceedings*, **54**,
7–27.

Hodgkin, A. L. (1983). Beginning: some
reminiscences of my early life. *Ann. Rev.
Physiol.* **45**, 1–16.

Hodgkin, A. L. (1988). Modulation of
ionic currents in vertebrate
photoreceptors. *The Helmerich Lecture.*
Proceedings of the Retina Research
Foundation Symposium. Vol. **1**, pp.
6–30.

Hodgkin, A. L. & Horowicz, P. (1957).
The differential action of hypertonic
solutions on the twitch and action
potential of a muscle fibre. *J. Physiol.*
136, 17–18P.

Hodgkin, A. L. & Horowicz, P. (1959).
Movements of Na and K in single muscle
fibres. *J. Physiol.* **145**, 405–432.

Hodgkin, A. L. & Huxley, A. F. (1939).
Action potentials recorded from inside a
nerve fibre, *Nature. Lond.* **144**, 710–711.

Hodgkin, A. L. & Huxley, A. F. (1945).
Resting and action potentials in single
nerve fibres. *J. Physiol.* **104**, 176–195.

Hodgkin, A. L. & Huxley, A. F. (1947).
Potassium leakage from an active nerve
fibre. *J. Physiol.* **106**, 341–367.

Hodgkin, A. L. & Huxley, A. F. (1950).
Ionic exchange and electrical activity in
nerve and muscle, *Abs. XVIII Int.
Physiol. Congress. Copenhagen, 1950*, pp.
36–38.

Hodgkin, A. L. & Huxley, A. F. (1952a).
Currents carried by sodium and
potassium ions through the membrane of
the giant axon of *Loligo. J. Physiol.* **116**,
449–472.

Hodgkin, A. L. & Huxley, A. F. (1952b).
The components of membrane
conductance in the giant axon of *Loligo.*
J. Physiol. **116**, 473–496.

Hodgkin, A. L. & Huxley, A. F. (1952c).
The dual effect of membrane potential on
sodium conductance in the giant axon of
Loligo. J. Physiol. **116**, 497–506.

Hodgkin, A. L. & Huxley, A. F. (1952d).
A quantitative description of membrane

current and its application to conduction
and excitation in nerve. *J. Physiol.* **117**,
500–544.

Hodgkin, A. L. & Huxley, A. F. (1952e). A
discussion on Excitation and Inhibition.
Propagation of electric signals along
giant nerve fibres. *Proc. Roy. Soc. B.*,
140, 177–183.

Hodgkin, A. L. & Huxley, A. F. (1952f).
Movement of sodium and potassium
ions during nervous activity. *Cold Spring
Harbor Symposium on Quantitative Biology*,
Vol. **17**, 43–52.

Hodgkin, A. L., Huxley, A. F. & Katz, B.
(1949) Ionic currents underlying activity
in the giant axon of the squid. *Arch. Sci.
physiol.* **3**, 129–150.

Hodgkin, A. L., Huxley, A. F. & Katz, B.
(1952). Measurement of current-voltage
relations in the giant axon of *Loligo.*
J. Physiol. **116**, 442–448.

Hodgkin, A. L. & Katz, B. (1949a). The
effect of sodium ions on the electrical
activity of the giant axon of the squid.
J. Physiol. **108**, 37–77.

Hodgkin, A. L. & Katz, B. (1949b). The
effect of temperature on the electrical
activity of the giant axon of the squid.
J. Physiol. **109**, 240–249.

Hodgkin, A. L. & Keynes, R, D. (1955a).
Active transport of cations in giant axons
from *Sepia* and *Loligo. J. Physiol.* **128**,
28–60.

Hodgkin, A. L. & Keynes, R. D. (1955b).
The potassium permeability of a giant
nerve fibre. *J. Physiol.* **128**, 61–88.

Hodgkin, A. L. & Keynes, R. D. (1956).
Experiments on the injection of
substances into squid giant axons by
means of a micro-syringe. *J. Physiol.*
131, 592–616.

Hodgkin, A. L. & Nakajima, S. (1972a).
The effect of diameter on the electrical
constants of frog skeletal muscle fibres.
J. Physiol. **221**, 105–120.

Hodgkin, A. L. & Nakajima, S. (1972b).
Analysis of the membrane capacity in
frog muscle. *J. Physiol.* **221**, 121–136.

Hodgkin, A. L. & Nunn, B. J. (1988).
Control of light-sensitive current in
salamander rods. *J. Physiol.* **403**, 439–471.

Hodgkin, A. L. & Rushton, W. A. H. (1946), The electrical constants of a crustacean nerve fibre. *Proc. Roy. Soc. B.*, **133**, 444–479.

Hodgkin, E. C. *et al.* (1979) *Arthur Lionel Forster Smith: chapters of biography.* Privately printed.

Hodgkin, G. L. (1934). Chapter V in *Keith Lucas*; edited by W. M. Fletcher and Alys Keith-Lucas, pp. 61–81. Cambridge. W. Heffer and Son.

Hodgkin, Lucy Violet (1921). *Life and letters of G. L. Hodgkin.* Privately printed.

Hodgkin, T. (1880). *Italy and her invaders.* Oxford University Press.

Huxley, A. F. (1957). Muscle structure and theories of contraction, *Progr. Biophys.* **7**, 255–318.

Huxley, A. F. (1959). Ion movements during nerve activity. *Ann. N. Y. Acad. Sci.* **81**, 221–246.

Huxley, A. F. & Niedergerke,. R. (1954). Interference microscopy of living muscle fibres, *Nature, Lond.*, **173**, 971–973.

Huxley, A. F. & Stämpfli R. (1949a). Evidence for saltatory conduction in peripheral myelinated nerve fibres. *J. Physiol.* **108**, 315–339.

Huxley, A. F. & Stämpfli R. (1949b). Saltatory transmission of the nervous impulse. *Arch. Sci. Physiol.* **3**, 435–448.

Huxley, A. F. & Stämpfli R. (1951a). Direct determination of membrane resting and action potential in single myelinated nerve fibres. *J. Physiol.* **112**, 476–495.

Huxley, A. F. & Stämpfli R. (1951b). Effect of potassium and sodium on resting and action potentials of single myelinated nerve fibres. *J. Physiol.* **112**, 496–508.

Huxley, A. F. & Taylor, R. E. (1958). Local activation of striated muscle fibres. *J. Physiol.* **144**, 426–441.

Huxley, H. E. & Hanson, J. (1954), Changes in the cross-striations of muscle during contraction and stretch and their structural interpretation. *Nature, Lond.*, **173**, 973–976.

Jenkins, Roy (1964). *Asquith.* London: Collins.

Jones. H. (1960). Herbert Wakefield Banks Skinner. *Biogr. Mem. Fellows R. Soc. Lond.* **6**, 259–268.

Jones,. R. V. (1978). *Most secret war.* London: Hamish Hamilton.

Katz, B. (1937). Experimental evidence for a non-conducted response of nerve to subthreshold stimulation. *Proc. Roy. Soc. B.*, **124**, 244–276.

Katz, B. (1947). The effect of electrolyte deficiency on the rate of conduction in a single nerve fibre. *J. Physiol.*. **106**, 411–417.

Katz, B. & Schmitt, O. H. (1940). Electric interaction between two adjacent nerve fibres. *J. Physiol.* **97**, 471–488.

Key, A. & Retzius, G. (1876). *Studien in der Anatomie des Nervensystems und des Bindesgewebes,* **2**, 102–112. Stockholm: Samson and Wallin.

Keynes, R. D. (1948). The leakage of radioactive potassium from stimulated nerve. *J. Physiol.* **107**, 35P.

Keynes, R. D. (1951a). The leakage of radioactive potassium from stimulated nerve. *J. Physiol.* **113**, 99–114.

Keynes, R. D. (1951b). The ionic movements during nervous activity. *J. Physiol.* **114**, 119–150.

Keynes, R. D. (1979). *The Beagle Record.* Cambridge: Cambridge University Press.

Keynes, R. D. (1989). 40 years of exploring the sodium channel: an autobiographical account. In *Mélanges de neurophysiologie à la memoire du Professeur Alexandre Marcel Monnier,* pp. 171–178. Privately printed.

Keynes, R. D. & Lewis, P. R. (1951). The sodium and potassium content of cephalopod nerve fibres. *J. Physiol.* **114**, 151–182.

Keynes, R. D. & Martins-Ferreira, H. (1953). Membrane potentials in the electroplates of the electric eel. *J. Physiol.* **119**, 315–351.

Keynes, R. D. & Rojas, E. (1974). Kinetics and steady-state properties of the charged system controlling sodium conductance in the squid giant axon. *J. Physiol.* **239**, 393–434.

Kilburn, T. & Piggott, L. S. (1978). Frederick Calland Williams. *Biogr. Mem. Fellows R. Soc. Lond.* **24**, 583–604.

Krogh, A. (1946). The active and passive exchange of inorganic ions through the surfaces of living cells and through living membranes generally. *Proc. R. Soc. B,* **133**, 140–200.

Kuhn, H. G. & Hartley, C. (1983). Derek Ainslie Jackson. *Biogr. Mem. Fellows R. Soc. Lond.* **29**, 269–296.

Leaf, A. & Renshaw, A. (1957). Ion transport and respiration of isolated frog skin. *Biochem. J.* **65**, 82–93.

Ling, G. & Gerard, R. W. (1949). The normal membrane potential of frog sartorius muscle. *J. cell. comp. Physiol.* **34**, 383–396.

Lipmann, F. (1941). Metabolic generation and utilization of phosphate bond energy. *Biochem. Enzymol.* **1**, 99–162.

Lorente de Nó, R. (1947). *A study of nerve physiology*, Vols. 1 and 2. In *Studies from the Rockefeller Institute for Medical Research*, Vols. 131 and 132. New York.

Lorente de Nó, R. (1950). The ineffectiveness of the connective tissue sheath of nerve as a diffusion barrier. *J. cell. comp. Physiol.* **35**, 195–240.

Lovell, A. C. B. (1975). Patrick Maynard Stuart Blackett, Baron Blackett, of Chelsea. *Biogr. Mem. Fellows R. Soc. Lond.* **21**, 1–115.

Lovell, A. C. B. (1990). *Astronomer by chance*. New York: Basic Books.

Marmont, G. (1949). Studies on the axon membrane. *J. cell. comp. Physiol,* **34**, 351–382.

Nastuk, W. L. & Hodgkin, A. L. (1950). The electrical activity of single muscle fibres. *J. cell. comp. Physiol.* **35**, 39–73.

Noda, M. *et al.* (1984). Primary structure of *Electrophorus electricus* sodium channel deduced from cDNA sequence. *Nature, Lond.* **312**, 121–127.

Neher, E. & Sakmann, B. (1976). Single-channel currents recorded from membrane of denervated frog muscle cells. *Nature, Lond.* **260**, 799–802.

Osterhout, W. J. V. (1931). Physiological studies of single plant cells. *Biol. Rev.* **6**, 369–411.

Osterhout, W. J. V. & Hill, S.E. (1930).

Salt bridges and negative variations. *J. gen. Physiol,* **13**, 547–552.

Overton, E. (1902). Beiträge zur allgemeinen Muskel- und Nervenphysiologie. *Pflüg. Arch. ges. Physiol.* **92**, 346–386.

Postan, M. M., Hay, D. & Scott, J.D. (1964). *History of second world war. Design and development of weapons: studies in Government and industrial organizations.* London: H.M. Stationery Office.

Pugsley, A. & Rowe, N. E. (1981). Barnes Nevile Wallis. *Biogr. Mem. Fellows R. Soc. Lond.* **27**, 603–627.

Pumphrey, R. J., Schmitt, O. H. & Young, J. Z. (1940). Correlation of local excitability with local physiological response in the giant axon of the squid (*Loligo*). *J. Physiol.* **98**, 47–72.

Ranvier, L. (1875). *Traité technique d'histologie*. Paris: Savy.

Ratcliffe, J. A. (1975). Physics in a university laboratory before and after World War II. *Proc. Roy. Soc. A,* **342**, 457–464.

Ridenour, L. N. (1947–1953). *MIT Radiation Laboratory Series, 1947–1953,* Ed. Ridenour. New York: McGraw-Hill.

Rowe, A. P. (1948). *One story of radar.* Cambridge: Cambridge University Press.

Rushton, W. A. H. (1932). A new observation on the excitation of nerve and muscle. *J. Physiol.* **75**, 16–17P.

Rushton, W. A. H. (1934). A physical analysis of the relation between threshold and interpolar length in the electric excitation of medullated nerve. *J. Physiol.* **82**, 332–352.

Rushton, W. A. H. (1937). Initiation of the propagated disturbance. *Proc. Roy. Soc. B,* **124**, 201–243.

Rushton, W. A. H. (1951). A theory of the effects of fibre size in medullated nerve. *J. Physiol.* **115**, 101–122.

Schaefer, H. (1936). Untersuchungen über den Muskelaktionsstrom. *Pflüg. Arch. ges. Physiol.* **237**, 329–355.

Skou, J. C. (1957). The influence of some

cations on an adenosine-triphosphatase from peripheral nerves. *Biochem. biophys. Acta*, **23**, 394–401.

Sharpey-Schafer, E. (1929). *The essentials of histology*. 12th ed. p. 175. London: Longmans, Green and Co.

Sigworth, F. J. & Neher, E. (1980). Single Na^+ channel currents observed in cultured rat muscle cells. *Nature, Lond.* **287**, 447–449.

Sloan, P. (ed.) (1938). *John Cornford: a memoir*. London: Jonathan Cape.

Smith, Mary F. *et al.* (1948). *Catharine Hodgkin*, ed. Mary Smith. Privately printed.

Somervell, J. (1924). *Isaac and Rachel Wilson: Quakers of Kendall, 1714–1785*. The Swarthmore Press Ltd.

Tasaki, I. & Takeuchi, T. (1941). Der am Ranvierschen Knoten entstehende Aktionsstrom und seine Bedeutung für die Erregungsleitung. *Pflüg. Arch. ges. Physiol*, **244**, 696–711.

Tasaki, I. & Takeuchi, T. (1942). Weitere Studien über den Aktionstrom der markhaltigen Nervenfaser und über die elektrosaltatorische Übertragung des Nervenimpulses. *Pflüg. Arch. ges. Physiol.* **245**, 764–782.

Thompson, D'Arcy Wentworth (1917). *On growth and form*. Cambridge: Cambridge University Press.

Trevelyan, G. M. (1943). *Trinity College: an historical sketch*. Cambridge: Trinity College.

Trevelyan, G. M. (1949). *An autobiography and other essays*. London: Longmans, Green & Co.

Trevelyan, G. M. (1954). *A layman's love of letters*. London: Longmans, Green & Co.

Trevelyan, G. M. (1951). Speech at Commemoration dinner. Annual Record (1951) Trinity College, Cambridge.

Ussing, H. H. (1949). The distinction by means of tracers between active transport and diffusion. *Acta physiol. scand.* **19**, 43–56.

INDEX

Bold-face numbers indicate illustrations.

Index

411

Index

Wilson, H.L. (grandfather), 18
Wilson, Mary F. (mother), _see_ Hodgkin,
    Mary F.
Wilson, Theodora (grandmother), 18–19
Window (metallic chaff used to screen
    bombers from radar), 197–200,
    206–9, 232
Winstanley, D., 50, 308
Wirth, Dr, 37, 39
Wirth, Frau, 34, 36, 38
Wirth, Marlene, 37–8
Wirth, Renate, 37–8
Wittgenstein, L., 72

Woodburn, James, 314–15
Woods Hole, 113–17, 119–20, 324–5
Woolf, L., 88
Worth Matravers, 153–5, 161, 168
Wray Castle, Lake Windermere, 32–3
Wright, Martin, 67

Young, F.G., 347
Young, J.Z., 66, 70, 119, 132, 254

zoology, 30, 32, 50, 55, 65–6, 70, 81
Zuckerman, Sir Solly (Lord Zuckerman),
    384

In the series *Labor and Social Change,*
edited by Paula Rayman and Carmen Sirianni

WITHDRAWN

✦ BLACK CORPORATE EXECUTIVES

BLACK CORPORATE EXECUTIVES ✦

The Making and Breaking
of a Black Middle Class

Sharon M. Collins

Temple University Press ✦ *Philadelphia*

Temple University Press, Philadelphia 19122
Copyright © 1997 by Sharon M. Collins. All rights reserved
Published 1997
Printed in the United States of America

♾ The paper used in this book meets the requirements of the American National Standard for Information Sciences — Permanence of Paper for Printed Library Materials, ANSI Z39.48-1984

Text design by Nighthawk Design

Library of Congress Cataloging-in-Publication Data

Collins, Sharon M., 1947–
    Black corporate executives : the making and breaking of a black middle class / Sharon M. Collins.
        p.   cm. — (Labor and social change)
    Includes bibliographical references and index.
    ISBN 1-56639-473-2 (cloth : alk. paper). — ISBN 1-56639-474-0 (pbk. : alk. paper)
        1. Afro-American executives — Illinois — Chicago — Case studies. 2. Afro-Americans — Employment — Case studies. 3. Discrimination in employment — United States — Case studies.   I. Title.   II. Series.
HD38.25.U6C65   1996
305.8'96073 — dc20   96-10735

To Annie, Fernie, Robyn, and Michael
and in memory of Fernando Collins, my father,
and Sheila Groot, my sister

CONTENTS

TABLES AND FIGURES

Tables

Figures

PREFACE

✦ The idea for this project arose in 1980 as I started to contemplate the relationship between economic opportunities for middle-class African Americans and federal government policies and legislation. This contemplation was triggered, in part, by emergent national attitudes that viewed race-based programs as ineffective — even harmful — for impoverished blacks and affirmative-action programs as unnecessary favoritism bestowed on a black middle class that could well compete without them. These attitudes congealed around Ronald Reagan's election to the White House and his campaign promise to cut back government bureaucracy, particularly social welfare programs that proliferated during the 1960s War on Poverty. Moreover, the Reagan White House openly challenged the policy and practice of affirmative action — a policy that in the previous decade had symbolized victory in the war on race-based employment discrimination.

William Wilson's 1978 book *The Declining Significance of Race* reflected some of these attitudes, through its title if not entirely its substance. A 1980 article by Carl Gershman and Kenneth Clark in the *New York Times* reflected these attitudes as well, articulating two sides of the debate on whether class or race effects blacks' economic opportunities. Increasingly, it seemed to me, editorials and new black spokespeople were stepping up to accuse traditional black civil rights leadership of exploiting the poor and manipulating federal policies and racial problems to gain narrow and self-serving resolutions.

But a more personal ingredient also provoked the idea for this project: my introduction to a new social stratum that included the black business elite of Chicago. This group of successful and well-known executives worked in high-paying jobs in Chicago's major corporations or were entrepreneurs who owned the largest — and possibly the only — black-owned company of its type in the nation. These people also integrated major companies during the mid-1960s and early 1970s when higher-paying jobs became accessible to blacks in the midst of civil rights upheaval.

The life-styles I observed appeared solidly upper middle class, even affluent, the executives imbued with conspicuous social privileges that flowed from their unique status. Yet, as these black businesspeople conferred among themselves about what they perceived to be the beginning of a new era of entrenchment in race relations, their tones conveyed fear. Although solidly assimilated into the social and economic mainstream of a major U.S. city, they feared the new public discourse and a federal agenda that threatened to dismantle the race-based programs that had assisted their rise.

Ultimately, then, this project was the culmination of my search to answer the question, What were these talented, educated, and economically successful people afraid of? (If William Wilson and others were right — if indeed class is more salient than race in predicting blacks' chances in the labor market — the changes in national policy should have little, if any, effect on them.) What I saw during this period, and in my own career trajectory, suggested to me a theory about blacks' job opportunity structure since the mid-1960s. I believe that race was a status that changed, albeit temporarily, from negative to positive during this period. This new positive status enabled some blacks to experience rapid upward economic mobility.

My theory evolved into an answer to my initial question. Although the people I observed were affluent, the system of jobs that supported their life-styles was a product of a particular historical period in which the federal government made efforts to improve the status of blacks. As federal policies and programs oriented toward blacks multiplied during the 1960s and 1970s, the number of professional and administrative jobs available to the black middle class multiplied dramatically. In 1980, however, the Reagan presidency, and its powerful challenge to the antibias legislation and welfare policies of the previous fifteen years, demonstrated that this period was ending. Black civil rights upheaval abated and, not coincidentally, so did federal expenditures and supports that created black economic opportunities. In short, when the government's approach to welfare expenditures and antibias policy altered, the economic base that enabled these people to rise economically and socially and to maintain their new life-styles was directly threatened.

A popular view among scholars and public policymakers is that macroeconomic and demographic trends stimulated growth in the African American middle class. I argue, however, that since the 1960s the African American middle class grew in a labor market in which government policy and intervention created job opportunities. The African American middle class is therefore a politically mediated, not a market-mediated, creature. In my view, job opportunities for middle-class blacks were enhanced by jobs, roles, and institutions created and expanded to alleviate black discontent and the social upheaval of the 1960s. Central to this thesis is the notion that the political foundations supporting blacks' class advancement puts the progress of African Americans in double jeopardy. The threat of downward mobility for African Americans in the middle class comes from changes both in the public-policy agenda and in the economy.

ACKNOWLEDGMENTS

✦ I have benefited from so much encouragement and advice that to thank each person by name I would need an acknowledgment as long as the manuscript itself. Therefore, I will say only that I appreciate each individual and have thanked every one of them in my private moments and, in many cases, in public as well.

There are people, however, to whom I owe a particularly large debt: My husband — one of the black businesspeople I refer to in the preface — and Dr. Arnold Feldman are among them. At the very outset, both helped me to articulate and formulate my ideas and to give them substance and logic. Arnold Feldman did this in his role as my professor at Northwestern University, while offering his unfailing dedication as my first full-fledged intellectual mentor. James Lowry, my husband, helped guide me with the perceptiveness of someone who well understood that his adult life was shaped in large part by the forces discussed in this book. Near the end of this project, Mildred Schwartz and John Johnstone — my colleagues at the University of Illinois, Chicago — stood by me, read my work, and cheered me on when I needed, but was unable to request, the structure and encouragement to complete it. To these people, and to all the others who helped to shape the ideas in this book, I offer my thanks.

✦ **BLACK CORPORATE EXECUTIVES**

Chapter 1

The Controversy over Race and Class

✦ Since at least the 1960s, the federal government has attempted to improve the economic and social status of blacks with legislation and policies to enhance and protect blacks' employment, voting, and housing rights. One result of government legislation, particularly laws designed to eradicate employment discrimination, was to allow significant numbers of skilled and college-educated blacks to enter the middle class through higher-paying professional and managerial jobs. Thus, blacks whose training and education positioned them to take advantage of federally mandated hiring practices began to compete in mainstream avenues to economic success and well-being (Farley 1984; Freeman 1976a). However, despite government intervention, the social and economic problems of many already disadvantaged blacks multiplied dramatically during the same period (Jaynes and Williams 1989).

Stark contrasts between the socioeconomic progress of contemporary blacks in the middle class and deteriorating conditions in the ghetto raise questions about what influences economic opportunities for black Americans. Researchers ask whether a new system of stratification in the labor market is evolving that responds to attributes associated more with class than with race. Here I take a different path to exploring the relationship between class and race by examining the role of political pressure and the careers of blacks in white corporate management.

I concentrate on relatively privileged blacks because my analysis of labor market trends since the 1960s demonstrates that characteristics associated with black advancement differ from those highlighted in existing research using aggregate data. In light of this analysis, I evaluate other research perspectives that perceive class, education, and skills, not race, as the deciding factors in the black experience. I also examine what those research perspectives imply for public policy.

The labor market analysis underpins an alternative paradigm that becomes the basis for interpreting the research presented here. In this paradigm, the middle class that sprang up among blacks benefiting from the civil rights movement is viewed as a politically mediated class, in this case, a class dependent on collective black protest, the growth of federal social programs and expenditures, and governmental antibias policy and intervention. Occupational gains made within the black middle class since the 1960s are fragile. Since the 1980s, these gains have been threatened as much by political changes in the African American community and by changes in federal policy as by macroeconomic trends.

Living the American Dream

Federal legislation to protect black citizenship rights, such as the Civil Rights Act of 1964, the Voting Rights Act of 1965, and Title VIII in the Civil Rights Act of 1968 (on housing), raised hopes that exclusionary systems that targeted blacks would erode. During the 1960s these hopes began to be realized in the labor market as employment practices changed and blacks' pattern of little or no occupational progress relative to whites finally altered. Before the 1960s, blacks' rate of entry into higher-paying occupations and white-collar jobs lagged far behind whites' and showed minimal progress. Except for periods of severe labor shortage during the two world wars, blacks' economic standing was severely limited by racial prejudice and may have even deteriorated during the Great Depression (Freeman 1976; Newman et al. 1978). Employment discrimination in the South restricted blacks primarily to field jobs in agriculture and to subservient positions, such as maids and cooks in private households. Even in the North, blacks were cast as inferior workers, excluded from lucrative industries and establishments and from jobs white workers found attractive. However, the enactment of modern civil rights legislation, in particular the Civil Rights Act of 1964, signaled a dramatic shift in blacks' ability to compete in the marketplace. Data for the 1960s and 1970s show marked increases in black-white male earning ratios (Smith and Welch 1977, 1978a; 1978b) and in the proportions of blacks in the professions (Freeman 1976a, 1976b, 1981). Moreover, occupational distributions for employed black men began to approximate those for employed white men (Farley 1977; Featherman and Hauser 1976; Hauser and Featherman 1974).

Advances in the African American Middle Class

Both E. Franklin Frazier (1957) and William Wilson (1978) define the black middle class in economic terms: blacks employed in white-collar, craftsmen, and foremen positions. By this definition, during the period of federal civil rights activity and the expansion of domestic assistance programs, blacks moved steadily into the U.S. middle class, joining the economic mainstream.

Between 1960 and 1980, blacks entered white-collar jobs at a rate much faster than did whites. The proportion of blacks in white-collar jobs increased 80 percent between 1960 and 1970, and 44 percent more between 1970 and 1979 (Figures 1 and 2). Overall, blacks more than doubled their proportion in white-collar jobs. By 1980, the proportion of blacks in white-collar jobs had increased a total of 124 percent, compared to only 25 percent for whites. At the other end of the occupational pyramid, the proportion of blacks in blue-collar jobs remained relatively constant, while the proportions in service and farm jobs declined sharply.

Blacks' upward mobility looked vastly different for men and women. African American women shifted from domestic and personal-service jobs into clerical and sales positions, but black men

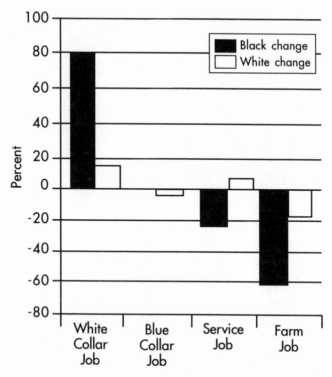

Figure 1. Changes in Employment by Race: 1960–1970

moved into professional jobs, including management and business-oriented professions such as accounting and law, for the first time in U.S. history. Thus, during the decades in which black initiatives set precedent for public policy, the occupational gap between skilled and highly educated black and white men began to shrink, although disparities remained (Featherman and Hauser 1976; Freeman 1976a; Smith and Welch 1977).

In 1960, only 5 percent of employed black men worked in better-paying white-collar occupations; by 1979, 11 percent did. Between 1960 and 1979, the proportion of black men in professional, technical, and managerial fields more than doubled. In contrast, the proportion of white men in the same fields increased only 18 percent (U.S. Bureau of the Census 1979, 1980). Although the economic

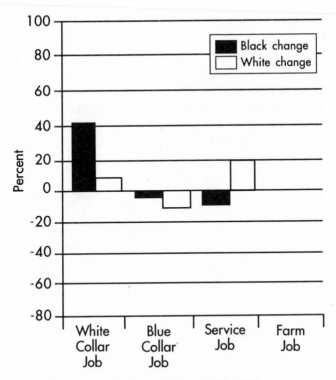

Figure 2. Changes in Employment by Race: 1970–1980

position of blacks still lags far behind that of whites, blacks — especially skilled or highly educated workers — have experienced substantial economic gains since the 1960s.

Deterioration in the Underclass

But as skilled and highly educated blacks reaped gains from civil rights protest and subsequent federal policies, the social and economic problems of the black underclass remained intransigent.[1] Life for blacks in inner-city ghettos became worse, not better, following the 1960s civil rights efforts. One in three blacks live below the poverty line, and blacks are about three times as likely as whites to be poor. Moreover, from the 1970s through the mid-1980s, the absolute and relative probability that blacks will be impoverished remained almost unchanged despite decades of governmental intervention to improve their socioeconomic status. Blacks with relatively less work experience and black female-headed households have fallen behind the general population during more than a decade of affirmative action and federal domestic assistance programs (Becker and Hills 1979; Freeman 1973; Mare and Winship 1980; Ross and Sawhill 1975).

The proportion of blacks who live below the poverty line, who are unemployed, or who have dropped out of the labor force remains discouragingly high (Farley and Bianchi 1983; Glasgow 1980; Ismail 1985; Parsons 1980). In short, economic conditions for unskilled blacks and blacks with low levels of education deteriorated at the same time economic opportunities for middle-class blacks flourished.

Perspectives on Black Economic Attainment

The bifurcated trends in blacks' access to labor markets have generated a heated debate on blacks' competitiveness in the U.S. economy (see Gershman and Clark 1980). Scholars such as Feagin and Sikes (1994), Zweigenhaft and Domhoff (1991), and Landry (1987) maintain that discrimination remains an obstacle to blacks' full economic participation, that institutional practices and individual prejudice erect barriers that still restrict blacks' economic chances, regardless of education, skills, and other class-related advantages. However, others argue convincingly that blacks' economic chances more likely are mediated by "nonracial" factors associated with class — family background and schooling — than by racial discrimination. For instance, Smith and Welch (1983, 1986) found that improvements in blacks' occupational status coincided with improvements in the quantity and quality of black education. Such findings tend to minimize the weight of race-based obstacles that restrict blacks' access to labor-market opportunities. Their underlying assumption is that race discrimination is more an individual anomaly or aberration than an entrenched characteristic of U.S. society. Other research connects disparity in occupational mobility to age and regional differences between black and white populations (Featherman and Hauser 1976; Hout 1984). Since education, region, and age are examples of "nonracial" explanations of blacks' differential attainment, such findings corroborate the claim that job opportunities for blacks are not anchored to racial discrimination, per se.

The Role of Economic Structure and Culture

The idea that racial barriers are eroding presumes that as blacks in the middle class attain quality educations and ascend occupational ladders, they assimilate into the economic mainstream. At the same time, barriers to black employment other than race — such as culture and macroeconomic changes — are seen to confront the underclass (Becker 1981, chap. 11; Murray 1984; Wilson 1978, 1987). For example, individuals who drop out, or are pushed out, of labor markets due to social pathology (Wilson 1987) and welfare incentives (Becker 1981; Murray 1984) are evidence of cultural deficits within inner-city black communities. More female-headed households and dependence on government welfare programs, along with reduced labor-force participation among adult black men, also illustrate dysfunctional ghetto features. These observations are not new. But unlike earlier theorists that portray dysfunctional ghetto features as a response to, not separate from, the conditions of the greater society, recent critics of inner-city culture argue that economic deterioration among blacks results from the self-imposed constraints of social disorganization and undeveloped human resources. This portrait of social disorganization among the black poor contrasts sharply with those of theorists — Kenneth Clark (1965), in particular — who linked ghetto pathology to racial prejudice and the pervasive structure of racial inequality. The current argument is that the dysfunctional components of ghetto life are no longer a symptom of contemporary racial discrimination and subordination (see Wilson 1987).

Even structural explanations for the massive deterioration of inner-city life dismiss racial discrimination as a background factor. A structural perspective attributes the plight of blacks in poor urban areas primarily to massive changes in the postindustrial economy (e.g. Harrington 1984; Wilson 1987). Wilson (1987) follows the lead of Kasarda (1980; 1986), for example, and views economic deterioration among blacks as an outgrowth of deindustrialization. Both scholars argue that rapid changes in urban industrial bases displace minority workers by creating a mismatch in which minority residents' educational backgrounds are no longer compatible with the incoming industries' needs. During the heyday of heavy industry, employment levels among blacks were high, especially in urban centers where blue-collar blacks found low-skilled but relatively high-wage jobs in industries like auto and steel manufacturing. But smokestack industries declined, and urban employment shifted into fast-growing service industries in which wages are polarized. Historically, blacks have been underrepresented in high-wage jobs in service industries. Conversely, low-skilled and blue-collar black workers have been concentrated disproportionately in poor-paying service jobs or quickly evaporating jobs in urban industries (Wilson 1987). In sum, the shift from goods- to service-producing industries and the accompanying decline in wages — rather than race per se — have dislocated black workers.

In this picture, scholars constructed a paradigm in which race is understood to play a diminished role in black employment relative to cultural and structural factors, which are seen increasingly to restrict opportunities available to less-advantaged blacks. Simultaneously, cultural and structural factors — in the form of valued education and skills and the expansion of skilled and high-paying jobs in service industries — benefit better-educated blacks in the middle class. Put another way, the same macroeconomic trends that cast working-class blacks into the underclass (e.g., shifts in technology and the increased numbers of white-collar jobs generated by service industries) also created new and better opportunities for other blacks to enter the middle class. In this scenario, dual forces converged to erode longstanding racial barriers to blacks with skills: the increased need for skilled workers brought on by an expanding service economy, on one hand, and educational improvements in the supply of black labor, on the other.

Structural and cultural perspectives, then, see the structure of occupations becoming increasingly color-blind. Not racial discrimination but black social dysfunction fed by welfare dependence, family background, and limited job skills explains race-related job inequality and low status. Conversely, the increased acquisition of marketable work skills and quality education for blacks already higher on the occupational ladder explains that segment's ascent to the middle class and emerging ability to successfully compete with whites.

Public-Policy Implications

The controversy over which factors influence economic opportunities for black Americans is not a theoretical exercise limited to academic arenas. The question also has public policy implications because the opposing ideologies justify divergent strategies for solving blacks' economic problems. In the 1960s, a politically liberal view that justified governmental protection and race-based policy as critical to black advancement prevailed among black intellectuals and leaders. However, since the 1980s, more conservative, essentially nonracial explanations of blacks' status in the economy have justified dismantling race-specific programs. For example, Wilson (1987), Glenn Loury (1985), Thomas Sowell (1983), and Carl Gershman (Gershman and Clark 1980) have argued that many contemporary problems of blacks fail to respond to race-based governmental interventions. Moreover, they argue, liberalism, black activism, and misguided government actions have fostered, rather than ameliorated, such problems. Wilson (1987), in particular, holds that two decades of race-based employment policy created opportunities for advantaged blacks but neglected to address training-based barriers to employment faced by disadvantaged blacks. In other words, the race-specific agenda tied to programs such as affirmative action created access to previously closed jobs, but only for those blacks who had the skills and education to compete for them.

Gershman (Gershman and Clark 1980), Loury (1985), and Wilson (1981) issue even more severe indictments of the political agenda they perceive as underpinning civil rights–related social policy. These authors paint black intellectuals and political leaders as self-serving advocates of race-specific government programs that are ill-suited to the needs of the poor. In their view, the liberalism of black leadership has led to remedial employment policies and programs that were exploited to gain jobs for advantaged blacks. Gershman and Loury also contend that governmental welfare programs, such as public assistance stipends, subsidized goods and services that fostered dependency among underclass blacks (also see Murray 1984; Sowell 1983; Williams 1982). In sum, race-specific programs and policies that have evolved since the 1960s have been under attack since the 1980s. In the 1990s, their role is challenged and their future remains in question. However, the growing consensus is that color-blind labor markets, not race-conscious policies, will protect and sustain black gains in jobs formerly closed to them.

A Different Paradigm of Race and Class

I differ with those who argue that blacks' gains in higher-paying jobs provide evidence that economic opportunities for blacks are linked to culture, education, and skills, not racial characteristics. I begin instead with the assumption that race remains important, even for the black middle class. However, I do not recapitulate earlier arguments that directly linked racial discrimination with blocked economic attainment. I believe that economic advances by the black middle class since the 1960s civil rights movement are considerable. At the same time, I view the black middle class of professional and managerial workers as a politically mediated class that fills a socially useful function. This class plays a unique role in the labor market and occupies a fragile position.

I also differ from researchers who take a narrow view of the effects of federal legislation on employment and depict equal opportunity employment legislation as the government's only method for spurring upward mobility among blacks. Such a narrow focus on regulatory policies obscures the indirect employment effects of government advocacy.[2] These indirect employment effects are job opportunities generated by an expanding network of government and private sector bureaucracy to increase the distribution of goods and services to volatile black constituencies. I view the black middle class as resulting both from public policy supports — such as federal legislation prohibiting employment discrimination, and the growth in federal social service bureaucracy — and from jobs created or reoriented to respond to blacks' civil rights–related demands and upheaval.

My conception of the black middle class is uncoupled from occupational categories. Rather, I define it by its relationship to a system of production. I hold that one by-product of race-conscious policies and programs was a new employment structure from which educated middle-class blacks have benefited. I suggest that better jobs for blacks may not be merely a factor of education or even the effect of affirmative action. Better jobs may depend also on substantive changes in the organization of jobs, a result of the need felt by the federal government and private employers to abate black upheaval and restore social order, which had been disrupted by civil rights activities.

Specifically, demands for black-oriented programs and employment policies altered the organization of jobs and institutions to distribute more services and financial resources to blacks. During the 1960s and 1970s, administrative functions and race-specific programs were created, expanded, or reinterpreted to respond to blacks' needs, which also increased the number of professional and administrative jobs available to middle-class blacks. More significantly, opportunities for good jobs in white-dominated settings emerged in a black-oriented delivery system. Blacks, therefore, remained "functionally segregated" in the labor market; black professionals in white institutions filled roles tied to the appeasement of blacks. They were not hired or promoted to meet the demands of total (predominantly white) constituencies. Executives I studied played a socially useful role in white companies during the 1960s and 1970s, but by the 1980s, they had become economically expendable.

In Chapter 3, I present evidence of functional segregation of blacks, showing first that during the 1970s, higher-paying jobs for blacks in the public sector were differentially concentrated in urban bureaucracies that serve disproportionately large concentrations of blacks. When federal funding expanded urban services targeting blacks, middle-class blacks were tracked into jobs that were developed to allocate additional resources to impoverished blacks and to restore social order. I also present evidence of functional segregation in private sector employment. For example, during the 1970s the first black-owned advertising agency was founded in Chicago. This agency was hired by white companies to reach out to black (but not white) consumer markets. Further, reorganization of work into racialized roles added a ghettoized component to blacks' gains in management during the 1960s and 1970s. Blacks were disproportionately moved into personnel departments and labor- and public-relations jobs (U.S. Bureau of the Census 1963, 1973) to administer corporate policies sensitive to blacks and, hence, lessen racial pressures on white corporate environments.

In this conception of how the black middle class grew during the 1960s and 1970s, even if the black middle class — more than the underclass — benefited economically from political demands, the benefits are mediated by blacks' political dependency. And, since the federal government and private employers do not respond to the black population unless it is disruptive, when pressures abate organizations will cut back race-conscious services and thereby dislocate people who became middle class by performing them. In this conception, the black middle class is the interim beneficiary of social policy to ameliorate black upheaval. The long-term beneficiaries are the larger systems of white elites.

these executives stayed in black-related jobs. Chapters 7 and 8 include the executives' perceptions of their own status changes within their companies. The story unfolds amidst a history that began with the expanding economy and race-conscious pressures of the 1960s and evolved into the increased economic and interest-group competition of the 1990s.

Some readers may be tempted to view this book as an indictment of race-based policies; that would be a misinterpretation. The people in this study are the most privileged, the ultimate beneficiaries of changes wrought by the civil rights movement. Having "made it" by almost any socioeconomic standard, they represent evidence that race-based policies have facilitated upward mobility for some segments of the black population. However, my intent here is to illustrate the resilience of segregating systems, even under social conditions designed to improve race relations. Public policy did indeed help some blacks grasp the American dream. Yet, now, twenty-five years later, many have hit the race-specific glass ceiling endemic to racialized roles. The same conditions that helped blacks succeed in corporate hierarchies also channeled them out of mainstream job functions and locked them into limited and fragile career paths.

The Investigation

To test my ideas about the labor market for blacks, I conducted two sets of in-depth interviews with the seventy-six highest-ranking black executives in Fortune 500 companies—the people whom researchers, public policy makers, and the general public refer to when they talk about black breakthroughs. I conducted the first interviews in 1986, the second in 1992. Conformity to corporate cultures, personal networks, and skill are key factors in individual mobility among high-ranking executives. Therefore, if any blacks can be said to have transcended racial barriers, it would be these executives. They are the ultimate black middle-class beneficiaries of the civil rights movement.

I considered blacks to be "top executives" if: (1) they were employed in a banking institution and held the title of comptroller, trust officer, vice-president (excluding "assistant" vice-president), president, or chief officer; or (2) they were employed in a nonfinancial institution and held the title of department manager, director, vice-president, or chief officer. I located these executives by using *Chicago Reporter* (1983; 1986) lists that cite the fifty-two largest white-owned industrials, utilities, retail companies, transportation companies, and banks in Chicago. I then asked knowledgeable persons familiar with the white corporate community in Chicago to identify black officers in these firms and other employees who might be able to provide names of higher-level black officers. I then asked these employees to identify blacks at the selected levels of management. Finally, I asked black executives who participated in the study to identify other top blacks in the selected Chicago firms. Eighty-seven executives were identified; seventy-six were interviewed.[3] Executive headhunters and business executives confirmed that I had defined and located nearly the entire population of senior level black executives employed in Chicago.

Although it is difficult to specify the exact number of blacks who met the study criteria, publications that feature the nation's top-ranking black executives, such as *Dollars and Sense* and *Black Enterprise*, routinely mention the people I interviewed.

In the mid-1980s, the interviewees held some of the more desirable and prestigious positions in Chicago corporations. Contrary to a popular notion that black women are greater beneficiaries of affirmative action than black men—in part because they fall into two federally regulated categories—only thirteen of the seventy-six executives were women. The joint effects of race and sex discrimination made black women largely absent in the higher-paying ranks of management and, indeed, the last to benefit from federal affirmative-action legislation (Leonard 1988).

Almost two-thirds of the men and half the women held the title of director or above, including three chief officers, thirty vice-presidents, and nineteen unit directors. (The total includes three people with the title "manager" whose rank within the organization was equivalent to director.) In addition to being the top people in Chicago in 1986, some of the interviewees were among the highest-ranking black executives in the country. Five of the men were the highest-ranking blacks in corporations nationwide. Almost half (thirty-two of seventy-six) were the highest-ranking blacks in their companies' nationwide management structure.

In the 1986 interviews, I explored whether or not these executives filled racialized roles that had emerged during the racially tumultuous civil rights period. Using resumés they had sent me before the formal interview, I conducted semistructured interviews in which I asked the executives to describe in detail each job they had held over the course of their careers. When a manager's description of a job or the nature of a job was unclear, or when it was not feasible to examine careers on a job-by-job basis, I asked, "During the late 1960s and early 1970s, social programs such as EEO [equal employment opportunity], manpower training, and community affairs were hot items in some corporations. Did you ever have a job in any of these areas?" Using their answers I differentiated the "racialized" and "mainstream" tasks for each job held by a manager. I considered jobs racialized if their description suggested a substantive or a symbolic connection to black communities, black issues, or civil rights agencies at any level of government. For example, one manager was hired by a major retailer in 1968 specifically to discover and elimi-

nate discriminatory employment practices. I coded this function "racialized" since it was designed to improve blacks' opportunities within the company at a time when the federal government increasingly was requiring such revisions. I labeled jobs "mainstream" if their descriptions revealed neither explicit nor implicit connections to blacks.

Fifty-one of these executives (67 percent) had held one or more jobs in a company in which they implemented corporate programs for, funneled corporate goods and services to, or advised the white corporate elite about black constituents. In both 1986 and 1992, I explored the possibility that racialized management positions lack job security because the need for racialized functions would vary in tandem with political conditions. I began my interviews with the hypothesis that since racialized jobs emerged in response to political pressures, they are likely to be treated as increasingly expendable as that pressure subsides. Between 1992 and 1994 I located and reinterviewed sixty-one of the original seventy-six executives to capture how they had fared under rapidly changing economic and politica conditions.[4] (I also talked with people in the business world famil both with Chicago's corporate executives and with various inte events not likely to be known outside the business community kept me up to date on Chicago's business news and helped r pret events that had implications for the study.) These f interviews provided a second and stronger test of my arg

In Chapters 2 and 3, I show that black mobility with class during the 1960s and 1970s was politically black social protest and federal government interv job opportunities for middle-class blacks during how this politically mediated system of econom a race-based job inequality that contributes fragility, particularly from the 1980s onwar

The remainder of the book draws on th 1986 and 1992. In Chapter 4 I look at th tions surrounding the entry of this su agers into major white corporations blacks' roles inside white corpor

Chapter 2

A Politically Mediated Opportunity Structure

✦ Affirmative action has come under siege recently not only as a politically unpopular program but as an ineffective policy for reducing inequality for targeted groups.[1] However, attempts to minimize and devalue the federal government's role in the upward mobility of blacks have not been decisive. Studies find that affirmative action and contract compliance significantly improved employment opportunities for blacks (Heckman and Payner 1989; Herring and Collins 1995; Leonard 1982). In addition, researchers who argue that federal intervention played a trivial role in blacks' mobility fail to consider the effects of employment opportunities that sprang directly from the proliferation of federally funded welfare and community-service organizations that coincided with blacks' protest for civil rights. (Civil rights activity in the 1960s only stimulated federal government efforts already underway on behalf of blacks; the first equal employment laws were established in the 1940s.)

Before the 1960s, women and people of color were the victims of employment discrimination (Burstein 1985; Farley 1984; Farley and Allen 1987; Jaynes and Williams 1989; Wilson 1978). In the effort to curtail such discrimination, since at least the 1940s executive orders have been issued and federal legislation enacted to prohibit discrimination against minorities and women workers.[2]

In the early 1940s, President Roosevelt issued executive orders declaring an end to discrimination in the federal civil service and creating the Fair Employment Practices Committee. The federal effort to prohibit discrimination continued during the 1950s when President Truman issued two executive orders to establish fair employment procedures within the federal government structure, to abolish discrimination in the armed forces, and to establish compliance procedures for government contractors. Given the entrenched nature of discriminatory employment patterns on the one hand and the escalation of black protest on the other, it is not surprising that the Civil Rights Act of 1964 was passed to further propel this legislative process.

While a variety of employment situations exist within the black middle class, patterns of blacks' upward mobility in the 1960s and 1970s were clearly not just the product of blacks' accumulation of skill and training. Members of the black middle class benefited from participating in labor markets shaped during the 1960s and 1970s in response to blacks' civil rights demands and the ensuing federal legislation. "Although blacks in the U.S. were better off absolutely in the 1960s than ever before in American history, their sense of deprivation relative to whites proved explosive" (Blumberg 1980). The demands of African Americans for greater investment in community and economic development, articulated by black leadership, set the ground rules for a social policy during the administrations of Kennedy, Johnson, and Nixon that generated growth in federal legislation and spending, inflating the labor market for blacks in both the public and private sectors. This inflation was the result of new bureaucracies and the enforcing of affirmative-action guidelines tied to federal expenditures.

Production's demand for labor can be conceptualized as an economic or policy response to consumers. Economic mandates are a response to consumer preferences mediated in the market. Policy mandates are a response to political pressure mediated through government. In the case of affirmative action, government enhanced the status of blacks by establishing employers' need for them, which in turn expanded labor-market demand. Employers sought college-educated blacks to fill higher-paying white-collar jobs, which translated into opportunities for new class position. Therefore, the black middle class that emerged during the 1960s and 1970s is a result of a policy-mediated, not a market-mediated, situation. This paradigm is meant not to rigidly characterize how the black middle class grew but to call attention to characteristics related to growth in the black middle class that have been overlooked.

The political impetus for black middle-class growth is embodied in four vehicles that implemented federal legislation and policies that enhanced the status of blacks: the Equal Employment Opportunity Commission, the Office of Federal Contract Compliance Programs, federal contract set-aside programs, and federally funded social welfare expansion.

Equal Employment Opportunity Commission

Title VII of the 1964 Civil Rights Act provided the muscle for federal attempts to eradicate employment bias in the United States, establishing the Equal Employment Opportunity Commission (EEOC) as the law's administrative agency. The EEOC required private firms with one hundred or more employees to report numbers of minority workers to the commission. Federal contractors with fifty or more employees and federal contractors or subcontractors selling goods or services worth at least $50,000 also are covered by the law. At the start, the EEOC was authorized to use "informal methods" to resolve complaints of job discrimination. But by 1972 the enforcement powers granted to the EEOC by Congress gave it the right to initiate civil suits in district courts. By 1973, the EEOC gained further leverage in the private sector by winning a consent decree from American Telephone and Telegraph (AT&T) agreeing to pay $38 million to workers discriminated against by the company (Hill 1977). At the same time, the EEOC created a system that tracked discrimination and "patterns and practices" charges against limited numbers of large employers (Purcell 1977). This demonstration of the EEOC's serious intent was a major impetus for corporate employers to seek black labor. Richard Freeman (1976:136) reported that "federally required programs involving job quotas that favor minorities have made minority hiring an explicit goal of major corporations. At IBM for example, every manager is told that his annual performance evaluation includes a report on his success in meeting Affirmative Action goals." In Heidrick and Struggles's (1979b) survey of Fortune 500 companies 52 percent who responded reported in 1979 that achievement of affirmative-action goals was a factor in reviewing management compensation. General Electric's policy sanctions and Sears's Mandatory Achievement of Goals program represent similar force-fed structural changes to increase black employment.

Comparing firms reporting to the EEOC with the total data reported by the National Commission for Manpower Policy for the years 1966 through 1974, Andrew Brimmer (1976) concluded that companies reporting under EEOC requirements opened jobs to blacks much faster than did nonreporting firms. Thus, in the eyes of white employers, the status of black workers was apparently enhanced. Other studies found that Title VII had a strong effect on the employment status of blacks in the private sector (Freeman 1973; Haworth, Gwartney, and Haworth 1975). Heckman (1976) found the effect of fair employment practices on employment status strongly significant. Purcell (1977) noted that at AT&T, the figures for blacks in second-level management and above increased 126 percent between 1972 and 1975 as a result of the company's different approach to employment.

Federal Contract Compliance

Additional federal government influence on white private sector employment policy is exercised through federal contracts, which also enhanced economic opportunities for blacks. Business enterprises with federal contracts or subcontracts of $50,000 or more were compelled to comply with federal hiring guidelines or face the withdrawal of federal money. Although loss of contract through noncompliance was a sanction of last resort and not well documented, 45 percent of all Fortune 500 companies reported that they had been threatened with ineligibility on compliance issues (Heidrick and Struggles 1979). I believe, therefore, that the threat alone was enough to change private-sector employment practices.

Comparative research during the 1970s on the impact of contract compliance in the white private sector found that government contractors hired significantly more black males than did nongovernment contractors. Ashenfelter and Heckman (1976) found that in the short run government contractors raised the employment of blacks relative to white males 3.3 percent more than did nongovernmental contractors. In the long run this effect was estimated to be 12.9 percent. Research also noted that more government contracts were awarded to less discriminatory firms (Heckman and Wolpin 1976) and that segregated white contractors were more likely to integrate than segregated white noncontractors (Ashenfelter and Heckman 1976).

More recent studies also have found that federal contract compliance programs significantly improved employment opportunities for blacks. Heckman and Payner (1989) found similar benefits in their study of the South Carolina textile industry when they isolated the specific effects of antidiscrimination laws from other programs and economic events that occurred at the same time. Leonard (1982) used establishment-level EEO-1 reports (filed by EEOC-covered employers) on more than sixteen million employees for 1974 and 1980 and establishment-level affirmative action compliance review reports for the period 1973 to 1981 to compare a matched sample of contractors with noncontractors. He found blacks' share of employment with government contractors grew significantly more than at noncontractor establishments, and that federal contract con.pliance programs substantially improved the employment opportunities for black men in particular. He argues that affirmative action supported the rise of the black middle class by increasing the demand for black labor in higher-paying, white-collar jobs relative to blue-collar, operative, and laborer occupations. Herring and Collins (1995) analyzed the 1990 General Social Survey and a 1992 survey of Chicago adults and also found affirmative action associated with higher occupational prestige for African Americans and with substantial incremental effects on the incomes of racial minorities.

Federal Procurement

The growth of the black middle class since the 1960s is also related to the government's own hiring efforts. For example, federal depart-

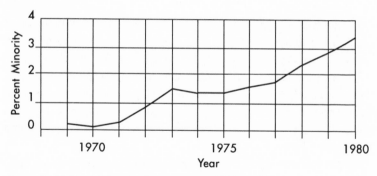

Figure 3. Federal Minority Procurement: 1969–1980

ments such as the Department of Commerce established specialized minority procurement procedures known as the contract set-aside program (U.S. Department of Commerce 1979). Before guidelines covered such transactions, government contracts for vast amounts of consumer goods and services were awarded almost exclusively to white-owned businesses. The set-aside program established the percent of government contracts that should go to minority-owned businesses, in an effort to assist businesses historically hindered by their inability to secure capital from white banks and to compete outside black communities. These programs helped to expand both the size and type of black firms by giving black entrepreneurs a chance to compete for sizable contracts in a protected setting.

In fiscal year 1969, minority business participants in federal procurement accounted for only .03 percent of total dollar awards, a figure that dramatically increased from 1969 to 1973, after President Nixon established the Office of Minority Business Enterprise. The gains are presented in Figure 3, which also shows how fluctuations in minority business procurement coincide with changes in White House administrations. The percentage of minority business participation in direct and indirect contracting decreased in 1974 and remained constant at approximately 1.4 percent of total procurement until 1976, a period when no policies addressed federal-wide goals for minority procurement. Any goals were local ones, if established at all.

In 1977, President Carter created the Interagency Council for Minority Business Enterprise as the principal instrument to coordinate federal purchases from minority firms; its palpable effect was to increase procurement from 1977 through 1980. Data from the U.S. Department of Commerce (1981b) indicate that as a result of contract set-asides, the amount of goods and services sold by minority firms to federal agencies grew from $3 million in 1969 to $3 billion in 1980.[3]

The probable effects of federal initiatives (and contract set-aside programs started on other levels of government) on the black business sector also show up in comparisons of gross receipts of majority- and minority-owned firms between 1969 and 1977. For example, between 1969 and 1972 minority business revenue increased by 57 percent, majority firms by 24.5 percent. During the same period, the proportion of black firms increased in almost every industry, as did the black percent of all firms' gross receipts (U.S. Department of Commerce 1981a). Here again are changes that suggest a relationship between the race-based economic policy during this period and the enhanced overall competitiveness of minority businesses.

Social Service Expansion

In addition to contract set-aside programs, new arenas for black employment also emerged as public welfare services and federally funded, black-run community organizations grew. According to William Wilson (1978) more than two thousand community action programs were formed during the War on Poverty. Some of these sprang from grass-roots concerns, others were government created.

Between the early 1960s and the middle 1970s, the public sector grew and social service bureaucracies, especially, proliferated. Estimates have been made that at least two-thirds of the recent growth in public employment can be attributed to social welfare programs. Brown and Erie (1981) connected the expanding educational bureaucracy to the programs of the Great Society era (e.g., the Elementary and Secondary Education Act, the inauguration of Head Start, and other preschool programs authorized by the Economic Opportunity Act). In addition, federal government increased funding to social programs in medical, public housing, and manpower-training areas, which also contributed to the growth of the public sector.[4]

This growth in federal bureaucracy occurred at least in part because blacks demanded that the poor had the right to receive welfare subsidies and that the government must provide them. The political origin of the War on Poverty has been confirmed by several authors (Donovan 1967; Levitan 1969; Piven and Cloward 1971; Yarmolinsky 1967). And, although federal welfare programs were not created exclusively for blacks, the public views them as attempts to meet black needs for more economic and social resources. One might, for example, view the War on Poverty as a response to the existence of various urban, not black, problems. But Piven and Cloward (1971:258–259) show that many urban problems occurred equally in rural areas, yet the Office of Economic Opportunity concentrated funds in large urban centers, where blacks are concentrated. Federal remedial programs affecting elementary and secondary education, such as Operation Head Start, were directed primarily at black children. In addition, the criteria for participating in dependency-related services such as the food stamp program and in welfare programs such as Aid to Families with Dependent Children were oriented toward blacks. In general, the expanding network of urban social services favored blacks because of their lower incomes. For example, inner-city blacks disproportionately tended to benefit from increased health, hospital, welfare, and public housing services.

Government was a significant employer of blacks even before the 1960s. But public settings seemed to strengthen that role when government at every level expanded the social service bureaucracy in that decade. In 1960, 13.3 percent of employed blacks worked for government; in 1970, 21.4 percent; and in 1979, 24 percent. Between 1960 and 1979, the proportion of employed blacks in government positions increased at twice the rate of whites. The relative chances of blacks and whites for being employed in government rose in these two decades from about equal probability to more than one-third more likely that blacks would be so employed. Robert Althauser's (1975) study of black and white graduates in the 1970s found that, indeed, black graduates tended to gravitate toward public sector jobs in larger proportions than white graduates.

As a result of these race-oriented programs, the black middle class made economic and social gains. As noted previously, the proportion of employed black men working in professional, technical, and managerial positions doubled between 1960 and 1979. Between 1960 and 1970 the probability of black as compared to white managers to be in government rose 67 percent, although the overall odds of employed blacks to be in government decreased slightly (U.S. Bureau of the Census 1963; 1973). Between 1960 and 1970, for example, the proportion of black managers and officials employed by government increased about 30 percent; in contrast, white proportions decreased by 15 percent. In 1960, 13 percent of white managers and 21 percent of black managers and administrators worked in government. By 1970, 27 percent of black managers but just 11 percent of white managers worked in government. Whether this increase is due to upgrading employed blacks or to an influx of black managers cannot be determined. Nevertheless, it seems evident that government directly contributed to blacks' enhanced economic opportunities, particularly in the higher-paying white-collar occupations. Between 1960 and 1982 the proportion of black men who were public employees rose to about twice the proportion of white men in managerial and administrative occupations (U.S. Bureau of the Census 1963, 1973; U.S. Bureau of Labor Statistics 1982).

The relationship between growth of the federal bureaucracy and the elevation of blacks' economic status is not a new one. During the era of the New Deal, for example, the proportion of blacks employed in the government more than doubled. A slightly higher proportion of blacks was represented in the government work force than the proportion of blacks in the population. According to Sitkoff (1978:76), "The New Deal . . . opened up professional opportunities on a grand scale for blacks in the federal government. Negroes worked in the New Deal as architects, economists, engineers, lawyers, librarians, office managers, and statisticians." It is not surprising, then, that increased government employment accounts for many of the significant changes in black employment patterns during the 1960s. Black accountants, engineers, lawyers, personnel workers, social scientists, and managers are all twice as likely as whites to be in government. Further, in 1980 close to half (47 percent) of employed black professionals and technicians worked in the government sector as compared to only 34 percent of whites employed in the same categories (U.S. Bureau of Labor Statistics 1982).

The demand for skilled blacks within the public sector is also reflected in statistics showing that in 1969, minorities held 7.8 percent of federal jobs at grades GS9 through GS11, while in 1975 this figure had almost doubled to 12 percent (Hampton 1977). The pay range for these positions was $20,000 to $32,000 in 1975 (Hampton 1977). Freeman's (1976a) estimate of black and white incomes for 1966 and 1969 shows that black, but not white, public employees earned more than their privately employed peers. Eccles (1975) reported that the ratios of black to white income in government were above the economywide average for college graduates.

Although the earning ratio between blacks in the public and private sectors decreased, in 1979 the median salaries of blacks in government continued to be higher than the median for blacks in private nonagricultural jobs by 25 percent (U.S. Bureau of the Census 1980a). According to Norman Seary, past executive director of the Voluntary Organization of Blacks in Government, 55 percent of all blacks in the federal work force in 1980 fit categories of professional, managerial, and technical workers earning $30,000 to $37,000 a year (Poole 1981).

The movement of middle-class blacks into traditionally closed market situations can be viewed optimistically as evidence that economic opportunities (i.e., jobs and income) associated with class position are mediated more by education and skills, and less restricted by barriers that are related to race. However, with few exceptions (Brown and Erie 1981; Collins 1983), existing studies fail to credit the activities of government in their analysis of factors that increased job opportunities for blacks. At this point, however, the relationship between federal government initiatives and blacks' economic mobility is clear.

The growth of the black middle class during the 1960s and 1970s, occurred in a peculiar way. Black advancement was greatest in the government and in the segments of private industry that responded to government equal employment initiatives. A picture emerges of black middle-class growth in which federal government policy has an important effect on the labor market for blacks.

Farley (1977) noted that gains made among blacks in the 1960s did not diminish in the economically stagnant 1970s. The significance of the "policy factor" is that it explains this phenomenon without removing the importance of race. Market demands in this period were not just for labor, but for black labor. Although the economy fluctuated, federal social policy oriented toward the improvement of blacks' social status fluctuated less.

Growth of this politically mediated labor market, however, contrasts with that in a free-market situation, which (hypothetically) employs labor in response to consumer rather than government preferences. The labor market of the 1960s and 1970s was essentially a product of federal dollars, which restructured market demands for black labor. The distinction is important if one assumes that race discrimination remains active, in which case the withdrawal of federal supports would erode this class position. In the 1990s, if blacks and whites were to lose jobs in the federal government in equal numbers, for instance, the impact on blacks would be substantially more dramatic. Despite optimism over the emergence of economic opportunities for blacks in the middle class since the 1960s, other authors have expressed this reservation. Freeman (1976a) predicted more than two decades ago that if government pressures for affirmative action weakened, so might the demand for black college personnel. The link between the growth of the black middle class and public policy highlights both the dependency and the fragility of the class position.

Chapter 3

Racialized Services
in the Workplace

◆ The 1980s ushered in the "Reagan revolution," a sharp reversal in the liberal democratic thinking that had dominated federal domestic social policy for fifty years. To understand just how vulnerable the black middle class is when federal policies change, it is necessary to define "racialized" services as they occur in policy- and in market-motivated situations. I define as racialized any services directed at, disproportionately used by, or concerned with blacks. Conversely, "mainstream" services are directed at, used by, or concerned with any consumer. For example, black-owned advertising firms are segregated in the market sector via links to black consumers. Administrators and professionals in public sector settings such as Chicago's Department of Human Services are in segregated situations because they are linked to dependent blacks.

I believe that, by the 1980s, most jobs that allowed the black middle-class to move ahead were disproportionately racialized, part of a system of tasks that would nullify blacks' potential for disruption. The idea of public and corporate substructures managed in response to local political pressures is not new (Piven and Cloward 1971, 1977). But I apply the idea here to argue that a black middle class of public and private sector functionaries was useful to white power structures during the 1960s and 1970s while at the same time occupying a precarious market position. The following data lend support to this idea.

When black political action disrupted the social and economic order during the 1960s, the response of white power structures was to develop social welfare, community development, and employment policies and programs to reinstate order. Two trends are notable in the occupational shifts made by black male college graduates between 1960 and 1970 (U.S. Bureau of the Census 1963, 1973). First, they made progress in almost every occupation outside the professions traditionally filled by blacks, such as social welfare and teaching. In 1950, only 61 percent of nonwhite male graduates were professionals or managers compared with 74 percent of all college men. By 1970, the ratio was 75 percent compared with 80 percent. In contrast, college-trained black women have traditionally had very similar job distributions to those of their white peers. In a second trend, college-educated black men moved into occupations where they could calm disruptive black elements of the population. Among selected professional occupations where employed black men with four or more years of college were underrepresented compared to whites in 1960, only two show gains that *exceed* the white proportion in 1970: personnel and labor relations, and public relations (U.S. Bureau of the Census 1963, 1973). The largest increases for black men in the 1960s occurred in personnel and labor relations jobs (400 percent) and public relations (300 percent); next in rank were social scientists (70 percent) and social welfare workers (67 percent). Without much more detail about what these work roles involved, the pattern appears consistent with my hypothesis. These gains are in occupations that have people-related, as opposed to technical, functions. If the white power structure during the 1960s was concerned about managing the upheaval among blacks, these are the kinds of jobs for which blacks would be in demand to manage or administer institutions, or organization units, where the white power structure mediates the needs of black consumers or the black work force. Significant portions of black middle-class income would thus rely on jobs dependent on heavy concentrations of blacks. Black service or product providers whose activities depend heavily on transactions with other blacks as clients have racialized jobs. Black supervisors overseeing a predominantly black work force also are racialized.

Racialized Roles in the Public Sector

Although few if any products, services, or policies in the public sector are consumed solely by blacks, some are consumed by much higher proportions of the black population than the white. Criteria for participation in dependency-related services such as food stamps and Aid to Families with Dependent Children were explicitly oriented toward blacks. Other services aimed at lower-income households differentially favor blacks. Low-income blacks tend to be overrepresented among those who use public facilities and services such as public housing, health and hospital care, corrections, and city transportation systems.

The legislative response to blacks' protest during the 1960s was to distribute subsidies to industries and portions of industries that dependent blacks use.[1] Besides pacifying the black underclass, these subsidies expanded the work settings and increased the level of incomes available to blacks in the middle class. Blacks working in racialized services were the indirect beneficiaries of the expansion of social services. At the same time, blacks employed in these services were intermediaries between white institutions and other blacks through racialized services, for social policy oriented toward blacks was funneled through them.

If my hypothesis about opportunities for black employment is correct, one would expect concentration in the services indicated above. Underrepresentation should occur in services oriented toward the more mainstream consumer. Public employment data from the U.S. Equal Employment Opportunity Commission (1980a) generally conformed to this expectation. In 1978 blacks made up only 20 percent of all city employees, nationwide. Yet blacks were 35 percent of city hospital workers, 34 percent of all city health workers, and, remarkably, 50 percent of all public welfare employees. Similar patterns existed on the state level. Although blacks made up only 14 percent of all public employees at the state level nationwide, in 1978 18 percent of all public welfare workers, 23 percent of all hospital workers, and 49 percent of all housing employees were black. Conversely, blacks made up only 11 percent of all financial administration workers, 7 percent of street and highway workers, and 5 percent of natural resource employees. On both state and local levels, sanitation and sewage workers represent the single category in which blacks were heavily concentrated in a mainstream rather than a racialized service. (This, however, accords with my idea of segregation by occupation. Sanitation services are useful for absorbing high proportions of unskilled black labor.) Comparing the proportion of all black and white workers in city and state government in 1978 showed that when job functions correspond to the expected disproportionately high use of a service or facility by blacks, black employees tend to be overrepresented. In cities, the black proportion employed in hospitals and public housing was more than twice that of whites, and about three times more in corrections. These figures create a striking picture in which blacks' economic opportunities were overconcentrated in race-oriented job functions.

A similar picture emerged on the state level. When job functions coincided with heavy black use of services, blacks were far more likely than whites to hold such jobs. Conversely, for jobs more likely to serve a general consumer, blacks had much less likelihood than whites to be employed. The single exception, again, was sanitation and sewage workers, where functions serve a general public. The extent to which this service was ghettoized in 1978 was remarkable. Blacks were eleven times more likely than whites to be employed in this function.

Within the context of this study, a better test of my hypothesis is to remove lower-status workers (clerical, blue-collar, and service) from the picture and focus on full-time professional, administrative, and appointed workers. (In this discussion I use the term *professional* to cover these categories.) The general picture remained the same (see Table 1).

Blacks were overrepresented in professional positions where race and economic status differentially defined consumer populations. On the city level, for example, black professionals were three times as likely as whites to be employed in public welfare and corrections, twice as likely to be employed in hospitals and health, and more

Table 1
Black and White Professionals, Administrators, and Officials: 1978
(percentage)

	City		State	
Agency	Black	White	Black	White
Public Welfare	15	3	22	15
Corrections	3	1	13	7
Hospitals	22	11	15	13
Health	7	4	10	11
Housing	5	3	1	.20
Community Development	4	4	1	.70
Sanitation and Sewage	3	3	*	*
Financial Administration	14	18	13	17
Police	4	12	4	2
Streets and Highways	2	5	3	10
Natural Resources	8	6	2	8
Utility and Transportation	3	8	.40	1
Fire	3	17	*	*
Employment Security	*	*	11	9
Other	6	4	4	6
Total†	99	99	99	100

Source: U.S. Equal Employment Opportunity Commission (1980a)

*Not reported, or less than 0.5 percent.

†Totals do not add up to 100 percent due to rounding and nonreported figures.

likely to be employed in housing. At the state level, black professionals were disproportionately concentrated in public welfare, corrections, and housing, although this service employed only a small
number of either race. Black professionals were also somewhat overrepresented in hospitals and employment security. It is not surprising
that they were only one and one-half times as likely as whites to be in
state-administered public welfare services, given that state welfare
offices, which tend to be scattered in small communities, are less
anchored than their urban counterparts to large concentrations of
black people. In a similar vein, state employment offices must meet
the needs of a widely dispersed population. In both cases there is
theoretically less implicit reason for blacks to be employed.

That black middle-class employment in the public sector was anchored to black populations is significant because it thus depended on vulnerable subsidies. Mainstream functions, such as police and fire, tend to depend on revenue generated through locally controlled city or state resources such as taxes or revolving funds. Racialized public services were heavily or entirely dependent upon revenue controlled at the federal level. This issue on its own would not be critical were it not that racialized programs addressed the needs of populations that were dependent and powerless (unless disruptive). Negotiations for funds take place in a political arena whose response to this population is unpredictable. Dependency-related services, like their recipients, were vulnerable to racial politics. Income from these services is essentially the "soft money" in government payrolls.

In the early years of the Reagan presidency the fifty-year trend of more and more federal dominance in social program areas was reversed. The Chicago Department of Human Services exemplifies the effects of subsequent shifts in federal funding on job stability. The department performed "public welfare" services for the city during the early 1980s, and its funding came from both city (corporate) and federal (program) revenue. Federal money supported the agency's social service arm, heavily used by dependent blacks. Corporate funds supported the agency's administrative functions. Almost all of the income for professional staff was generated by federal money, whereas only the top levels of management were supported by corporate funds. In a *Chicago Tribune* article, "Minority City Workers First to Go in Reagan's Aid Cuts," that appeared on 14 December 1981 local officials explained that "the policy of placing minority workers on federally funded payrolls ensures that most of the city employees to be laid off because of budget cuts will be black and Hispanic." Of the 400 employees laid off at the Chicago Department of Human Services during that time period, approximately 75.9 percent were black. Similarly, of the 186 workers laid off from the Department of Health, almost 40 percent were black. In contrast, the Streets and Sanitation Department, where funding is generated more locally, showed few effects from the federal cutbacks. The

Chicago Sun Times reported on 29 March 1981, under the headline "Minority, Female Job Gains Periled," that this department was only 20 percent black in 1981.

General employment opportunities for professionals can be measured by comparing their distribution in various public functions. For example, city housing services employed a larger proportion of all city professionals (3 percent) than did state housing (.2 percent). Conversely, city health services employed a smaller proportion of all professionals (4 percent) than did health services at the state level (11 percent). Comparing city with state functions that tend to perform racialized services, such as public welfare, the smaller the percentage of total employed, the greater the odds that blacks will be employed. For example, in 1982, states employed 15 percent of all full-time professionals in welfare functions; cities employed only 5 percent. Yet the proportion of blacks to whites was almost four times greater at the local level. The same pattern held true for hospital, health, and corrections.

An apparent explanation for this employment pattern is that blacks tend to find jobs where the black population is concentrated and where they exercise political power. In many large cities, then as now, blacks and other minorities make up over one-half of the total population. At the same time however, cities in general are more revenue dependent than are states. Therefore, in terms of both occupation and government level, blacks are concentrated in fiscally vulnerable situations.

The link between black employment in the public sector and black populations increases this vulnerability. Public jobs are responsive to both community demands and political pressures. For example, the *Chicago Sun Times* reported on 12 December 1994 in an article headed "Hispanics Condemn Lack of School Jobs" that between 1989 and 1994 African Americans in Chicago's public sector lost ground in policy-making city jobs, while whites, Hispanics, and Asians made sizable gains: The proportion of African Americans decreased by 13 percent, as whites, Hispanics, and Asians gained by 28 percent, 85 percent, and 82 percent respectively, although the base point for Asian Americans was low. One explanation for this shift is that other urban groups have begun using the political strategies that during the 1960s and 1970s produced results for African Americans. That is, they apply political pressure to increase access to, or to protect, institutional privilege. In 1994, for example, Hispanic activist groups in Chicago demanded that more Hispanics be hired as administrators and teachers in the city's public schools, necessarily replacing African Americans then disproportionately represented in this job pool. Similarly, in Los Angeles, Hispanic activist groups targeted jobs in the U.S. Post Office, a job resource that in the not-too-distant past contributed to the economic stability of blacks and to blacks' ability to be middle class.

Similar activity can be found among other ethnic groups (Ramos 1994). In a politically mediated opportunity structure, public jobs are resources that can be oriented toward one group or another to satisfy demands. In the 1980s and the 1990s, protests organized by more politically active coalitions of labor work to the detriment of now mollified African Americans.

Segregated employment patterns to appease black demands were also suggested in employment data on federal-level jobs in 1980. The departments of Housing and Urban Development, Health and Human Services, and Community Services Administration showed heavy concentrations of black employees (U.S. Equal Employment Opportunity Commission 1980a). For example, of the Community Services Administration's nine hundred employees, 44 percent were black (U.S. Equal Employment Opportunity Commission 1980b). This is precisely the type of agency that sprang up to absorb black demands for increased social and economic resources in the sixties; it was abolished by 1982. These data do not tell us if blacks were already concentrated in these types of public services and moved up into professional and administrative positions, or if they were pulled into these positions as part of employers' response to 1960s political activism. Although the data and calculations suffer from weaknesses, the analysis tends to support the argument that, by the 1980s, much of the gain made in the middle class in professional and managerial jobs occurred in public settings where policy absorbed black demands.

Some obvious questions unanswered by these data eventually led me to collect the qualitative data on which most of this book is based. For instance, were blacks concentrated in black-oriented services and functions because they preferred these to mainstream opportunities or because these represented the job avenues with least resistance to blacks' entry? In addition, these data, in contrast to the qualitative data presented in Chapter 5, do not allow us a close enough look at the workplace to measure racialized pockets, even within mainstream functions.

Racialized Roles in the Black-Owned Business Sector

Government policy inspired growth in the black business sector during the 1960s and 1970s, as well as in the public sector. Data available also suggest that black entrepreneurs held racialized rather than mainstream market niches by the late 1970s. The great preponderance of black-owned business (98 percent) remained concentrated in retail and selected services where, historically, black businesses have been forced to market their wares almost exclusively to black consumers; a sizable proportion of all minority business is based either on federal procurement or on sales to the minority community (U.S. Department of Commerce 1979). A 2 April 1982 *Chicago Tribune* article, "Report Urges Business Aid for Minorities," typical of media coverage at the time, reported that minority business owners pushed for federal aid to help black firms penetrate general (i.e., white) consumer markets, an indication that black businesses were relegated to the limited growth markets in inner-city neighborhoods and tied to public aid income, as they remain today. Thus racialized, they are useful for serving low-income blacks in potentially volatile environments. In addition, they are indirectly linked to policy fluctuation which adversely affects disadvantaged blacks.

Even when federal policy on black business development stimulated expansion among nontraditional service businesses during the 1970s, the hypothesis should still hold that black entrepreneurs concentrate in areas where they hold racialized and functional positions related to black populations. Black businesses represented in the *Chicago United Compendium of Professional Services* (1980) not only typify nontraditional firms in 1980 but were publicly acknowledged successes. To be included in the 1980 compendium was one measure of professional and business competence. One could reasonably expect these successful and nontraditional firms to be the most likely to join those serving the mainstream consumer market. Businesses listed there further illustrate the intersection of black business functions and market demands.

The compendium, published in 1980, lists blacks in three advertising firms, five architecture and engineering firms, six management consultation firms, three certified public accounting firms, four law firms, and seven personal services firms. All except two firms listed were established between 1965 and 1979. Most, therefore, are firms in which blacks were able to capitalize on race-related market incentives. The two exceptions were an advertising firm established in 1950 and an accounting firm established in 1939. The following excerpts describe these firms and illustrate the type of market they cornered:

> [Advertising firm] offers the following services: marketing planning, marketing research, media planning and buying, creative services, print and broadcast.

> [Advertising firm] is a full-service Chicago based advertising agency primarily concerned with the black consumer market. Selected Clients: Coca-Cola (U.S.A.) and McDonalds.

> [Architect and engineer], general experience covers: urban renewal clearance, housing programs, conversion projects, municipal planning. Selected Clients: City of Gary Indiana and Public Building Commission of Chicago.

> [Management consultant] . . . provides management consultant and public relations services . . . on employment practices and affirmative action programs for minorities and women. Selected Clients: People's Gas and Joint Legislative Administrative Committee on Public Aid.

[Certified public account firm] has been under contract with the City of Chicago to perform pre-grant award audits — as well as financial audits of agencies and grantors under federally funded programs. Selected Clients: City of Chicago, Office of the City Controller and Commonwealth Edison.

[Legal services] . . . land transactions, secured commercial transactions, the defense of property claims collections, contract compliance, casualty claims and labor-management problems. Selected Clients: United States Department of Housing and Urban Development and Mammoth Life and Accident Insurance Company.

[Personnel service] . . . specialization is minority recruitment in all areas of employment. Selected Clients: Honeywell and International Harvester.

The first thing that is apparent from the descriptions is that these firms were performing services comparable to those offered by their majority competitors. That is, the advertising firms do market research, and the engineers plan building sites. Not only do these firms show clientele from the black private sector (Mammoth Life), but they show relationships with Coca-Cola, Honeywell, and People's Gas as well. And second, most were depending on racialized services to draw their mainstream consumers. For example, the advertising firm capitalizes on its black consumer market, while other types of firms address the corporate response to equal employment opportunity in hiring and promotion. These were the services that were most expected to attract business from the majority corporate world.

Since the compendium offered only short descriptions of business functions, I conducted telephone interviews in 1982 with one professional representative from each black firm listed. I asked general questions about whether the firm did business with any sector not mentioned among "selected clients," and what kinds of services were performed for clients. For example, if all clients listed in the compendium were from the white corporate sector, I asked whether business was done with public and black private clients as well. I

Table 2

Nontraditional Black Firms in Racialized and General Services

	Service	
Client	Racialized*	General†
Public	A E M M M M C C L L L	E E E C P
White Private	A A A M M M M L L P P P P	E E C C C L P P
Black Private	M C C C L L L L P P P A	
Total	36	13

Note: The following symbols are used to indicate type of firm: A = Advertising ($N=3$); C = Certified Public Accounting ($N=3$); E = Engineering and Architecture ($N=5$); L = Law ($N=4$); M = Management Consulting ($N=5$); P = Personnel Services ($N=6$).

*The racialized category includes the services directed at black consumer/manpower needs.

† The general category includes services directed at a general population/market.

also asked for examples of the kinds of work a firm performed in each sector. Based on the services that were described, I then decided whether or not the service was generally black related. For example, I placed accounting firms involved in establishing general accounting systems in the Zenith Corporation in the mainstream category of services, and accounting firms establishing systems for certain types of federal contracts in the racialized category of services. One firm was establishing an accounting system for a California regional office to be used specifically for Comprehensive Employment and Training Act (CETA) funds. CETA program funds, of course, targeted blacks. Table 2 shows the sector distribution of these firms' clients and whether or not they were related to racialized (R) or mainstream (G) services. Type of service was categorized for each sector in which the firm was involved. If a sector involved services in both categories (R and G), the firm appears in both. All firms doing business with other black firms were automatically classified as racialized, along with their other sector involvement.

Even discounting firms' dependence on the black market, Table 2 illustrates that they show up in racialized business services more often than not. Of course, this is an overly simplified categorization.

I did not ask businesses how many contracts they received from where, to do what, and with what profit margins. Nevertheless the table is a beginning effort to show that the strongest foothold outside black markets occurred in "black specialties" offered to white institutions. Black law firms negotiated on behalf of their white corporate client when affirmative action or federal contract compliance was at issue, or when white firms were entering contract negotiations with black proprietors or land owners, or when insurance claims for the black insured needed settling. Black personnel service and management consulting firms were almost exclusively involved in black manpower development and executive search for white corporations. Black management consulting firms also helped in the political negotiations of corporations establishing services in the black community. Black advertising firms helped white corporations exclusively to penetrate black consumer markets. Roles were race specified even among firms that dealt with white corporate structures in mainstream functions. Engineering firms with business in the white sector entered into these relationships as subcontractors, based on the corporate need to hire minority firms to compete for federal contracts. Although these functions show up in the General category, it can be debated whether or not they belong there.

Even in the public sector, firms were drawn into performing racialized services. Personnel service firms provided workers for such ghettoized agencies as the Office of Manpower, and law firms were hired for contract compliance and labor-management issues in such racialized services as the Department of Housing and Urban Development. Certified public accountant firms performed pregrant audits and general audits for racialized institutions such as Cook County Hospital. Management consulting firms provided technical assistance primarily to ghettoized units such as the Department of Human Services or the Office of Minority Business Development. Engineers provided professional services to ghettoized manpower sites such as the Northshore Sanitary District. These firms were relatively successful products of moral and legislative commitments made in the past few decades to establish greater exchange between white corporations and black enterprise. Yet even here black businesses appeared to be useful only when they served in symbolic or intermediary positions between white structures and black clients. The degree to which these firms competed in the corporate world on the basis of company profit, as opposed to policy, predicts the softness of their market position. Past growth in government-let contracts, coupled with stipulations for black subcontractors, helped form a mutually beneficial relationship. However, when public policy changed and profit considerations encompassed broader issues as the 1980s progressed, this alliance probably changed.

Racialized Roles in the White Private Sector

Salaried blacks employed in the white private sector showed similar relationships to those in the public sector in the late 1970s and early 1980s. A 1979 survey of Fortune 500 companies reported that more than 29 percent of black executives versus 3 percent of white were in personnel specialties (Heidrick and Struggles 1979b). These are positions in the corporate structure typically responsible for affirmative-action plans and implementation. An example of how companies used such positions lies in the Coca-Cola and Heublein negotiations with People United to Save Humanity's (PUSH) trade association in 1982. In 1971, Jesse Jackson launched PUSH covenant negotiations to expand opportunities for black individuals and business. Some of these negotiations took place with black community affairs officers who were viewed by PUSH representatives as major stumbling blocks to PUSH objectives for these corporations. In general, not only were these positions functional in dealing with the problem of blacks, they were an effective way for corporate entities to address government hiring policies while minimizing black power. Personnel, labor relations, and public relations officials operate outside the strategic planning or production areas that typically lead to power within the corporation. They are the "soft money" within the business world.

Over 22 percent of black executives in Fortune 500 companies in 1979 were in manufacturing, a category that nationwide accounted for the second largest industrial concentration of black labor (Heidrick and Struggles 1979b; U.S. Bureau of Labor Statistics 1982). One might suppose that these executives were useful in managing large numbers of blacks, while helping the companies meet affirmative action requirements (Fernandez 1981). At the same time they were concentrated in the "soft" sector of the economy in an arena of jobs that experienced a steep decline. As public policy is reversed, the implicit need for blacks in these kinds of positions may also diminish.

Taken together, these data tend to support the idea that in the 1960s and 1970s the black middle class became economically dependent on governmental policy and programs instituted to deal with black unrest. In the late 1970s, black public sector employees were concentrated in the portions of government operations that legitimized and subsidized black underclass dependency, such as social welfare institutions. In the black private sector, black-owned businesses remained concentrated in economically underdeveloped areas or dependent on policy-supported joint ventures with white firms or in protected and racialized consumer markets. Salaried blacks in the private sector seem to have been concentrated in intermediary positions between white corporations and black consumer, manpower, or policy issues.

The racial disparity in the structure of professional opportunities available to blacks is clear. In the years surrounding social upheaval among blacks, members of the black middle class, no longer restricted by occupation or income, remained segregated. In this system of segregation they were concentrated in institutions dependent on government subsidy and in race-oriented services. Thus, by 1980 race was strongly implicated in blacks' economic position in the U.S. economy, despite the obvious ascent of blacks into middle-class positions.

Chapter 4

Tan Territories, Urban Upheaval, and the New Black Professionals

Question: How did you find your job in personnel management for [a major white-owned insurance company]?

Answer: It was all luck. I was at an NAACP convention. They saw me and offered me a job right there on the spot [in 1966]. Over the next three to four years they hired twenty-five or thirty people like that. If you were black and walked by they'd just about grab you off the streets.

—From an interview with a senior investment analyst and director employed in a venture capital division of a firm in the insurance industry

✦ A radical alteration occurred in the hiring practices of white-owned companies around the middle 1960s. Until then, black employment patterns in major industries reflected the U.S. cultural norm of de facto segregation and discrimination against black workers (Burstein 1985; Farley 1984; Jaynes and Williams 1989; Wilson 1978). Blacks disproportionately filled a job ghetto comprised of low-paying, physically demanding, and subservient blue-collar and service work for white private industry (McKersie n.d.; Wilson 1978).

In Chicago, blacks tended not to be placed in the principal office of multiestablishment firms; rarely did they fill white-collar jobs, such as sales and clerical jobs, and they were almost totally absent in professional and managerial jobs in white-dominated settings (McKersie n.d.). In Chicago in 1965, blacks were completely excluded from policy-making positions (Chicago Urban League 1977).

From about the mid-1960s onward, however, major white-owned corporations began to hire blacks into white-collar jobs — in particular, into previously closed professional and managerial positions. The term *previously closed* is borrowed from Richard Freeman (1976a:146) to characterize a particular change in the structure of jobs available to the black middle class, one in which blacks increased their proportion in business-related professions because they increasingly were recruited into white-owned, white-oriented economic settings. This juncture of breakthroughs (blacks' entry into new types of jobs and into new job settings) meant new job markets for the black middle class. During the 1960s and 1970s in the managerial world a new elite of blacks emerged as executives in major white corporations. In 1968, for instance, International Business Systems in New York hired into the engineering department the company's first black professional. A white New York corporate official candidly noted in a *Harvard Business Review* article in 1970, "I think we have reason to be a bit defensive about our past record. Thirty-five years ago we used to look for Germans and Swiss, even for the lowest level jobs. Then during World War II, we began to hire the Irish. By the time fighting broke out in Korea, we were taking Italians and Jews. Now we're actively recruiting blacks and Puerto Ricans" (Cohn 1975:11). Between 1960 and 1983, the proportion of black male college graduates employed in managerial jobs increased from 7 percent to 18 percent (Freeman 1976b; U.S. Bureau of Labor Statistics 1983). The executives I interviewed were at the forefront of this trend in the private sector. One director of personnel reported that when a major food manufacturer hired him in 1968 he was the only black college graduate in the Cincinnati corporate offices. Moreover, he was the first black ever in the company's management training program. Another personnel manager for a major national manufacturer reported that, in 1970, he became the first black promoted to manager in the Chicago location of the company. One ranking man was the first black professional a Fortune 50 company employed and, in 1968, the first to work in the Chicago corporate office. A fourth man, a high-level executive who works for a well-known consumer goods company, was one of the first five blacks the company employed in sales. He was hired in 1966, among the first of a large number of blacks hired by the company over the next six months.

One-half of the executives (thirty-six of seventy-three) I interviewed were the first blacks to integrate the professional and managerial job structure of a major white corporation at some level. Each entered the white private sector as the first black professional or manager hired by a company, either in the company's regional, state, city, or district offices or at specific company sites, such as corporate headquarters, plants, divisions, or departments. Even those who were not, technically, "firsts" were either one of only two or three blacks in a company's professional-managerial job structure, or part of the first cohort of blacks hired in the company's initial effort to recruit blacks into its professional-managerial ranks.

Some researchers explain the improved market for blacks by linking black employment to cyclic factors, pointing out that blacks' position in the labor market improves during economic expansion and declines during times of recession (Jaynes and Williams 1989; Levitan et al. 1975). Yet gains made by black workers in the professional-managerial job strata during the 1960s, under conditions of economic expansion, also increased during periods of recession in the 1970s (Farley 1977; Freeman 1976b). This phenomenon suggests that factors other than the economy play a crucial role in black attainment and provide countercyclic supports that sustain blacks' gains.

A second explanation suggests that improvements in the quality and quantity of education have been key to blacks' access to broader job opportunities (Smith and Welch 1986). However, they cannot solely account for the sudden inclusion of black Americans into higher-paying white-collar occupations in the 1960s. The quality of black education had increased from the 1940s onward (Jaynes and Williams 1989), and states and the federal government increased antidiscriminatory employment measures during the same postwar period. Moreover, before the 1960s, even similar educational achievement did not produce equality in black, relative to white, occupational attainment (McKersie n.d.); higher levels of black education actually accentuated some race-based economic differences (Siegel 1965; Thurow 1969).

These macroeconomic and human-capital explanations account for the presence of a pool of qualified black workers available to take advantage of the new opportunities, but they do not explain why those opportunities occurred when they did. Political interventions completed the equation and opened the job market. Human-capital theory views market dynamics as neutral, while I view the expansion of the black middle class during the 1960s and 1970s as a market response to corporate and state interventions intended to abate insurgency from below — a politically mediated phenomenon.

Job Markets for College-Educated Blacks

Contrary to popular public perception, until the 1960s, the most profound barriers to economic equality for blacks in the labor market were erected against college-educated and skilled blacks competing at the top of the economic hierarchy (Freeman 1976a). For instance, until the 1960s the absolute gap separating the earnings of black and white men actually increased as their amount of schooling increased (Farley 1984). One study indicates that racial differences in earnings in 1960 were almost five times as great for black men with college degrees as for black men who dropped out of elementary schools (Siegal 1965). Another classic study found that the earnings gap between college-educated black men and their white counterparts was about two and one-half times greater than the gap between black and white men with only eight years of education (Thurow 1969). A third highly influential study showed that the more years of schooling blacks had, the larger the gap between their purchasing power and that of comparably trained whites (Freeman 1976a). As late as 1959, the average earnings for nonwhite men with college educations was 20 percent lower than those of white male high school graduates (Freeman 1976a).

Not only pay discrimination but job restrictions severely limited the life chances and economic opportunities of college-educated blacks. For instance, before 1960 it was easier for unskilled and less-educated black workers to find jobs on the bottom rungs of the occupational ladder than for black college graduates to work in a white-dominated setting in more prestigious professional jobs. An illustration of this relative disparity appears in ethnographic accounts of discrimination in federal government employment at the end of World War II. Although statistical descriptions indicate that blacks in federal government jobs were not affected when white veterans returned home, the small number who held professional jobs in white-dominated government agencies were most likely to have their jobs threatened (Newman et al. 1978). In other words, blacks lost the small gains they had made during the war toward equality in federal employment and upward mobility. In contrast, blacks in low-status service jobs and black professionals in segregated settings fared better. For example, blacks did not lose out to returning white veterans who were low-status job holders, such as custodians, maintenance people, or laborers; they maintained gains made during the war in professional jobs in environments largely run and staffed by blacks, such as Freedman's Hospital (Newman et al. 1978; U.S. Senate 1954:204). In the 1950s, when more than one-third of employed nonwhite men were unskilled laborers or farm hands, the etiquette of workplace race relations held that college-educated blacks could not supervise whites or work immediately alongside them, nor could they work in any setting that was considered "refined" (Newman et al. 1978:33).

Overt racial discrimination and economic disincentives pervaded the job market in 1960. Consequently, in Chicago, African American men with enough education and experience to hold jobs equal to those of white men actually stood 15 percent lower on an economic index of occupational position (McKersie n.d.). Nationwide, college-educated black men were grossly underrepresented in the broad range of high-paying and high-status professions and virtually excluded from the business-related professions and the field of management. Only about 7 percent of nonwhite male college graduates in 1960 were managers compared to 18 percent of college-educated white men (Freeman 1976b:11). Blacks constituted less than 1 percent of accountants and engineers and just 1 percent of lawyers working in the United States. The absence of opportunities for college-educated blacks in the managerial and business-related professions is underscored by the failure of any major white corporation to make recruitment visits to historically black institutions in 1960 (Freeman 1976:35).[1] In this way, black college graduates were excluded from the pipeline that feeds the entry-level managerial and professional positions in large corporations. Effectively locked out of good jobs in white-owned industry, college-educated blacks were confined to a limited number of occupations that served the black community, such as physicians, teachers, and self-employed business owners (Drake and Cayton 1962; Frazier 1957).

During the mid-1960s, after a history of virtually total exclusion from better-paying jobs in white businesses, highly educated blacks were actively recruited by white corporations and experienced a striking transformation in the employment market demand. The longstanding pattern of decline in the income of educated blacks compared to whites was reversed. The ratio of black-to-white income rose most rapidly for managers. Between 1966 and 1970, the fraction of black men working as managers or professionals increased substantially, by 1970 almost doubling the 1960 numbers. The number of white corporations conducting employee searches at predominantly black colleges jumped dramatically after 1965 (Freeman 1976:35). In 1965, about thirty visits by white corporations to black college campuses took place. By 1970, the average black college campus received ten times that many visits. The same black colleges that corporations had shunned before the 1960s became an important source of black manpower recruitment in 1970.

Pre-1960s Job Markets

Although before the 1960s, skill and educational qualifications of black applicants had been important issues for white employers, placing high on these characteristics was not enough to eliminate employment discrimination. The individuals I interviewed were prepared by virtue of their education to start in professional and managerial jobs in private industry. Most of them (89 percent) had at least a bachelors degree when they entered the labor market. Over one-third (38 percent) had earned advanced degrees.

These proportions closely parallel the level of education of white male senior-level executives; 94 percent of top executives in Fortune 500 companies surveyed in 1986 had bachelors degrees; 42 percent had graduate degrees (Korn/Ferry 1986). Moreover, the level of education of my interviewees is well above the median level (about one year of college) for salaried male managers in 1960 (U.S. Bureau of the Census 1963). Slightly more than one-half of the black graduates I interviewed received their degrees from a predominantly white college or university. Yet a college education (as most other qualifications) failed to open corporate doors to prestigious and higher-paying jobs for those entering the labor market before the mid-1960s, compared to the jobs they held afterwards. Twenty-four of the twenty-nine (86 percent) who entered the labor market before 1965 had at least a four-year college degree, yet fewer than half got their first jobs in a white-owned establishment. Those who did work in the white private sector were concentrated in low-paying or segregated white-collar jobs, such as clerical and black-oriented sales occupations.

Private Sector Employment

The skewed midwestern labor market for blacks in the 1960s matched the national pattern, reflecting racial discrimination and the exclusion of blacks. In Chicago, as in the rest of the nation, blacks were placed in inconspicuous positions vis-à-vis the white public and were underrepresented in sales and clerical jobs. Their occupational position was much lower than that of similarly educated whites, and they were underrepresented in high-paying and high-status jobs (McKersie n.d.).

One case illustrating the underemployment faced by educated blacks is that of a marketing manager with a bachelor's degree in chemistry who expected to work with a major chemical company when he entered the midwest labor market in 1957. Instead, he says, "It was tremendously difficult to find that kind of job. . . . I had huge difficulties getting work. My first job was as a stock clerk." Although factors not related to race may explain his search difficulties, they occurred during a time that chemicals were a growth industry in the United States (Urban League 1961). His report is consistent with research indicating that blacks were underrepresented in the faster-growing industries in Chicago between 1950 and 1966, particularly trade, finance, insurance, and real estate (McKersie n.d.). Moreover, while his job opportunities in the white private sector were initially limited, the scope of his opportunities later broadened dramatically. After languishing in an entry-level technical job in a chemical company, he was recruited by a major clothier into a management trainee slot in 1967. In 1986 he was a vice-president and marketing manager in a major Chicago-based firm within the wholesale food industry.

A second college-educated man, a regional vice-president of sales for a major consumer goods company, also believed he experienced underemployment. After his graduation from a white midwestern college, he went to work in the marketing department of a large oil company. He reported, "I was supposed to be doing a lot of their market research and stuff like that. But I was just a clerk really, and there really wasn't much of a career opportunity." As he put it, "They would tell me, when openings came up, that, you know, their other [i.e., white] managers had to have those slots. So I just didn't see a future at [the company] being that I was black." In this case, also, perhaps he was limited not because he was black, but because of nonracial characteristics. If so, however, such characteristics did not preclude several major companies from offering him chances to enter management training positions after 1965. Moreover, he did not languish at the lower levels but moved rapidly up the job hierarchy once he went to work for his current employer.

Other illustrations of white-collar occupational segregation and truncated job opportunities come from college-educated blacks who entered sales fields in Chicago before the mid-1960s. In 1960, the median level of education for whites in sales was about twelve years;

three of the five executives I interviewed had a college education when they began their careers in sales before 1965.

In the late 1950s and early 1960s, consumer goods and wholesale trade companies adopted a marketing strategy in which they racially segmented sales areas into white and "special" markets known as "tan" territories. The terms are euphemisms for geographic areas dominated by black consumers. The practice of hiring blacks into predominantly white sales organizations and then steering them into black-oriented sales jobs was, and still is, common. A vice-president and regional sales manager for a Fortune 500 industrial said of his employer that "you would find that in Harlem and the Bronx, Brooklyn, you would basically have black sales reps, or Hispanic."

The first series of private sector jobs held by one man was limited to sales in black communities, although he had a bachelor's degree from a private white university. Between 1959 and 1964 he worked for a major alcohol beverage company and for a major tobacco company. Next, he found work in a black-owned hair care products company. It was not until 1968 that a white-owned company finally recruited him for a "crossover" job in general market (white) territories. This individual's career included time as a professional baseball player, and he described how routes into the white private sector were open only to a very limited group of blacks previous to anti-discrimination legislation. "Black [athletes] were the first in all of these [sales jobs], and we were selling to the black community. If you were a black athlete, then you had some high visibility and that generally was how you got into a major corporation."

Confining blacks to tan territories was a way for companies to better compete for the black consumer dollar without alienating white colleagues and customers. Although tan territories were the first markets that companies outside the South opened up to black salespeople, they did not offer jobs from which blacks would ordinarily move up in a company or get training in the merchandising business. Black territories were considered poor sales targets because the preponderance of black consumers had limited disposable income. The unanticipated consequence of such discriminatory marketing and employment policies was the discovery of the economic potential and the name-brand loyalty of the black community. Thus today, many companies aim different marketing strategies at black, white, Hispanic, and other special target groups. And companies have adopted some of the approaches developed for black consumers by black special market people to general markets and advertising (Davis and Watson 1982:18).

Professional Jobs and Government Contractors

Some of the people I spoke with did break into the professional mainstream in white companies before the 1960s. However, these opportunities were available only to those holding advanced degrees. Moreover, even within this educationally elite group, job opportunities suffered from racially based limitations. Almost exclusively, good job offers came from companies serving as federal contractors that therefore were sensitive to government oversight.

From the 1940s onward, the federal government stipulated that private industry receiving defense-related government contracts must address employment discrimination. One executive I interviewed had an MBA in finance from a relatively prestigious white university on the West Coast. He indicated that when he entered the labor force in 1962, federal government contract compliance laws induced large government contractors to hire highly educated blacks. And, indeed, between 1960 and 1966, blacks increased their employment in prominent companies in Chicago, most of them federal contractors with Plans for Progress (McKersie n.d.). While this seems like a positive reminiscence, he went on to say, "I wanted to be in the investment banking community, but there was no opportunity there at all. I finally settled on a job as an accountant at [an aerospace firm]. The [company] stacked the roster, they wanted lots of graduate degrees, and they wanted minorities, despite the fact that there were not the obvious or blatant kind of affirmative regulations [at that time]. And I was not doing work [at] the level of an MBA."

Although this executive was able to break into a traditionally closed profession, he also believed that he began his career as an overqualified but underutilized accountant. There may be some statistical support for his beliefs, since in 1960 only about 9 percent of all accountants and auditors had five or more years of college education. In 1986 he was the only black CEO of a major insurance company, a subsidiary of a Fortune 500 conglomerate.

A second example is striking because the man I interviewed was one of only a handful of blacks who had a graduate degree in engineering in 1960. In 1986, he was vice-president of operations for a large electronics and communications firm.

> I must have gone on forty-seven interviews. I heard that I was too old, I was too young, I was overqualified, I was underqualified. I was everything except the fact that nobody wanted a black engineer. So when I came to this [defense contractor], they were the first corporation that really offered me a job as a full-blown engineer. [However] I worried about where I could go [in the company]. There was one other black engineer in the whole corporation and he had been around a long time. This man had graduated from the University of Illinois with a mechanical engineering degree when blacks couldn't even eat on campus. And all he was was an assistant manager.

Government Employment

Fewer than half of the twenty-nine people I interviewed who entered the labor market before 1965 worked initially in the white private sector. This is consistent with the fact that government was a significant resource for black employment prior to the 1960s. In 1960, for example, just 13 percent of white managers, but 21 percent of black managers and administrators, worked in government (U.S. Bureau of the Census 1963). The proportion of black managers and officials was 62 percent greater than that of whites (U.S. Bureau of the Census 1963). Before the mid-1960s, then, these highly educated men found their jobs in government agencies, black businesses, and black nonprofit agencies. The experience of a vice-president for sales for a Fortune 500 retailer is typical of that era. He began his career as a caseworker for the Chicago Housing Authority; in 1986 he was one of the highest-ranking black executives in the United States.

> When I came out of the army, I made some inquiries into job possibilities prior to discharge. And, in fact, a whole group of us had sent out a number of resumés to companies. I was in an unusual outfit. It was [made up] mostly of college graduates. And, I guess of about 2,600 people in that [outfit] maybe 15 percent were black. We had sent out a number of resumés to corporations, to social work concerns, to foundations, to colleges. The responses we got [were] very interesting. The responses [that blacks] got from companies concerning management trainee positions were nonexistent, you know. White companies did not respond to black resumés. Or, they'd say that they didn't have anything open that you'd be interested in. It was interesting because there was a definite difference in the response of these companies between black and white resumés.

This man said that he came to the same conclusion "that educated blacks came to [then], and prior to that. The best opportunities [were] with governmental agencies and quasi-governmental agencies." When white corporations in the Midwest discouraged his application for management training jobs, this executive used his college degree to find a public sector job in social work. Five years later, in 1965, he was courted by a white company to enter management training.

His personal experience, and even his memory of these events, may be biased. However, others told similar tales about government employment. A director of affirmative action who worked for the same Midwest-based company said, "I was a teacher; my parents were teachers. At that time that's pretty much what you were going to do if you were able to go to college. You were going to be a teacher or a social worker." In 1961 she began working as a government employee; in 1966, she was recruited for a management training program in white private industry.

The hiring policies that excluded these people from good jobs in corporations in the Midwest also were reported by educated blacks who sought work in other parts of the country, although the prevalence of these conditions is not explicitly documented. For instance, a current vice-president of manufacturing who entered the labor market in 1964 in Arkansas noted:

> There were just not many opportunities. So, you catch the first thing you can get. And, at the time I was looking, they were just starting this [poverty] program. It had nothing to do with my background and nothing to do with my interests at the time. I had a degree in math, and I was looking to get into research. But, again, in that part of the country, and during those times, you either had to work for the government, or . . . that's it.

In 1966, he answered an advertisement in a white daily newspaper and found an entry level training position with his current employer, a company in the food-manufacturing industry.

In summary, before the mid-1960s corporate America made few professional and managerial jobs available to college-educated blacks, as the reports here reflect. After the mid-1960s, these people were in much greater demand in the private sector; they entered high-paying and prestigious jobs in white settings that had rigorously shunned them only a few years previously. Between 1950 and 1966 the occupational position of black compared to white men actually declined in Chicago (McKersie n.d.). But after the mid-1960s, at least for some better-educated blacks, the employment outlook changed dramatically. As one executive summarized it, "The same employment agency that turned me away in 1960 helped me find a job in the private sector in 1966."

Political Pressures on Private Employers

Despite their high levels of education, the people that I interviewed experienced their race before 1965 as a criterion for exclusion from good jobs in private industry. Further, data show that, given education and age qualifications, black men in Chicago faced somewhat more job exclusion than was generally the case for blacks in other central cities (McKersie n.d.). The Chicago experience illustrates the transcendent social fact of the time period, economic exclusion. After 1965, however, race became less of a barrier. Fifty-four of the seventy-six people I interviewed (71 percent) attributed this shift, at least in part, to the political pressures that shaped the 1960s and influenced this hiring period.

By the mid-1960s, federal employment legislation and an increasingly militant black community pressured the white business community to employ blacks. Grass-roots groups led by black ministers and other community leaders demanded that major employers start programs to expand black employment and business procurement opportunities (Cohn 1975). At the same time, urban rioting appeared to jeopardize not only the stores and plants but the sales markets of white businesses. Moreover, Title VII of the 1964 Civil Rights Act gave private citizens the right to sue over employment discrimination and made legal prohibitions against job discrimination enforceable for the first time in history (Hill 1977). In the 1960s, the federal government stipulated that major employers should correct employment inequality and by the early 1970s required federal contractors to submit affirmative action plans for hiring and promoting blacks and other specified minorities. The people I interviewed believe that new job opportunities emerged because of this federal affirmative action legislation and because of community-based political pressures, including urban violence.

Federal Legislation

By the time Title VII was enacted by Congress it had become increasingly evident that meaningful enforcement of a fair-employment law would require active federal intervention in hiring (Hill 1977). The premise of the law was that, based on past performance, major employers were likely to maintain the very barriers to equal employment opportunity that in the 1960s jeopardized the well-being of the entire nation (U.S. Commission on Civil Rights

1969). The employment protocol of the American Can Company in the 1960s illustrates the potential of Title VII to improve the socioeconomic status of blacks through federal intervention. One facility of the company, which was a large federal contractor, drew its employees in 1969 from a rural area in Alabama where the population was about 57 percent black. Yet, the facility perpetuated racial inequality by employing blacks in just 7 percent (108) of its 1,550 jobs, and in only "several" skilled positions (U.S. Commission on Civil Rights 1969). The company also maintained other systems of inequality, such as owning a town in which totally segregated rental housing was maintained for company employees. In this town only 8 of the 123 company-owned houses that were rented to blacks had running water and indoor toilets. The contractor and subcontractor clause of Title VII enabled the government to intervene with the threat to withdraw federal contracts. As discussed in Chapter 2, the power of such a threat could alter exclusionary practices in this and similar employment settings. In 1969, an estimated one-third of the nation's labor force was employed by federal contractors, who made up a sizable proportion of the largest industrial employers (U.S. Commission on Civil Rights 1969). Accordingly, Title VII had the power to ameliorate a significant source of race-based social inequality.

The two categories of employers that fall within the regulatory sphere of Title VII are (1) companies with one hundred or more employees and federal contractors with fifty or more employees and (2) federal contractors or subcontractors selling goods or services worth at least $50,000. Consequently, large segments of corporate America came under pressure in the 1960s when they became targets of civil litigation and the courts ordered broad remedies when enforcing the central concepts of the law. In 1965, for instance, the Illinois Fair Employment Practices Commission ordered the Motorola Corporation to either stop employment testing altogether or replace their current biased test with one that was nondiscriminatory (Hill 1977). Although the monetary award for damages was eventually rescinded, the ruling received much publicity and set an important legal precedent (Hill 1977); indeed, employers' use of tests has been successfully challenged.[2] In 1966, in *Hall v. Werthan Bag Corp.* (251 F. Supp. 184, 186 [M.D. Tenn 1966]), the district court upheld the right to bring class action suits and thereby rejected efforts of employers to protect institutionalized forms of discrimination. In 1968, a federal district court held in *Quarles v. Philip Morris, Inc.* (279 F. Supp. 505 [E.D. Va. 1968]), that Philip Morris's assignment of black employees to departments with limited advancement potential was discriminatory. The court required the company to adopt a plan of interdepartmental transfer and promotion to eliminate this disadvantage.

Corporate America also came under pressure in the 1960s as the federal government's contract compliance programs increasingly held federal contractors accountable for following aggressive hiring plans, known as affirmative action. In 1967, the Office of Federal Contract Compliance (OFCC) began to impose affirmative-action standards on designated federal projects, such as the construction of the Bay Area Rapid Transit system in San Francisco (U.S. Commission on Civil Rights 1969). In 1969, the OFCC for the first time commenced proceedings to debar contractors for noncompliance (U.S. Commission on Civil Rights 1969). By the early 1970s, the federal government reached the apex of its commitment to implement programs to correct employment inequality. In 1970, Order No. 4 required contractors to set goals and timetables for hiring minorities. In 1972, Revised Order No. 4 specified in detail how contractors should comply with hiring obligations or face losing federal contracts. Major employers also came under investigation and litigation by the EEOC and settled job discrimination suits for millions of dollars in damages (Hill 1977; Lydenberg et al. 1986).

The judicial interpretations and enforcement of Title VII, the federal scrutiny of major employers, and the threat of losing sizable federal contracts all proved powerful incentives for corporations to correct longstanding employment inequalities (Ashenfelter and Heckman 1976; Freeman 1973; Leonard 1984; Vroman 1974). Spurred by them, the corporate world suddenly found ways to open job opportunities to a new echelon of skilled and educated blacks. For instance, the percentage of minorities among Philip Morris's officials and managers rose from no blacks in 1969 to 5.7 percent minority in 1972. In 1985, 15.4 percent of the company's officials and managers were minorities (Lydenberg et al. 1986).

In 1972, General Electric was among the first high-tech companies to create support programs to increase the numbers of minorities in engineering. General Electric also began a summer job-training program for minorities and funded minority scholarships at colleges and universities. Continental Bank in Chicago set up the Inroads program to provide summer jobs for black college students to familiarize them with the banking industry. A Northwestern student sponsored by the program became the first black in the bank's bond department, according to an article in the *Chicago Reporter* in May 1974, "Few in Top Jobs." The Chicago office of Arthur Andersen & Company, a Big 8 accounting firm, started recruiting at black universities in 1965; the Chicago office of Touche Ross first recruited at black schools in 1967, according to an October 1975 *Chicago Reporter* piece, "In the Red on Blacks." Three-quarters of the people I interviewed who entered managerial fields after 1965 (twenty-seven of thirty-six) reported that their employment was a result of a company effort designed to hire blacks. A senior investment analyst, whose first job in a white company was in the financial industry, describes his recruitment in 1972 this way:

> After they found out who I was they recruited me. I just happened to be out at [the company] and put in an application. Once they found out who I was they came after me. [Question: What do you mean by who you were?] Black. They saw I was black. They were looking for a black. They had put the word out in the [black] community. One of the people they had put the word out to in the black community was a person that I knew very well, and I had been listing that person as a reference on my applications.

One striking example of this flurry of recruitment came from a technical director of research for a Chicago communications company. He reported that in 1971, when he had just completed his doctorate in engineering, a major corporation was so interested in him that they not only offered him nine different jobs but even proposed sending him back to law school so he could work on their discrimination suits.

It was within this revolutionized hiring context that the employment opportunities of the executives I interviewed were greatly enhanced after 1965, reflecting the much broader market demand for black labor. Almost three-quarters of them (fifty-four of seventy-six) accepted their first job in the white private sector between 1965 and 1974, when federal implementation of Title VII was at its peak. A white supervisor at a large Chicago-based retailer told one executive with a college degree in education during his job interview, "We'd love to have your type." When I asked what he meant by "your type," the executive told me that, in 1972, "they needed a nigger. . . . They told me soon after I was hired that they were interested at that point in upping their numbers for the affirmative-action program. Because they had products that, if they were going to have to sell to the government, they had to have an affirmative-action program." After 1965, first jobs for these interviewees came much more often from the private sector than from government and black businesses. The rate of entry into the professional-managerial strata of private sector jobs also jumped dramatically. About twice as many post-1965 entrants as pre-1965 entrants had access to the white private sector via white-oriented supervisory, managerial, and management-training jobs.

During the 1960s and 1970s, these African Americans increasingly were hired into higher-paying and highly visible jobs that gave them authority in their interactions with whites. In the private sector, they were hired into sales tied to total consumer sectors, rather than to black sectors, and into management and trainee positions in which they supervised whites. An executive employed in the communications industry described the effect of government legislation on his personal mobility and the reduction of racial disparities in supervisory positions in his company. He explained that his transition into management in 1969 was, at least in part, a result of political pressure stemming from government affirmative-action legislation. He reported that the executive director of the company was strongly influenced by affirmative action. He told his department heads and directors, essentially, "We don't have any black supervision in this location. I want you to go out and, at your next opportunity, find a black who's qualified to be promoted to supervisor, and promote him." So somehow I was located in the company. I had done reasonably well, and they came down from [Columbus, Ohio] and interviewed me and offered me the supervisory job. I was promoted at a time the affirmative-action policies were really getting off the ground, and people were resisting that notion all across the company.

An impression reported by a second executive who is employed by a major food company is consistent with this theme. He, too, believes that he owes his transition into management to governmental pressures in the background. "In 1973, I think that their objective was to get a black in a management position and preferably to get one who was qualified. I think that was part of the project I was serving when I was promoted." Overall, these alterations in the pattern of black employment are substantively consistent with research based on aggregate data on black middle-class growth (Farley 1977; Freeman 1976a). The economic status of the black middle class rose as college-educated blacks were hired to fill new work roles. Government intervention via affirmative-action and contract compliance programs played a crucial role in advancing many of these individuals, and it did so by legitimating, not overriding, objective criteria that entitled blacks to equal employment.

The Rise of Black Militancy

The federal government was only one, albeit powerful, ally in blacks' struggle for more social and economic resources during this period. Preceded by a decade of black-white confrontations in the South, civil rights legislation was accompanied by the escalation of black activism and civil disorder in northern urban areas (Bloom 1987; Sitkoff 1981). The executives I interviewed believed that a combination of these escalating social factors also contributed to the expansion of their job opportunities, particularly in the 1960s. That these middle-class blacks consciously connected their gains to disruption in the lower classes points to a race-conscious connection among blacks that transcends class differences. Moreover, it differs from the response of the white middle class, which tended to distance itself from white political action groups such as the Students for a Democratic Society (SDS) and the Weathermen. Upon entering managerial fields in the white private sector, these black executives recognized that organized black boycotts and the ideological militancy of black organizations were instrumental in creating access to these jobs.

An example of this perspective comes from a manager of community affairs who told me that when black political organizations such as the Student Nonviolent Coordinating Committee (SNCC) were becoming more active in the North, his employer was unprepared, both culturally and administratively, to respond to approaches from these groups asking for jobs and contributions. In addition, this manager reported that the emergence of black militancy and the call for "black power" encouraged existing black staff in the company to coalesce and become increasingly insistent that the company demonstrate social sensitivity to the needs of the black community. As the manager summarized the late 1960s context:

> Those were things [that] were in the works at that time [and] the leadership of [the company] looks around here and it's all white men that you see. There wasn't a black face in any one of these [Chicago corporate] offices. They had a few blacks in some cubical . . . but no one that was to be a manager. So here was a tremendous opportunity to make this nigger a manager. [Question: This nigger? Do you mean you?] Sure.

In 1970, this man was the first black to be hired into the company's corporate offices with the title "manager"; he became the manager of community relations.

By the 1960s, blacks' campaign for equal rights was based on activist strategies reflecting the rise both of more aggressive black ideologies and of mass-based, grass-roots organizations (Bloom 1987). The narrow and more conservative, legalistic approach to winning racial equality taken by the National Association for the Advancement of Colored People (NAACP) during the 1940s and 1950s gave way to direct action and confrontational strategies by newer black organizations that reflected both the growing cohesiveness among blacks and the rise of black militancy.[3] In the South, changing attitudes helped in 1957 to create the Southern Christian Leadership Conference (SCLC) and, in 1960, the more militant SNCC. Based in the North, the Congress on Racial Equality (CORE) was known for its broad-based appeal to blacks and its unorthodox, militant methods of reform. These organizations used nonviolent and mass collective activities to disrupt the status quo and pressure white institutions they sought to change (Bloom 1987).

During the 1960s, protest demonstrations transcended issues relevant only to the South, and the movement focused on a broader range of economic issues (Bloom 1987; Newman et al. 1978). As militancy increased, black demands for better housing, education, and jobs spread across the entire nation.[4] And, as black protest shifted from the South to the North, sit-ins, shop-ins, and other forms of demonstration highlighted the need of blacks for broader economic opportunities. In the mid-1960s, for example, the NAACP started a nationwide campaign to open previously closed jobs to blacks, particularly higher-paying supervisory, professional, and managerial positions, and to admit blacks to specialized company training programs. CORE held demonstrations that, among other issues, pressed for black jobs in retail businesses, banks, and the construction industry (Bloom 1987:196). In short, blacks protested the unequal allocation of economic resources and demanded a black fair share of corporate contributions, jobs, and black-owned business procurement opportunities. The *Harvard Business Review* notes that Fortune 500 companies were now approached by the more militant minority-group organizations such as CORE, SNCC, and the Welfare Rights Organization (Cohn 1975:10).

Black Economic Boycotts

By the mid-1960s, southern business had already experienced the impact of boycotts in which black consumers withheld their patronage from white businesses (Bloom 1987). Given its success in the South, the use of black consumer power as a tool to negotiate for jobs and for black economic development in the North proliferated in the 1960s. In Chicago, for example, Jesse Jackson in 1967 organized Chicago's black ministers to support Operation Breadbasket, the economic arm of the SCLC, to find jobs for blacks in bakeries, milk companies, and other firms with products that had heavy black patronage (Colton 1989). In 1971, Jackson launched PUSH for the same purpose: to use black consumer power to increase the number of black jobs and black business procurement opportunities in white corporations (Jackson 1979).

Although the impact of black economic boycotts on the southern social structure is well documented (Bloom 1987; Sitkoff 1981), their effect on black jobs in the North appears not to be precisely known. However, economic boycotts by blacks arguably would provide enough motivation for companies to get involved in equal employment opportunity programs. Retail outlets and consumer goods manufacturers are particularly vulnerable to a black boycott of products (Haynes 1968). For example, a 1961 study by Roper and Associates commissioned by Pepsi-Cola found blacks drank more soft drinks than whites and were much more likely than whites to drink Coca-Cola. In response to this survey, Pepsi-Cola launched an advertising and public relations campaign to woo black consumers, which netted Pepsi-Cola almost $100 million in additional annual profit (Gibson 1978).

An argument for corporate responsiveness is also implied in the recollections of a black personnel director at a food manufacturer with high name-brand recognition. He noted that in the late 1960s there was an "extraordinary awareness on [his employer's] part that there was a black consumer market out there that [the company] needed to appeal to." During his tenure, he recalled, the company hired a black consulting firm that presented evidence to senior management of a large segment of black consumers who "never heard of mayonnaise, but who eat [brand name] like crazy . . . who have never heard of cheddar cheese, but eat [brand name] like crazy."[5] This man went on to say that, as a result of this heightened awareness of buying patterns, the company started a black marketing program. He said, "I hated that title, [but] that's what we called it at the time." The threat posed by economic boycotts is a consistent theme in the tales of corporate recruitment. A director of community affairs and district personnel manager described his management trainee position in 1966 as a response to Operation Breadbasket's threats of a black consumer boycott against a chain of food stores in Chicago: "The store had about 32 operations in the black community and nobody in management. They had a big black consumer base. So Jesse came in and got a [hiring] covenant. I got hired as a part of that covenant. Blacks started coming into the business structure at that time. I would say [the store] put at least twenty to twenty-five blacks into management training at that time."

A similar story was told by a man who had worked in the New York clothing industry. This vice-president, whose first job was assistant manager in a clothing store in 1966, was hired when the NAACP organized black consumers to take action against the retailer. Until that confrontation, he said, "it was well known in the black community that blacks could not get good jobs there," and blacks were absent within the management structure of the company. When he was hired, he explained, "managers were, you know, the guy would be the manager of the porters, or the elevator starters, or the kitchen. But there were no front-line blacks in buying or merchandising responsibilities. And black people spent a lot of money in [the store]. So that's how I got in — because of the pressures of the NAACP picketing."

A man who entered the white private sector in Ohio echoed his story. This director of human resources for sales operations explained that he was the first black college graduate recruited in 1967 as a management trainee in a Cincinnati-based consumer goods company because of the actions of the SCLC. Although his hire might have been merely a coincidence of timing, research indicates that black demonstrations against a single employer often prompted other employers to hire blacks as well (Meier 1967). He recalled, "Some of the companies were being boycotted . . . by SCLC and a coalition of ministers in the Cincinnati area. They had gone [after] companies like Kroger and Procter & Gamble and other highly visible consumer companies. Before they hired me there weren't any blacks and there wasn't a management training program. I was it."

By 1982, Jesse Jackson had signed more than a dozen moral (i.e., not legally binding) covenants with major corporations such as Coca-Cola, Seven-Up, and General Foods (Jackson 1979). Such willingness on the part of consumer goods companies to sign trade agreements, and the insights of the managers I interviewed, point to a sensitivity in the retail business to its market share and, therefore, its consumer base. In markets where blacks constituted a "sizable enough" proportion of sales, there is reason to believe that top management would respond to boycotts by increasing the number of blacks visible to consumers and upgrade some blacks to jobs in management. Of course, data on black employment comparing firms by the size and strength of their black consumer market would be necessary to support this point.

The Urban Crisis

Black urban unrest also appears to have helped motivate major employers to start programs to expand black job and economic opportunities. For example, one executive reported that his employer, a leading food company, faced tremendous pressure during the riots from black groups in different sectors of the country. Black organizations were demanding that the company create black ownership as the price of doing business in black communities. As a result, in 1968 the company set up an urban affairs program, headed by this executive, that generated two black-owned franchises in Chicago and one in Cleveland, firsts in the company's history.

Between 1965 and 1970 one-third of 247 urban-based Fortune 500 companies started programs that would expand black economic opportunities, "principally to help discourage boycotts, violence, and other threats to company well-being" (Cohn 1975). Piven and Cloward (1971) offer evidence that the private sector participated heavily in a strategy of appeasement by providing a multitude of services and technical assistance to black ghettos via community action, manpower-training grants, and urban redevelopment programs. To preserve a profit-making environment, U.S. businesses responded to the riots and the shock waves generated by the 1968 Kerner Commission's report by starting black economic development programs as one solution to the urban crisis (Cohn 1975; *Fortune* 1968; Henderson 1968).

The series of long hot summers of urban riots began in the mid-1960s and continued until the end of the decade.[6] With half of the black population in the United States compressed within the ghettos in large northern cities (Bloom 1987), private industry faced pressures in addition to those from federal legislation, potential profit loss, and bad public relations due to economic boycotts. The people I interviewed believed that companies also opened jobs in fear of the destruction of physical assets, such as company plants, located in or near volatile black neighborhoods. They attributed their enhanced opportunities for professional and managerial job opportunities during this time to the spiraling outbreaks of black riots in central cities. Two-thirds of thirty-five informants entering the white private sector after 1965 in managerial fields described their companies' vulnerability to black consumer boycotts and to urban upheaval. For example, a manager of community affairs for a public utility who was hired following riots in Chicago shared the following perspective:

> I'm not too sure that if we could we wouldn't [have taken] the cable out of the ground and off the poles [in] . . . north Lawndale which is black, and south Lawndale which is mainly Hispanic. Some of the employment programs [were] motivated [by] a desire to do good . . . but some of it was to develop good public relations as a buffer around the assets that the company has in [those neighborhoods].

Typically, literature on urban rioting looks at the relationship between racial insurgency and the expansion of social welfare and government programs for the black poor (Isaac and Kelly 1981; Piven and Cloward 1971, 1977). In contrast, my research suggests that riots helped to create white-collar jobs and economic programs for blacks in the private sector. Although a tendency exists to associate urban programs with the black disadvantaged and the hard-core unemployed, these programs just as often were outreach efforts by major employers to hire and economically upgrade a more advantaged black population. Urban-based Fortune 500 companies had more minority and urban affairs programs to recruit and train blacks who were both skilled and qualified to fill jobs than programs to employ or train the hard-core poor and unemployed (Cohn 1975:11): twenty-one companies made special efforts to recruit blacks directly into managerial and executive positions, while seventy-one others set up management-training programs for blacks already on the company payroll. In reaction to urban riots, companies in every industry group set up these new programs (Cohn 1975). A new corporate consciousness is also suggested in the account of a vice-president of sales who was the first black ever hired into management in a huge retail company. Significantly, he was hired in 1969, immediately following a period of rioting in East Cleveland during which a nearby competitor burned down: "Our store was not far from East Cleveland. All of a sudden headquarters called [and] said, 'We want you [for the management] job.' The store I was put in . . . was 50 percent black. So then it became very clear. [Question: That your employment was connected to the rioting?] That's right." (My question may seem to have loaded the answer, but by that point in the interview the man's condescension toward the actions of his employer had become quite apparent. I interpreted his truncated explanation as a shorthand used because he viewed me, a black woman, as an insider who would understand what he meant by "it became very clear.")

His interpretation is consistent with research reported in the *Harvard Business Review* (Cohn 1975). He viewed the increased black visibility in management as a public relations effort forced upon the company to stave off the destruction of its assets. It is also consistent with more recent observations of the effect of the 1993 Los

Angeles riots; this one from the *Los Angeles Times* is typical (Harris 1994):

> Two years ago, Joe Naphier, then a human resources manager on a McDonnell Douglas commercial airline project, stood on the executive floor with company President Robert Hood and watched Los Angeles burn. The two men talked. Two months later, Naphier, who had been with the company for 14 years and was then the only black general manager, was named to head a diversity committee. Four months later, he was promoted to ombudsman, a position last held by a company vice president. He got a nice raise as the first black face ever in that sensitive position . . . and has no problems telling you that the riots are the reason he's there. The paradox of riots emerges. Riots failed to benefit the social and economically disenfranchised, rather — in this instance at least — it helped a well-placed few.

Rioting has been interpreted as an attempt to draw attention to the needs of poor blacks, yet its result in the 1960s was the destruction of urban ghettos that never got rebuilt. Social policy responses to alleviate black upheaval in urban areas benefit blacks who were less, rather than more, impoverished by their environment — in this part of his argument Wilson (1978, 1981) is precisely right. For the poor, jobs are still hard to come by, the black family is still in trouble, and the life chances of residents of urban ghettos look dim and are darkening. The consequences of rioting seem much less ambiguous for the better-off, such as the group I interviewed. They are part of a new black middle class working in professional and managerial jobs once traditionally set aside for whites in private sector institutions.

It is interesting to recall, nevertheless, that middle-class blacks consciously connect their gains to disruption in the lower classes. A racial awareness exists within the middle class of a political connection that overrides some ideologies based on class divisions. Moreover, a deeper irony may be revealed over time. Various coalitions of middle-class blacks may, indeed, be self-serving, but blacks in the middle class are not necessarily a permanent, nor even the ultimate, beneficiary of this new social arrangement. Blacks benefited from a political climate and racially oriented policies that produced mechanisms (i.e., roles) to address black needs. Ultimately, however, these racially sensitive mandates alleviated political pressures on white corporations. Without such pressures, what will happen to these black personnel?

Chapter 5

Race Tracks and Mainstream Careers

✦ Tracking African American managers into racialized jobs was a strategy white companies developed during the 1960s and 1970s, when these executives' value became tinged with race-conscious political purposes. For example, ten of seventeen managers (59 percent) that I interviewed, all with highly technical skills as accountants, engineers, chemists, and so on, were asked to fill such jobs. Twenty-two of the forty-five people (49 percent) in my interview group who became affirmative-action and urban affairs managers started in line areas but were recruited for racialized jobs, twelve of them (55 percent) by senior level white management, usually either senior vice-presidents or chief executive officers. Nine (41 percent) turned down the first offer and were approached a second time by top management. Eleven (50 percent) were given salary increases, more prestigious job titles, and promises of future rewards. The push in companies to fill new administrative roles in employment and social policy areas cut across both personal preference and previous work experience. A midlevel manager in his forties comments: "It was during the early 1970s, and there weren't very many people around that could do anything for minorities. . . . I mean, . . . all the companies were really scrambling. All you saw was minorities functioning in [affirmative action and urban affairs] and it doesn't take much brain power to figure out that that's where most of us were going to end up." A white senior vice-president of human resources noted that top management deployed people from line jobs into affirmative-action jobs to signal the rank and file that the company was serious in its commitment. Transferring an experienced line manager into affirmative action increased the credibility of a collateral role and enhanced its effectiveness. From one perspective it would seem that employers either ignored these executives' education and experience or used it against them.

These executives are among those who rose to the top by managing affirmative-action, urban affairs, manpower-training, and technical assistance programs. Twenty-six of the seventy-six black managers I interviewed spent their entire careers in racialized jobs outside the corporate mainstream. Another twenty-five held one or more racialized jobs but eventually moved permanently into the mainstream. Only twenty-five had careers made up only of white corporate mainstream jobs.

This high concentration in racialized jobs is consistent with the notion of a politically mediated black middle class. It indicates the nature of pressures on corporations and shows the incentives for getting blacks into these areas. Conversely, this high concentration calls into question notions of a color-blind allocation of labor and of a color-blind market demand.

To obtain a rough comparative measure, I conducted an informal survey of top white executives by asking twenty CEOs of major Chicago companies if they ever held affirmative-action or urban affairs jobs. (I asked about these jobs specifically because they typify racialized jobs.) Some seemed startled by the question, and only one reported having worked in either area, a CEO whose tasks in urban affairs fell in a different category from those performed by my black interviewees. Although this man represented the company on several citywide committees to improve race relations, his job, unlike the black executives I interviewed, was a part-time and temporary assignment, not a full-time and permanent position. The results of my informal survey suggested that among the managerial elite in Chicago, blacks are likely to have held racialized jobs, but whites are not. Moreover, just 5.9 percent of 698 respondents to Korn/Ferry's (1986) survey of corporate vice-presidents, senior vice-presidents, executive vice-presidents, chief financial officers, and group vice-presidents held positions in personnel or public relations departments, where companies tend to house affirmative-action and urban affairs jobs. More typically, the track to top jobs in companies includes profit-oriented positions such as sales, operations, and, more recently, finance (Korn/Ferry 1990).

African Americans in Mainstream Careers

Mainstream careers are grounded entirely in jobs with goals oriented to general (i.e., predominantly white) constituencies, not jobs produced in response to black protest and subsequent social policy. The career of a forty-year-old vice-president and regional sales manager for a Fortune 500 company in the manufacturing and retail food industry illustrates a mainstream work history. The executive holds a two-year college degree in natural sciences. When he entered the private sector in 1960 as a market researcher for a Fortune 500 East Coast oil company, his job involved marketing to the total (predominantly white) consumer market, not to "special" (predominantly black) markets. In 1968, he accepted a position as a salesman with his current employer, and, even in the midst of the civil rights era, he was never assigned to a black territory. He moved up through the sales hierarchy from salesman to sales manager, zone manager, district manager, area manager, division manager, and, eventually, to his present job in the company. Throughout his ascent, he was never responsible for a predominantly black sales force or for strategic marketing to the black community when he managed geographical areas. His current employer once offered him an affirmative-action job in personnel, but he declined the offer because of negative experiences in a similar, but unpaid, role thrust upon him by his first employer. Moreover, he perceived affirmative-action jobs as lacking power in the company.

Another illustration of a mainstream career executive is a highly recruited woman with an MBA from the University of Chicago who was a vice-president of investor relations. She entered the white private sector in 1968 and worked her way up in banking through a series of financial assignments. In 1984, she was recruited for an assistant vice-president and director's position with a leading food manufacturer in Chicago where, she said, she "could be a part of the management team." In 1985, she again was recruited, this time by her final employer in the private sector, and became a full vice-president and company officer. By 1993, she had left there to start her own business. This woman was never asked to implement, nor did she ever manage, programs related to blacks. (I asked, for instance, if she had participated in any bank program designed to give financial advice to black organizations, if she had ever consulted primarily with black consumers or investors, or if she had administered any Small Business Administration minority business start-up programs sponsored by the bank during her tenure.)

Mainstream executive careers represent the affirmative-action ideal, as opportunities for talented blacks to compete for power and prestige in business bureaucracies. Yet mainstream African American executives stand out as the exceptions. How were they able to avoid racialized assignments? One explanation is that their employers viewed them as too well trained to shift out of the mainstream. The banking executive's MBA from a prestigious school undoubtedly made her a unique commodity in 1968 relative to other black job candidates. Yet level of education has not proven a good predictor of career track. Executives I interviewed at each level — bachelor's, master's, and doctoral — had at least a fifty-fifty chance of getting a racialized job.

Another possible scenario is that companies filled affirmative-action and urban affairs jobs on an ad hoc basis. That is, when the need to develop programs arose, a company first looked in-house for black candidates to fill those positions. Conversely, if racialized jobs were already filled, mainstream people were more likely to stay mainstream. Each of the three firms the MBA worked in had affirmative-action and urban affairs programs in place when she came on board. Ten of twenty-five in mainstream competition also reported such programs in place when they were hired. In addition, the vice-president and regional sales manager offered his opinion that no pressure was exerted on him to manage affirmative action because another black professional in the company subsequently agreed to take the offer.

However, blacks' ability to move up in the mainstream of a company did not mean their careers evolved free of the influences of job discrimination (see Fernandez 1981; Jones 1986). Nevertheless, these two African American executives and others who built mainstream careers were extremely successful relative to most African Americans (and most whites). Both the MBA and the vice-president and regional sales manager made six-figure salaries. Both were officers in their respective companies in the 1980s.

The African American Mobility Trap

Most of the executives I interviewed (51 of 76) moved into and stayed in, or moved through, racialized jobs that also created barriers to corporate mobility. For example, a 46-year-old man who I'll refer to as the frustrated manager was initially hired by a major steel company in the 1960s for a job administering a federally funded in-house program for disadvantaged youth. Funded by the U.S. Department of Labor, the program was designed to train, or retrain, predominantly black Chicago youth in skills that would qualify them to work in Chicago's white private sector. The frustrated manager, who was a social worker employed by the city of Chicago before he was hired by the steel company, said the company identified him as a candidate for this position through his work with inner-city youth and gang members. He recalled being told during the initial interview with the personnel manager that his active ties to Chicago's inner-city youth, and the implication these ties had for program development, was a key reason the company was interested in hiring him. His response: "I told [the personnel manager] that I don't want a nigger job and I don't want to be dead ended. That's the job I didn't take. But I saw some value in the manpower training because it was an inroad for minorities and females."

The frustrated manager defined a "nigger job" as corporate "positions preidentified for blacks only. Those jobs have high-ranking titles and are highly visible but do not have any power in a company. Those jobs are not with the mainstream of [a] company [so they] would [not] turn into any kind of career with the company. [In contrast] real jobs [were positions by which] good performers could rise [in a company]." In other words, people in "real jobs," but not "nigger jobs," performed valued corporate functions and were thereby able to move up the corporate ladder. This man seemed to be ambitious and intense; he punctuated his recollections about his job interview by pounding his fist on the boardroom table. He recounted his career with pride in his voice, but also a hint of ambivalence about how far he had been able to go. He believed that because manpower-training programs served the disadvantaged (i.e., black) population, they created a need for skilled black labor, which gave him a chance to move into the white private sector in an administrative slot and to develop, he said, a "different kind of work experience." But, although he was interested in the job, he negotiated, he said, "up front so that [the company] would not dead end" him. He did not want this to be one of those "jobs in companies [where blacks are left] to die on the vine."

He also observed that corporate jobs administering manpower-training programs, which sprang up in the wake of urban riots, were similar to the training programs themselves: Both were vehicles corporations used to bring blacks into the private work force. However, he took the job because he perceived it not as a dead-end position but as a valid step on a career path leading to personnel director. He aspired to a career in personnel over other functions such as production, because he believed that "black people [at that time] had no involvement in managing [other] parts of a corporation." He also said, "I'm a people person and I knew I wanted to be a director of personnel. This was a way to get there, [and] that was agreed upon in that interview with [the personnel manager]."

It is doubtful the personnel director of the steel company in this case seriously viewed the manpower-training job as a route to a director of personnel job. Although the company explicitly recruited this manager because of his networks in the black community, implicitly management may have chosen him because they needed to hire a black person. The company was based in a riot-torn community, and the manpower-training program was part of an effort to improve the company's poor record of employment and training of African Americans. According to this man the company at the time employed only one other black professional. Apparently no one on board would, or in the company's judgment could, fill the position of manpower program director.

After the frustrated manager had the manpower-training program for two years, "very successfully," the company created a job for him as a community relations representative, part of a move "to institutionalize the [minority] manpower-training program within the organization" after federal funding ended. Two and a half years later, he moved into a second newly created job, community relations director, a promotion in both title and salary. Still proudly, he attributed his career mobility to his success in employing residents from the surrounding black community.

Although, according to the frustrated manager, the company was willing to promote him and increase his salary because he was meeting a need, his promotion did not move him into a mainstream personnel career. When asked what his last promotion meant in light of his original career goals, he admitted he was aware, even then, that his future in the company might be limited. "You have a little step-ladder . . . a logical progression [of positions] you have to go through if you are to ever become a personnel director. I wasn't doing any of that. As far as I could see, the company wanted black folks to be my only responsibility." The manager reminded his superiors that his career goals lay elsewhere, but he was not deployed into mainstream personnel. He therefore viewed his movement within the company as promotions "in place," evidence that he "was not really experiencing mobility in [the corporate] structure." He also believed his ability to accomplish the job he was in was limited because of conflicts with the personnel director. He said, "My problem [in developing a good minority recruitment program] was not one of identifying qualified minorities, but of stopping discrimination among those with authority." An important part of his job, he pointed out, was to identify, and attempt to correct, irregular hiring patterns for which the regional director of personnel was ultimately responsible. "[Whites] were hired . . . who did not have high school diplomas, who could not speak English [and] could not pass any of those battery of tests that they give to the blacks or the women." When the program to improve minority hiring "worked out well, [the director of personnel] was the person who was embarrassed. He was the person who was made to change."

Believing his in-house options were limited, the frustrated manager volunteered to be a loaned executive on a citywide corporate project engaged in community and economic development. He felt this route "represented more training and [potential] mobility." In the early 1970s, after leading seminars on community relations for the steel company, he was courted for a personnel job in ten other companies. He decided to start over with a new computer firm and try, once again, for a director position. Yet, ironically, "community work" was involved even in his subsequent jobs in the computer field because those companies also were "not doing well in the recruitment of minorities."

I was unable to reinterview this man in 1992. However, the person who originally referred him to me reported that he had been transferred to California, was "still doing community affairs, and was ready for an early retirement." Secondhand reports are often suspect, but the two are friends, work for the same company, and talk with one another often. Our mutual acquaintance also said that their recent conversations about their current careers increasingly focus on personal disappointment and missed opportunities.

The frustrated manager, like one-third of the executives I interviewed, gained status in the private sector by filling corporate positions linked to black constituencies. At the same time, filling race-based roles effectively locked him out of conventional routes up the corporate ladder. Several facets of his career in the steel company converged to keep him out of the mainstream. First, he performed well in a position that, at that time, was valued in the company. Top management rewarded him with higher salaries and kept him assigned to that area. Consequently, he was excluded from experiences that would broaden his mastery of more generalized personnel functions. The lack of generalized experiences further undermined the legitimacy of his claims for promotion to a mainstream personnel job. Moreover, when he identified hiring biases in the company, intentionally or not, he criticized and alienated an important potential mentor, the regional personnel director. Finally, if he habitually used the word "nigger" in a conservative corporate environment it may have further minimized his chances for promotion. The provocative nature of that racial epithet could cause his image in the company to suffer and increase his chances of being left behind. He used the word often during our interview, and I wondered at the time if he used it with the personnel director.

Career-Enhancing Strategies

During the 1960s and 1970s, the executives studied here either began their careers in mainstream jobs or wanted to exchange racialized jobs for mainstream assignments. About half who wanted to move into the mainstream (25 of 51) achieved their goal. By 1986, executives who had left racialized areas had been out about nine years on average. They were distinguished from those who remained in racialized jobs by falling into one of two categories. One group was able to decipher the rules of the game by seeking mentors and other sources of information about meaningful career routes. They used that knowledge to generate career-enhancing moves, which included requesting alternative assignments. In the second category are workers for whom racialized jobs inadvertently became a career springboard toward solid — mainstream — ground.

Reading the System

Fourteen of the twenty-five who escaped the racialized sphere during the 1960s and 1970s simply asked to be reassigned. In contrast, only seven of the twenty-six who stayed in racialized jobs during this time asked to be reassigned. Virtually all who requested reassignment perceived both the trend in corporations to have black managers fill affirmative-action and urban affairs jobs and their potential to limit opportunities in the long run of corporate life. The group's collective consciousness is summed up in an observation about people who turn this type of job, made by a vice-president at a major electronics firm. This man started one affirmative-action program for a company and turned down several similar job offers from other companies. Referring to his stint in affirmative action, he said: "They would send me to some of these conferences [and] . . . you'd walk in and there would be a room full of blacks. . . . And I met titles, . . . directors and you name it, of equal employment opportunity. It was a terrible misuse at that time of some black talent. There were some black people in those jobs that were rather skilled, much like myself."

He recognized such job ghettoization as a race-related mobility trap for black managers. "During the 1960s and 1970s blacks in these jobs had fancy titles, but basically they were in dead-end positions," he said. Black executives approached to take such jobs faced a career dilemma. They believed they had to be committed team players to get ahead in the company, and at the same time they feared that if they succeeded in affirmative action or urban affairs, top management might never transfer them.

A forty-six-year-old sales vice-president expressed the first side of the dilemma when he explained, "In this company you don't turn down requests when they come from a senior vice-president, and especially when they look like a promotion." When he was asked to fill affirmative-action positions, however, he said he also "just wanted it to be up front" that he "didn't intend to keep that job forever," that he thought it necessary to set time limits on his transfer. Expressing the second side of the dilemma, he said, "You can do those [affirmative-action] jobs too well [or top management feels] . . . this is where you need to be." The sales vice-president mentioned earlier concurred with this observation. "All we had to do was to look at blacks around us to come to that conclusion."

Although they risked appearing recalcitrant, managers headed for mainstream careers stipulated time limits when taking racialized jobs and, once in these jobs, assertively requested reassignment. Many were eventually mainstreamed. The sales vice-president, for instance, transferred into an affirmative-action job from floor sales, but also set the stage for his mainstream reentry by negotiating a one-year limit on this placement before accepting the position. The company honored its promise and moved him out of affirmative action and into a buyer's position.

A current vice-president of human resources for a clothing manufacturer also requested a transfer from affirmative action because he recognized the limits of the job. This man transferred into affirmative action in 1967 when he worked for a federal contractor-operator of a huge munitions plant in southern Indiana. Although he was trained in and working as a research chemist, the personnel director approached him about transferring to personnel. He was the only black professional in an operating environment that was facing intense federal scrutiny. At the time he was approached, "the federal government had the big push on government contractors to do more for affirmative action," he explained. The plant was particularly vulnerable because it was wholly supported by millions of dollars in federal contracts. "A lot of the federal investigators were black," he observed, "and . . . so if anybody had a chance of staving off all kinds of repercussions, then a black probably had a better chance."

The job came with a salary this manager characterized as "a pretty good deal." Yet he agonized over taking the offer because he would be leaving "the laboratory," he said, for a job he "wasn't trained for."

To his surprise, personnel proved to be his true niche, and he went on to develop the plant's first affirmative-action program. But in 1970, he was worried about the vulnerability of his career track. He reasoned that affirmative action, as a field, was transient and did not offer technical skills that would allow him to branch out in a company. Affirmative action "is the kind of field where . . . a few laws might change, but the concept doesn't. Once you know those [laws], there's not an awful lot more to learn." Like the frustrated steel company manager, he realized that this avenue could diminish his ability to manage non-race-related personnel areas.

He also expressed the dilemma of becoming too successful. "My biggest concern was I was going to end up becoming the guy who handles all the EEO problems in the corporation. I'm thinking, [even] back in the early 1970s, . . . that if something ever happened to affirmative action, where it wasn't popular anymore, I wouldn't have any other marketable skills." Consequently, in 1972 he asked for, and moved laterally into, the mainstream job of personnel manager.

Twelve of the fourteen executives who asked to be reassigned to nonracialized positions pointed to the role of mentors and, more significantly, to their own fact-finding efforts as fundamental elements of their eventual transition into mainstream jobs. Those who stayed in racialized jobs read the business environment narrowly by observing blacks around them. But those who exited these jobs were more cognizant of the overall structure of a particular corporate hierarchy. They educated themselves, or found mentors, and assessed the corporate environment early in their careers, which enabled them to identify career-enhancing moves in their firm. For instance, a vice-president in a communications firm said that two years into his job, he knew "you can spend all the time you want . . . in personnel, and public relations, and that kind of stuff. But you aren't going to be a vice-president of this company, or president of this company without [going through] operations." Seventeen of twenty-five people who left racialized jobs specifically analyzed which routes led to the executive suites in their company. A vice-president in the communications industry told me, with a hint of condescension at having to state what to him appeared obvious, "All you had to do was look at who ran the company and see what areas they came from. I just observed who was sitting where. I looked at those yellow bulletins because they announced organization changes and because they also give personal bios."

Using Racialized Jobs as Springboards to the Mainstream

Executives I interviewed who had turned functionally segregated assignments to their advantage often found senior management mentors who trained and propelled them into core corporate positions. A vice-president of operations who has done postgraduate work in physics and engineering, for example, between 1968 and 1972 was an equal employment opportunity manager. The employee relations director at his first firm approached him to set up the company's affirmative-action program. He had been with the company eight years and, he said, "I wanted to get into management. That was the first and only opportunity that I felt I was going to get."

He said he believes he was approached for this role because "there was some concern that if you put the wrong guy in there that he would just raise all kinds of hell. And what safer guy could you get than someone who's sitting in the engineering department?" (Also, the company had only one or two blacks in professional jobs, and none in personnel.)

The company may have chosen this man merely because he seemed to have the requisite interests. During his off-hours he volunteered with black community agencies and often attempted to convince the company to donate funds to community projects in black areas. The firm, an aerospace company that subsisted on federal contracts, was extremely vulnerable to federal oversight. But in 1968, when the federal government's approach to contract compliance was relatively untested, the company was still unclear about affirmative-action mandates and its own direction in developing programs. As this executive put it, "Nobody knew how to do it, but everybody knew it would have to be good."

The company's dependence on federal contracts and its consequent emphasis on compliance with affirmative action made his job highly valued and anointed him with a status he knew no other black in the company had. He received "a job with a manager title, exposure to the company's inner works, and visibility to the corporation's top people." Yet he approached the job as if it were a stepping-stone rather than a permanent stopping place — a chance, as he put it, to "let me get my nose somewhere" and "to get something for myself." As he developed the affirmative-action program in conjunction with a senior vice-president in the company, he did get something for himself — a powerful and active mentor. Up to that point, he said, "essentially I lacked what most blacks lack . . . sponsorship. I was totally on my own." He credited the sponsorship of this executive vice-president with his ability to turn his career around. "He was a white guy who . . . got to know me. Supported a lot of things that I wanted to do. And said, 'You know you've got a lot of capability, and it's a waste to keep you here in EEO. So I want to send you back to [Massachusetts Institute of Technology] because I think when you come back we can get you ready for a senior management job.' "

After administering affirmative action for five years, this man entered MIT's Sloan School of Management executive-training program. Soon afterward his mentor retired. When he returned from MIT, he expected the company to reward his achievements. "I was looking for a position where I could eventually do something, where I had some power." But offers for that type of position were not forthcoming. "I had forgotten I was black." Moreover, his mentor's protection and advocacy were gone. Despite his postgraduate work and success as an MIT-Sloan executive fellow, the offers that he received from his employer and, he stressed, from "so many other companies were to direct affirmative-action programs."

He said he heard about "all those guys I'd been with [at Sloan], . . . and their promotions." He believed that, at minimum, he should have been promoted to company director. After a lengthy pause, he said "[I] was screaming inside my head; I was hurt terribly by the [job offers]." But in spite of his initial disappointment, the training at Sloan (and, I would argue, the existence of federal anti-bias legislation) helped him to redirect his career. He told his employer, "No, hell no," he said. "I wouldn't even be interested in a job making decisions between black and white cars, let alone black and white people. I'm through with that. I've done my share." He demanded a line job, he said, because he "was no longer so naive to be seduced by title or salary." He subsequently became a project manager in his original company, staying with the firm for the next two years.

There is more bad news, however, and it is grounded in a by-now-familiar dilemma. In his efforts to achieve affirmative action as the company's EEO manager, the interviewee had alienated the man who later stepped into the executive vice-president position vacated by his mentor; he had "embarrassed" this person and "shoved something" he did not want "down his throat," he said. When he became a project manager, this adversary became his direct supervisor. His opportunities within the company stagnated once again.

One way to interpret the careers discussed thus far is to view them as blacks' struggles to succeed in a world where they are anomalies. In this context, blacks' unique status worked both for these people and against them. They were approached for racialized jobs because their visibility brought them to the attention of senior management. Such jobs were both potential springboards for, and hazards to, entering mainstream corporate training and competition. With an eye on both factors, they negotiated conditions for taking affirmative-action and urban affairs positions. In the case of the MIT-Sloan graduate, the racialized job was the vehicle that gave him a new chance — a mentor — and the year of additional specialized and prestigious training. Yet playing racialized roles also pigeonholed this man, inviting new affirmative-action job offers, creating a powerful enemy, and robbing him of a new mentor. Ultimately, however, occupying a racialized job created a window of opportunity, which this man used well. He eventually got other offers and went on to become a vice-president in a major firm.

In his first jobs, from 1963 to 1968, a vice-president and central region manager in a consumer goods industry he developed special markets successfully for two Chicago companies.

Although these special markets directly addressed each firm's economic initiatives, they were not a springboard from which blacks were expected to gain organizational power. Nevertheless, by 1968 this man's outstanding sales and performance records prompted senior management to create a new position for him as a special-market sales manager. His upgrade, while earned, was considered a radical innovation for this employer. His new position made him the second black manager in the company; the black executive who predated him ran urban affairs. He said that during this time, "I always wanted to break into the mainstream market but . . . I didn't even try. I just tried to do the best I could in that particular area." In 1973, five years after becoming special-market manager, his aspirations to participate in the mainstream came to fruition when the vice-president of sales promoted him to central division manager. From that point on, he moved quickly and steadily up the mainstream corporate ladder.

When asked what he thought influenced his ability to enlarge his racialized career, he recalled learning "a host of bottom-line functions" that eventually led him out of special markets. "I thought it was positive," he said. "It allowed me to learn the business — distribution, pricing, taxation, expenditures, promotional activities." His optimistic view of his opportunities, however, obscures the hiring discrimination that influenced his assignment to sales territories dominated by black consumers. He gives his early placement the gloss of an apprenticeship when he would have received similar — or possibly broader and better — training had he been given general-market assignments. But he was making his assessment based on what was possible for him at the time. When he was hired by this Chicago company, white companies simply did not offer blacks sales jobs in white-dominated geographic areas.

On the other hand, unlike the case of the frustrated manager in the steel industry, the same vision and skills this man demonstrated in special markets were recognized as valuable to the broader, mainstream sales territory. Although the timing of his recognition may have been an effect of governmental activity or some other race-related corporate considerations, his special-market assignment spotlighted his performance and prompted upper management to risk moving a black employee into a traditional (white) sales area. As he put it, "The vice-president of sales felt that I was good enough. He brought me in and really started teaching me the business."

The career path of a finance company's senior vice-president is a third example of converting a racialized job into a mainstream trajectory leading up the corporate hierarchy. In 1970, this man was transferred out of an entry level position and into a management role in the company's guaranteed loan program designed to assist small and minority businesses. The position had many pitfalls because it involved screening and lending capital to the most difficult, and economically vulnerable, customer base — owners of small black businesses. Thus when this manager generated profits, his success was both surprising and noteworthy. "We pulled it off," he said. "We made some good loans, and some folks became quite wealthy because of [them]. We had some real winners. We were doing so well that we got a pretty good reputation." As a result of his performance in lending, he was promoted into a commercial area he characterized as a "big hitter" and "for whites only."

When asked to give some background on this promotion, he said he viewed it as a natural progression in line with the job he had. His response assumes that his performance running the loan program created the perception among top management that he had the necessary skills to fill the higher position. Significantly, at about this time, a higher-placed executive took an interest in him and eventually became his lifelong mentor. In a tangible way, his racialized job in a profit-generating area gained him visibility and recognition that led to a permanent position in a formerly all-white domain.

Both this man and the special-markets manager discussed earlier became the first black members of all-white senior management teams. In both cases, their initial, and racialized, assignments became launching pads for mainstream careers. This man's client and business skills were perceived by top management as transferable to and, more important, better exploited in a mainstream profit-driven area. In addition, each had a mentor who played a crucial and active role in his ascent. Also, both mentors stepped forward after the workers excelled despite the limitations perceived to be inherent in their racialized assignments.

Golden Handcuffs and Social Obligations

Like the frustrated manager in the steel company, eighteen of the twenty-six workers who stayed in racialized jobs during the 1960s and 1970s aspired to mainstream positions at varying points in their private sector careers. Why, then, did only seven of the twenty-six request reassignment into core corporate functions? One theme among people who stayed in racialized jobs was that their ambitions were shaped, and sometimes thwarted, by their racial identity and a sense of racial solidarity. A second theme was that their ambition, compounded by their lack of practical knowledge about constructing corporate careers, made them easily seduced by racialized jobs and the corporate perquisites that often accompanied them. One manager who worked in sales before he took on affirmative-action responsibilities reminisced in a voice filled with irony that the move "was supposed to be an honor."

Individual Activism and Group Commitment

Racialized jobs struck a familiar chord with the executives who stayed in them, because they addressed social and political issues that permeated blacks' existence. The problems of black people were problems these men and women both lived and intellectually understood. Most were part of the first wave of blacks in the white private sector to benefit from civil rights pressures. They acknowledged their debt to the decades of struggle and expressed a strong obligation to "give something back" to the black community. Thus, when offered a job intended to assist blacks, some welcomed the jobs as their chance to help channel previously withheld resources such as jobs and contracts back into the black community. Once harnessed to the job by a commitment that linked their sense of group obligation with the administrative interests of a company, they became pigeonholed and entrenched in a racialized area. The recollections of a fifty-five-year-old director of affirmative action reflect this dilemma.

He began his private sector career in catalog operations management for a retail company in the mid-1960s. In 1969, in response to his complaints about the lack of management opportunities for blacks in the firm, the CEO asked him to set up the equal employment opportunity section. In 1992, he was still head of affirmative action for his third employer. Noting the disparity between where his career started and where it had gone, I asked him why he did not move back into a mainstream function. He mentioned a variety of factors.

I was trained for the operations end of the business but stayed in the [affirmative-action] end because I felt . . . I had an obligation to do something [for other black people] about that time, you know. Prior to that I really wasn't doing anything other than helping myself. But then in the 1960s people became really motivated . . . "black is beautiful" and the whole concept. And [the job] provided me with an opportunity to make a social contribution and also to continue to be successful in business.

At that time, he idealistically viewed affirmative action as beneficial to the business world because it changed the business environment. He also viewed it as a business contribution that generated resources and opportunities for less-advantaged black people. With pride he recalled that he was a member of one of the first Chicago coalitions to aggressively "address the issue of hardcore unemployment." He also offered what he considered a critical distinction between his generation of black managers and the succeeding generations of black business people: The new black managers "do not have the same sense of social responsibility that we had, . . . [they don't feel] an obligation to those back home" — that is, to other black people.

Paradoxically, he also believes that taking on the affirmative-action position was a misstep in his career. "If I had to go back and do it all over again," he said, "I would not stay in affirmative action. Them that brings in the dollars is where the most opportunity is. I advise my sons, . . . stay out of the staff functions, although those functions are very necessary." He went on to name people who took different routes, whom he views as "making it."

This manager's dilemma raises the question of whether, in the context of white corporate culture, some blacks' sense of group solidarity and responsibility worked against their own aspirations and mainstream development. Indeed, those who did permanently transfer out of racialized areas viewed such jobs as stigmatized in the corporate culture. Moreover, to hasten their departures, some quickly pawned their jobs off on black replacements who, they said, "cared more," or who had "closer ties to the black community."

Others who stayed in racialized jobs explicitly commented on the trade-offs they made between individual aspirations and community commitment in accepting racialized jobs. The case of a forty-five-year-old director of an urban investment program for a major consumer company provides one illustration. Before entering the private sector, this man spent a brief period as a professional basketball player and learned that he enjoyed interacting with people. Consequently, when he was hired in the early 1970s by a large insurance agency he requested a position in personnel. From that point his jobs and experiences evolved smoothly in mainstream personnel functions, and he let it be known that his eventual goal was to become a regional personnel vice-president.

Several years had passed when a senior vice-president in the company offered him a position as division personnel manager. Although the title does not signal a racial component, the job was, in fact, a specialty position focusing on strengthening the company's affirmative-action program. Like the frustrated manager in the steel company, this man knew that to progress up the corporate ladder in personnel, he needed to remain in and successfully carry out a series of mainstream functions. However, he took the racialized job offer: "I felt that if I had to make a choice, affirmative action for me was the most important thing. Because the rest [of the company's personnel functions] were going to take care of themselves." He subsequently moved in and out of racialized positions and occupied one when I first interviewed him. In his attempt to balance his community commitment and his career goals he repeatedly made career decisions he knew worked against his career interests. He knew, he said, that "people would perceive me as someone who was not being developed," and that, while he deeply desired it, he "probably would never become a regional vice-president." After a brief pause, he added, "Because I'm black, the company did probably use me to a certain extent." Yet after he made that comment, his race consciousness seemed to lead him to a different and intriguing conclusion: "But I didn't . . . mind using myself, if I could get more . . . black folks into the company. So it's hard for me to know what the truth is. Who's using who, and how much?"

He wasn't alone in his thinking. Fifteen of twenty-five executives who stayed in racialized areas also mentioned their sense of obligation to use their presence in the company to make a difference and to give something back to the black community. For example, an urban affairs director explained that staying in his job was his duty and his "way of civil righting back then." A director of corporate contributions explained his position by saying that "community affairs had been done in a little different way . . . by two other people who were white community-activist types. They were doing the job as agents of the company. And, while I'm sure people would say that I was doing the same thing, I didn't see it that way. I saw it more as the other direction. And [I think that I made] some breakthroughs in that area." A manager of government affairs noted, "We had a small group of about eight of us that met. We were all community relations managers for major companies. . . . We'd meet informally for lunch because we felt that our role was to facilitate some progress in the community and we couldn't do that if we didn't talk together and make a solid front."

Golden Handcuffs

But while a commitment to racial group solidarity was one motivation to stay in racialized jobs, less altruistic and more materialistic incentives to stay existed as well. *Golden handcuffs* refers to situations in which persons cannot take advantage of alternative opportunities because the cost of lost benefits would be too high. The power of golden handcuffs was summed up by a consultant who in 1968 moved into affirmative action from his job in accounting. "I sent the company signals," he said. . . . "I wanted to go back [into accounting] . . . but they do what it takes to keep you satisfied, salary increases and what not. So I stayed with this job." In other words, the material and psychic rewards associated with these jobs eventually proved to be traps.

Those in racialized jobs were greatly rewarded because they handled new and unpredictable contingencies facing companies. For example, more than 80 percent of first racialized jobs were created when these people filled them. Because these functions were new, top management looked to them for guidance and direction in shaping these areas. In turn, the ability to provide guidance gave these managers a unique status in companies. Managers who filled these jobs were on a first-name basis with corporate CEOs as well as with black nationalists from the streets, and their ability to walk between these conflicting groups made some of them favorite sons in corporations in the civil rights era.

In contrast to those who left racialized jobs, perceiving them as a barrier to their quest for assimilation, people who stayed in racialized jobs perceived them as their chance to become key players in shaping the goals of mainstream corporate life. It is a view similar to that held by the vice-president in the electronics firm when he described the EEO job as his chance to make it in the company and become a manager. The distinction, however, is that he saw the job as a stepping-stone; people who stayed in racialized jobs perceived them as powerful.

Among these was a fifty-two-year-old man who in 1986 worked for the same private sector employer that recruited him from his job at the Urban League in 1967. In both his first job as manpower coordinator and the job as district personnel manager that he held when I initially interviewed him, his function in the company was, he said, "always something relating to the black community." He "examined issues in the community," was "used as a negotiator," and generally "fixed problems with blacks" whenever they existed. Initially, these assignments fit in well with the overall direction he had envisioned for his career.

> My experience, both professional and in life, prepared me for that role in the company. That's what I do best. And I don't want to, didn't expect to, do anything more. The way I cajole myself, perhaps, is that somebody was going to do that job. And I felt that I would do that job with a sense of the community in mind. Not just because it would be a good job to earn money, and get you in and among people of ilk. Although it does. [Question: What do you mean by people of ilk?] I'm talking about people with money, and influence, and class.

He then paused and chuckled, as if reflecting on other incentives that influenced his career deliberations. Perhaps he was struck, as I was, by his use of the term "cajole" as an apparent synonym for "fool" and wondered if altruism was the only motivation shaping his decisions. His next comment revealed an equally significant reason he agreed to take the role.

> I made more money than I had ever dreamed of. Ordinarily, . . . I probably wouldn't have very much to do with people of ilk. [Not] if I was in some other kind of job. . . . Socially, that is not my place. You see, I represent a billion-dollar business here when I'm out there. And [there are] . . . not many places out there that are not open . . . to me. And that was one of the things that I discovered here. That's why I stayed.

Others echoed his sentiments, saying they stayed in racialized jobs in part because of their perceived benefits. During the 1960s and 1970s, they viewed such jobs as offering faster advancement, greater freedom and authority, and higher visibility and access to white corporate power brokers than the mainstream jobs of their black contemporaries. This new and distinctive social status also gave them an aura of power and prestige in the eyes of their black peers.

When I asked the affirmative-action director who advised his sons to stay out of staff functions if he ever tried to move back into the mainstream after he took on equal employment opportunity functions, to my surprise he said that he turned down a buyer's job offered to him by a vice-president in merchandising with his first employer. "I was stubborn at that point," he said. "No, I didn't want that." Given that buyers were key people in that organization and that the job was a stepping-stone to higher-paying positions, his refusal signals the attractiveness of racialized positions in companies during the civil rights era.

> Remember now, this [equal opportunity] stuff was exciting and there's a trap that you get into. Those of us who are in this kind of area talk about it all the time. It's kind of a golden handcuffs trap. We used to go on the convention circuit around the country . . . the Urban League and the NAACP, promoting our individual corporations. We were visible. We were representing the

company. We had big budgets. I mean, you know, you go to every convention. And [you can] get yourself two or three suites and entertain all the delegates. You could spend $15,000 or $20,000 at a convention. I never had that kind of money to spend, to sign a check, so it was very attractive.

To fully appreciate this manager's perspective, we should also remember that the economic rewards and social status that accompanied racialized positions were unimaginable luxuries to most blacks — in this or any employment sector — in the years preceding federal fair employment legislation. Those who stayed in racialized jobs were ambitious men who saw themselves doing the best they could, given the limited job possibilities blacks historically had in white companies. They weighed the jobs' perquisites against the career stagnation common among the handful of blacks who had previously attained management roles and remained trapped in low-level positions.

In the 1960s and 1970s, these people thought racialized jobs were their best opportunities for social and economic advancement. As a fifty-three-year-old director of corporate contributions, then in his twenty-third year in a racialized job, told me, "That was the place for us to be." Many now see the downside of that decision. With the benefit of hindsight, the affirmative-action director explained, "I believe that had I stayed in operations [I would have] continued to move up, and that's where the clout is. But the opportunity just wasn't there [for blacks] when I first started with that company." After a slight pause he added, somewhat ruefully, "Things changed, and it is now."

Comments on Mentorship and Role Models

Role models and mentors may have changed the course of these men's careers. The white corporate milieu and collegial relationships with whites were new and mysterious to many people I interviewed. Only twenty-six of seventy-six came from families where at least one parent was a professional, managerial, or sales person. Moreover, only two of the twenty-six had a parent, close relative, or friend who had professional work experience in a major corporation. Half had graduated from all-black colleges or universities. Since these executives, like most blacks, had not been exposed to the white corporate world, they had no one to help them decipher the rules of the game. Thus, historical restrictions on blacks' access to white corporate culture played a role in shaping their managerial career preferences.

Those who stayed in racialized jobs were as ambitious as those who got out of them; indeed — and ironically — ambition was a large part of the reason they stayed where they were. During the civil rights era, racialized jobs made educated, ambitious blacks company stars. Paradoxically, the most attractive features of the jobs, such as starting titles and salary, freedom, and visibility, for some of them also diluted their desire to move into the companies' mainstream areas. But with black role models, aggressive mentors, or more knowledge about company hierarchies, would those who stayed have made different career choices? Did they need mentors and role models to perceive alternative career options as truly possible? The district personnel manager who spoke of representing a billion-dollar business may, in retrospect, be satisfied with his career choices, but he now also understands that visibility among white business elites was not the same as power. The longtime affirmative-action manager now knows that, for corporate success in the long run, social commitment must come second to business decisions, important only when they support a company's profit-generating function.

When confronted with racialized job offers, these executives lacked the experience, role models, and mentorship to assist them in reading company culture. And, in the absence of support, blacks who remained in racialized jobs in Chicago corporations turned to each other for help in making career decisions. The affirmative-action director said, "It was a case of the blind leading the blind. I was stupid. I remember the CEO saying . . . 'We want you to take this beautiful job. It's going to pay you all this money. It's going to make you a star.'"

A sense of disappointment comes through as these people look back on their careers. Middle age, regardless of race, is a period when people review their lives; some regret past choices, even if they are by objective standards successful. Among people who were firsts in history, expectations about making a difference and achieving economic success may run even higher than the norm, and their disappointment that much greater.

Chapter 6

Peacekeepers, Crisis Managers, and Conciliators

> I can think of at least three or four instances where [the company] would have gotten [very] stringent sanctions from the EEOC and I interceded and basically negotiated [the] sanctions down . . . one or two levels. I would get informal agreements that the company could comply with. I [also] kept them out of trouble with the minority employees, who were pushing for better participation in management, and promotions, and salary increases. And I did it without selling anybody out. When I came in, there was talk of boycotts and everything by . . . certain groups of employees. In fact they did call Reverend Jesse Jackson in a couple of times. Jesse Jackson and I . . . worked things out. Yeah, . . . I kept [the corporate] hind end out of trouble a couple of times, at least.
>
> —Urban affairs director in the consumer goods industry

✦ Black executives remained in racialized jobs because white corporate leadership needed them there. The executives reaped economic rewards and enhanced social status, while their employers received a reliable work force to "absorb" black demands and government regulations. Companies who had to adjust to new social demands looked to these executives to (1) help companies conform to federal legislation, (2) reduce pressure from the external black community, (3) maintain racial harmony inside the company, and (4) help companies increase or protect their share of the black consumer market.

99

Conforming to Legislation

When Title VII of the 1964 Civil Rights Act was enacted, companies stepped up efforts to comply with federal employment regulations. After 1965 major corporations sought black manpower from historically neglected sources (Freeman 1976:176). That is, white employers began to recruit from previously ignored, or little-known, black colleges. They also began researching, developing, and implementing affirmative-action programs. The EEOC reported that it had advised 6,517 private sector employers in 1974 on recognizing and correcting discriminatory employment practices. The commission noted that the startling increase in its caseload resulted from an increase in federal compliance activity and in court decisions, as described in Chapters 3 and 4 (U.S. Equal Employment Opportunity Commission 1978). As corporate accommodation to federal regulations spread, the need for a cadre of managers to help companies alter discriminatory personnel practices also grew. In personnel management, for example, blacks assisted white companies in the hands-on recruitment and training of blacks. Sixty of the seventy-six executives I interviewed reported that they were instrumental in either starting or expanding minority recruitment programs to meet federal hiring guidelines. These executives also paved the way for the entry and promotion of blacks inside white-dominated institutions to meet affirmative-action requirements and protect government contracts. Indeed, twenty-six of the fifty-one (51 percent) who filled racialized jobs during the 1960s and 1970s were instrumental in helping top management detect exclusionary hiring practices and reorient personnel departments to conform to government requirements. Moreover, twelve of the twenty-six (about 46 percent) continued to specialize in this area.

The career of the head of affirmative action for a multinational diversified retailer reflects this trend. In 1966, this executive became the first black professional woman ever hired by the company. In 1970, she became the company's first black in personnel at any level, including file clerk and typist. At the time she was hired, the firm — a huge supplier of commercial goods to the U.S. Department of Defense and therefore accountable to compliance regulations — had in place an explicit set of procedures for excluding black applicants from all but the most menial of jobs. This woman started in sales in Gary, Indiana, but her status as the only black woman professional in the company gained her the attention of top management. After a series of interviews with the company's CEO, she was handpicked by him to go to Chicago and "clean up" personnel practices.

> The CEO made it a point to know me personally once I came to the company. I would say, being the first black woman [professional], he watched how I handled myself for a long time. Once he saw I was competent, and got along with [white] people, and didn't cause a lot of trouble, we had a series of talks. He told me he wanted me to be his ambassador in personnel. We discussed that possibility for a long time. He didn't want me to do anything rash, but he wanted [me to be his] eyes and ears in that department.

Under the company's double standards for pay and benefits based on race, blacks received less compensation and fewer or no job benefits. Moreover, the company used a code to channel black applicants into "suitable" jobs, such as janitorial work, and to designate which of two sets of skills tests would be used to evaluate job applicants. After the executive reported these procedures to the CEO, most of the most blatant were discontinued. She then designed and implemented the company's first affirmative-action program, including goals and timetables.

By reducing her firm's vulnerability to contract compliance reviews, this woman helped protect millions of dollars in contract awards. "There was no way the company was going to take a chance on losing those [federal] contracts," she said. "Even if we had contracts held up for two days, and it's worth millions, you are losing thousands of dollars [a day]." Ultimately, she was promoted to company director of affirmative action for helping the company avoid the costs and disruptions accompanying governmental sanctions.

This executive was important to the company not only for helping it conform to federal requirements, but also for her ability to act on behalf of top management. While not attempting to radically restructure the system, she managed to "keep them out of trouble."

Being a trusted company person carries with it costs as well as benefits. A recurring theme in the interviews was that affirmative-action jobs were easy to get into, but often hard to get out of, particularly if the executives had earned the company's trust. That dynamic was at work in the career of a forty-three-year-old group director of human resources interviewed in 1986. Starting out in the private sector in mainstream personnel, he moved quickly into an affirmative-action job, although the field was never his career goal. His reasons for moving into the racialized job path resembled those of others: He felt the pull of his black identity and an attendant obligation to help the black community. His ambition, whetted by corporate perquisites and financial rewards that come with performing a valued job, also played a role in his decision. But what began as a mutually beneficial duet became a tug of war when he tried to get out of his racialized position. Ultimately, he failed. His ability to read what the company wanted to accomplish and his success in creating an effective program conspired against him. In other words, this human resource director was a loyal company man and a highly competent black professional, qualities that made him more valuable in affirmative action than in the mainstream of the company. Finally, his salaried job offers in the private sector became defined by that function.

Tracing this executive's career in detail provides an object lesson in the dangers of doing a racialized job too well. He started in 1966 working in one of the old poverty programs, the Opportunity Industrialization Center, a skills-training center for the hard-core unemployed. He then moved into the private sector as a (mainstream) employment manager for a nationally known bakery, where he stayed for the next three years. In 1970, a white headhunter approached him about a personnel job in his current employer's Illinois plant. The opening was in a large, rapidly growing department that seemed to offer greater chances for advancement than the bakery, and he became optimistic about getting the job when the headhunter told him he was "exactly what the company was looking for." During his job interview, the personnel director said that the company needed "more black managers" and that "the opportunities are really there if you have any skills and ability." The company was, and still is, a big government contractor, and this man was the first black in the company's history to become part of plant management.

Between 1970 and 1972, the Illinois complex grew from 450 to 2,200 employees and expanded from one to three plants. When the dust settled, a general personnel manager was at the top of the personnel hierarchy, and reporting to him were three plant personnel managers and a newly created employment and training manager. Until this point, this man had been happy as a plant personnel manager and, given a choice, would have stayed in that job. Instead, the company brought in three outside people to staff the human resources program and slotted him as the employment and training manager — a move that may have been connected to the racialized functions associated with that newly created position. The Office of Federal Contract Compliance Programs (OFCCP) was most active in that period, and this man reported that he was heavily involved with increasing the proportion of employees who were minorities. "The employment function controlled the affirmative-action program for the location," he said. "That was my job. I had to ensure that we had black applicants to go into the jobs that we filled. And to be sure we filled them in a ratio that was at least in compliance with federal regulations. As far as the programming of it was concerned, it was my responsibility."

The complex was the company's largest and highest-profile facility. Between that visibility and the acknowledged difficulty of maintaining compliance in a rapidly increasing work force, the manager's results in employee relations and affirmative-action successes won him positive attention from company leadership. One year later, the company created a national affirmative-action operation and offered this manager the job. He attributed the offer to his demonstrated success and the company's need for a nationally coordinated program. I suspect that the company also saw that having a black representative run this office not only signaled the company's good intent but, given the tenor of the times, was a political necessity.

After noting that in the early 1970s most affirmative-action managers were black, he said, "I had become the [company's] highest-ranking black personnel person in the country. I guess from that standpoint they looked at me as a likely suspect. I guess I shouldn't say suspect. They came and said, 'We've got this new slot. We'd like you to come to Chicago and talk to us about it.' In a series of three interviews I was offered the job, and I took it." He had reservations about accepting the job, for he believed that "sometimes the company wants to pigeonhole you," and his objective "was not to be in affirmative action for the rest of my life."

He recognized the job's attractions, however. For instance, he believed that taking the job was "probably the fastest way to hop slots." Also, he felt an obligation to "give back" to other blacks and help the black community. After hearing what the job was designed to do, he said, "I [didn't] know if I trusted anybody else to have the commitment to the job that I thought I would have." Thus, at the time, the offer appeared to be a golden opportunity: a chance to advance his career and to play a significant role increasing opportunities for others in the black community. Nevertheless, before accepting the job he exacted the promise that he would remain in it no longer than three years, and he asked, he said, "about other jobs I could go to from here. This was [not] to be a forever job."

The company was true to its promise. In 1976, he was transferred from affirmative action into a mainstream position as a personnel manager at the company's headquarters. But even with an affirmative-action program in place, in about 1978 "[the company] still didn't have any black vice-presidents and black account managers, and black marketing managers, or female, or Hispanic, or anything else." He said that "top management was very aware that PUSH [(Jesse Jackson's People United to Save Humanity) was] out there . . . doing covenants" with big companies. Specifically, "much of PUSH's involvement was brought about by black employees' discontent . . . black people inside [these companies] . . . were disgruntled and complaining [that] somebody's got to do something about [white] people."

Since the company was based in Chicago, which is also the head-quarters for Jackson's organization, it was no surprise to this man when the senior vice-president of human resources approached him with an offer. The request, if indeed the senior vice-president was merely asking, was for him to develop an internal compliance pro-gram and do what was necessary for the company to avoid bad public relations and maintain black consumer alliances.

> My perspective was I wanted to contribute, but you can't con-tinue planning all your life, for affirmative action anyway. I've since learned that you can. I actually turned the job down when it was first offered. And [the senior vice president of human resources] came back again and said, "You know, we could go outside and we could probably bring somebody in, but . . . we would have lost some opportunities for effectiveness. We'd be in a position . . . to give you assurance that you're not going to be stuck in the job if you will accept that responsibility."

In that position, his value to the company would be based not solely on his ability to do the job but also in the fact that he was both black and a known and dependable commodity in the company. Indeed, the company's repeated attempts to coax him into the jobs suggests that top management was highly comfortable with him. His unique value to the company was further underscored by top management's offer of a significant salary increase, which he described as "worthy of making the move."

A corporate manager might turn down his superiors' request once and still be seen as a team player. But to remain competitive in most corporate environments, one would certainly not decline twice. He took the job because he believed he could eventually leave it with the blessings of his superior and, conversely, because he could not turn it down without endangering his career.

Two years after accepting the job, after, he said, "an internal analysis, and after we avoided the charges, and the compliance reviews, and the threats of being brought into litigation," the company rewarded his performance by mainstreaming him. In a move that signaled the company's commitment to his ultimate career development in human resource management, he became a manager of compensation. The company continued to groom him by further expanding his experience and, one year later, moving him laterally to regional personnel manager.

Regional personnel manager was typically a feeder job into the group director job, yet his career took another detour in 1983 into an urban affairs/EEO position.

> We talked about the fact that I'd paid my dues on the job, and that there is only so much that you can do before you begin to repeat yourself, and a number of other things. They went off and came back a second time and said, "How can we make this job appealing enough for you to reconsider?" I asked for quite a bit — and I [told them] I'm not a career affirmative-action person. And unless I can look forward to [being] one of the group's personnel heads, then I'm not interested in that job.

He remained as urban affairs director for almost three years before the company made the director of human resources job available in 1985.

A reasonable interpretation of this man's career in 1986 was that he had paid his dues in affirmative action and received his payoff, an upper level midmanagement mainstream job. However, the company was sold to a new parent company in the late 1980s, and the new leadership's assessment of him greatly differed from that of his former superiors, he said. Under the former management, he repeatedly was slotted into affirmative-action jobs because he was trusted by senior management and considered a loyal team player. But the new senior managers looked at his record and viewed him as an affirmative-action manager, a regulatory person, and someone who might prove adversarial.

Even outside that company, potential employers defined his skills by his detours, even though he did not present himself as a career affirmative-action person. By 1992, he had left the corporate world, feeling forced out, and started his own business. One reading of this career trajectory is that reentering the mainstream after performing racialized assignments may depend on a series of fortunate circumstances that happen to befall an executive.

Reducing Pressure from External Black Groups

After the mid-1960s, corporations found strategies other than affirmative-action programs to adjust to the political climate. Community relations, job training, and other urban affairs programs, essentially appeasement functions that provided service and technical assistance to black ghettos, were now deemed urgent because of the fear of racial violence.

Between 1965 and 1970, at least eighty-two Fortune 500 companies started some kind of black economic development and employment program to "principally help discourage boycotts, violence, and other threats to company well-being." In 1970, 201 of 247 city-based Fortune 500 companies reported having urban affairs programs, but only four such programs existed before 1965 (Cohn 1975). Thus, at least 197 of the 500 largest companies in the United States started urban affairs programs within the five-year period characterized by civil disorder, which included the most costly outbreaks of urban riots in the country's history.

Initially, white managers filled urban affairs jobs (Cohn 1975). But increasing the presence of African Americans allowed corporations to initiate programs to appease black communities without siphoning off white employees from other areas of the company. Moreover, the political activism of both middle- and lower-class blacks, including the grass-roots demand for black representation, necessitated naming blacks to represent the company and direct these programs. Seventeen of the fifty-one executives in this study who held racialized jobs during the 1960s and 1970s functioned to reduce pressure from black groups that were external to the company. Eleven of these seventeen (65 percent) stayed in them for most of their career.

Within the context of the times, workers filling external relations jobs were useful for easing tensions between the white corporate world and black America. In one case, an executive I interviewed moved into a community relations function in 1967 when, he said, the city was "undergoing a little upheaval, a little racial unrest, maybe, for want of a better term." Community groups were staging sit-ins that escalated to near riots in the East Coast town where company headquarters were located. This executive was approached by his mentor because "there was nobody [black] there," referring to the corporate offices. (This situation is reminiscent of the frustrated manager in Chapter 5 who was employed to coordinate the manpower-training program for a steel company after management began to perceive urban riot as a threat to the company's assets. In both cases, company headquarters — and thus huge capital investments — were located in riot-torn neighborhoods.)

Another case further explicates executives' value in buttressing vulnerable companies in urban areas. An executive who filled the urban affairs position for a midwestern bank in the late 1960s, after one of the "long hot summers" of race riots, established tutorial and summer employment programs, although banking is not a seasonal employer. His goal, he said, was to show "responsiveness on the part of the institution," thereby decreasing the bank's vulnerability in the case of future riots. To emphasize this point, he explained somewhat sardonically that the bank created the urban affairs function after "the natives jumped out and started to burn down the city. The bank saw what riots could do, and they had assets to protect." Unlike businesses that retreated to the relative peace of white suburbia, banks could not reasonably move their headquarters out of the city, he explained. For example, proximity of the bank to the Federal Reserve enabled the institution to keep money longer, since the time it took to transfer money was minimal. Bank officials weighed the cost of potential riot-related property damage against the benefits of an inner-city location. Their solution: to create an urban affairs department, viewed "by top management as an investment in the city and . . . in its own security," he said.

He described his job in the late 1960s and early 1970s as a "consultant to the top house," meaning he met directly with the top managers, he said, "to change the way the bank did business with blacks." Both he and the bank's management team believed that he could "keep them out of trouble" since he held membership in the group by which the bank felt threatened. His success in developing the bank's relationship with the black community was well rewarded. In the early 1970s, the bank's board of directors voted to make him a vice-president and director, and thus the highest-ranking African American executive in one of the country's largest banks.

In a similar situation, a black executive for a steel company was promoted to project manager in 1974, after a major construction contract worth millions of dollars came under attack by the *Chicago Defender,* an African American newspaper. Both the African American media and community that surrounded the construction site publicly criticized the company, pointing to its complete lack of African American suppliers and contractors. As this manager summarized it, "I was one of the few blacks [in the company] and I was pointed out. [The company wanted me to get] them out of trouble . . . because they were in big trouble. [Blacks] were threatening to picket the bank [that provided the loan for the project] . . . and [the company] was in trouble [because of] the minority community which was threatening the project." This man could easily maneuver between the corporate and black communities. Born and raised in a segregated black community, he had also earned a bachelor's degree from Dartmouth, an elite white college, and wanted to succeed in the white world.

Other executives among those I interviewed who held external relations jobs described their roles as corporate ambassadorships. They plugged corporations into black civil rights and social service organizations and, in general, represented white companies in black-dominated settings whenever necessary. A director of urban affairs, for example, said that in 1971, one of the reasons they hired me [as their director of community affairs] was because I knew everybody in town. I came up working with all the grass-roots organizations from all the neighborhoods. [The role was to] make [the company] look good. I did what they needed done to look good in the community. They utilized me in that fashion. For eleven years I was just their spook who sat by the door, and I understood that. Certainly I was, and I charged them well for it.

Each of the executives in this category was skilled at brokering the interests of their companies and successfully "absorbing" the tensions between white companies and urban black constituencies. The manager recruited by his mentor was able to appease protesters and reinstate order in the company headquarters town by developing a series of black-oriented outreach programs and becoming visible to the black community. The steel manager's impact on the community went further; since the corporation was involved in rebuilding the area after the riots, he also was able to improve job training and employment chances for blacks, at least temporarily. Finally, the man at the midwestern bank developed vehicles to "promote the bank's visibility and good name" and thereby created a reservoir of goodwill in the black community.

Racialized jobs were a sort of black ambassadorship, with black executives negotiating with African American activists to avoid pressure and bad publicity that threatened corporate operations and profits. Color and culture were integral to jobs to "cool out" and "keep a lid on" blacks and to manage other sources of political pressure. When blacks demanded race-based reparations during the 1960s and early 1970s, they also demanded "maximum feasible participation." This included having black representatives at all levels of an organization to convey a black perspective. As one executive I interviewed put it, "Blacks would be surrounding a company saying, 'We don't want to talk to some flunky or some white guy, we want to talk to a black vice-president.'"

Furthermore, according to one business and civic leader in Chicago, high-profile operations such as mortgage pools, private initiative councils (PICs), and other federally funded, privately run skills-training programs were started after the riots "to create economic incentives and create a feeling that things will get better for people that have been left out of prosperity."[1] The executives who administered these programs played key roles in eliciting feelings of social investment among black people.

In contrast to mainstream external relations in which corporate concern over blacks' goodwill and well-being was all but absent, racialized versions of these jobs were links to previously neglected populations, forged to appease blacks and discourage racial violence (Cohn 1975).

Maintaining Racial Harmony inside Companies

Racialized roles also neutralized race-related pressures inside corporations, where black executives mediated antagonistic, and potentially explosive, black-white relationships. Thirty of the fifty-one racialized managers I interviewed described racially volatile labor relations between black workers and white management. As in the case of external pressures, in internal conflicts they interceded as interpreters and conciliators, explaining black political action, interceding in racial conflicts, and, in general, ameliorating the racial tensions that played out inside the work environment.

They were able, for example, to bridge the widening gap between black labor's discontent and the apprehensions and bewilderment of top management. One executive was asked by the CEO of a major oil company to become management's agent in a racially sensitive work situation, explaining and then quieting blacks' organized resistance to company policy. In 1968 the company had a flourishing ghetto of lower-paid jobs occupied by African Americans who were becoming increasingly vocal in their grievances. Complaints stemmed from their systematic segregation and concentration into a single department at company headquarters and into selected low-paying occupations. Black workers who had, in general, been docile about their job stagnation became militant. When, both individually and collectively, they stationed themselves outside executive offices to demand racial justice in job competition, this man — the only black supervisor employed in the building — was selected by top management to address the situation.

> I was called [from the accounts department] into the president's office, and two senior vice-presidents were there. They didn't understand what was going on in [the] department, and they called me in to explain it. Now I wasn't the organizer of this thing. But I guess because I was the senior black in that department they asked me. That was the first meeting. The second meeting they called me in, not to explain it, but to stop it. And I guess I've been the company's affirmative-action person from that day on.

The need to contain the potentially damaging effects of the workers' political activities on the company's day-to-day operations broadened the value of this worker in the eyes of top management. Once he shifted out of account services, he became a well-paid expert, advising senior executives on the new etiquette of race relations. Given the degree of black activism at the time, he was more valuable to the company as a "cultural consultant" than in his former supervisory role. While his job as intermediary was not one that he had envisioned for himself, it would have been to his disadvantage to decline the assignment. Tensions in the company were running high. "I was in a position where somebody upstairs asked me to take this on. I couldn't turn down the top man's request without a very good reason," he said. Revealing the uniqueness of his status, and his value in the eyes of top management, he said, "They wanted me, [and] I told them what it would take to get me. I was at the top of my game where I was, so they were going to have to pay."

As in the case of the human resources director mentioned earlier, this worker was valued not solely for his color but also because top management was comfortable having him act on its behalf. Why else would he have, as he put it, "got what I wanted"? He said, "I got more money with nobody reporting to me than I got with ten people reporting to me, [a] $25,000 [pay] increase." He could not have gotten this raise solely on the basis of race, although the company felt it needed a black person badly in this role. An unknown quantity would be too risky a gamble in this volatile environment. The company wanted someone trustworthy, and just any black person would not do.

Given the scrutiny of top management and the level of racial animosity that existed in the company, one wonders how this manager would have fared in the company had he failed. He didn't fail and ended up, as he put it, "staying in affirmative-action forever, [because] they do what it takes to keep you satisfied."

Another manager's case is strikingly similar. A black director of testing in a mainstream technical job, he was deployed into personnel management to work at reducing racial tensions inside the company. As in the corporation cited earlier, job opportunities for black workers in this company were restricted to clerical activities. One-third of employees were black, and most were "secretaries and [office] administrative types." Ghettoized into a low-paying job sector, black employees developed highly disruptive strategies in protest. As in the previous case, senior management told the executive "that there was a lot of dissatisfaction in the ranks, [and] there was a definite need for someone [black] to be visible in personnel." The issue, of course, was not that "someone" be visible, since a white personnel manager was already in place. Rather, it was that the "someone" be black. He acknowledged this when he said his job was to "make blacks in the company feel they were being related to, . . . to present a positive image to the staff, [and] to show them they could make it." In short, this African American manager could offer what his white counterpart could not, a presence both politically meaningful and pacifying to blacks.

Protecting Product Markets

Given the volatility that prevailed among blacks in the 1960s and 1970s, retail companies in a competitive marketplace exploited and protected their existing market shares by hiring black salespeople to court and keep black consumers. Ten of the fifty-one executives in racialized jobs were hired or promoted specifically into sales slots oriented toward a black consumer constituency. For example, a vice-president who began his career as a sales agent for an insurance company recalled that he was hired in 1971 because his employer "saw that the black market in Rockford [Illinois] was . . . untapped." Consequently, the company "was really looking for someone to take [on sales] in the black community." This man lacked previous sales experience, and sales was not a career path he had planned to take. However, the company developed a race-specific need, so he took advantage of the opportunity.

Another executive, who worked for a major retail company in Washington, D.C., noted that he and twelve other blacks were hired by the company at about the same time, and that all were placed in stores in the inner city.

Historically, consumption patterns generally have not been considered race related. But beginning in the mid-1960s, black advertising agencies convinced companies that blacks related to distinctive cultural symbols, and market research firms showed that black consumers had unique buying patterns. Using blacks in sales was one way to project an image that would attract black consumers. Moreover, although political pressures have since lessened, consumer goods companies still use race to enhance their image among black consumers. For example, McDonalds has a series of television advertisements that was created by a black-owned advertising firm, features black rap artists, and is directed to black audiences.

Using blacks in sales is primarily a response to markets, but during the 1960s and 1970s, black executives also assumed racialized roles as intermediaries between white corporations and black buyers in response to race-based political pressures. Companies during that period were threatened with boycotts that could have withdrawn large segments of buyers from their existing consumer base. The rise of one executive employed by an educational testing and supplies company is an example of how corporations responded to such pressures. In an era when blacks were demanding more representation and community control, a change in a New Orleans school district's management put blacks in key buying positions. This man was promoted, in 1965, from an entry level customer representative to manager of community relations, after his company was threatened with losing a lucrative market in the school district. He stated flatly that he was promoted because "the company was abruptly shut out, [and] they needed somebody black to represent [them]" to this new sales constituency. In other words, he was assigned to a highly visible sales position because a black presence in that role would project a message of corporate responsiveness to black concerns and racial sensitivity. Black buyers for the school district could then feel good about patronizing the company.

For similar reasons, several managers were assigned to administer politically inspired, but economically motivated, business programs. For instance, after black consumers joined ranks with civil rights advocates to boycott products, set-aside programs were established by companies to allocate a percentage of their purchases to minority businesses. One executive oriented a major retailer's urban affairs program to black business development after the company was threatened with boycotts.

Relative to his white peers in the company, this man had moved rapidly up the mainstream corporate ladder. A counter clerk in 1964, within three years he had moved through a series of supervisory slots to become a store manager. At the age of twenty-three, he was making $23,000 a year: a $10,000 salary and a $13,000 performance-based bonus.[2] "Running the stores was a piece of cake. I was dazzled by the money and very self-confident," he said. Company officials soon began considering him for middle management. An advisor told him to apply for a promotion because he had both the talent and the skills to take on more responsibility. In 1967, he moved from store management to area supervisor. Coming up through the ranks was the typical route for the company's top executives and, therefore, "the way to go," he said.

Less than a year after his promotion, however, he was asked to apply for a job creating an urban affairs program for the company. The circumstances that led to this request were relatively straightforward. The company viewed blacks as a sizable proportion of its customer base, and civil rights activists had confronted the company with specific demands backed by the threat of a nationwide boycott. The selection process differed significantly from that used to fill other jobs at the same level. It included interviews with the head of personnel, a senior vice-president, the head of licensing, the corporate legal council, and the president of the company, "real heavy hitters," as he said. He was offered, and accepted, the job, which was "to work with the licensee department and [come] up with minority candidates around the country to become [store owners]."

Within this operations-driven corporation, a manager with demonstrated talent for business operations, particularly one who had a senior management mentor, would generally be considered a serious contender for a top level mainstream position. Slotting such an employee into urban affairs might appear to be a frivolous use of talent. But in 1968, with no other blacks working at the company's corporate offices (save for one janitor), the company was vulnerable to racial protest. From this perspective, deploying a black midlevel manager, a known commodity in the company, into corporate urban affairs was a rational business decision.

In their book on black executives, Davis and Watson (1982) report that the first blacks hired into management during the 1960s believed they were valued by companies as much for color and image as for any other characteristics. In the case of this midlevel manager, using a black to negotiate with black activists could defuse charges of racism that community leaders had leveled at the company. It also cast a member of the activists' constituency as a representative of the company, thus creating a symbol of progressive interests, while minimizing differences. In this manager's case, as in others', race was at least as highly valued as any other qualification, including skill in performing a midmanagement mainstream function, and was the basis for his assignment to black business development and community relations.

While this manager's job seems straightforward — avert a boycott by creating more opportunities for blacks to be business owners — it was actually fraught with political pitfalls. He had to perform under the scrutiny of both black and company (white) constituencies with strong, and conflicting, class and race interests. He had to broker the interests of white top management, white store licensees, black candidates for store licenses, and black civil rights activists and appease whites in the company who resented what they thought of as blacks' special treatment. He was responsible for the conflicting tasks of building relationships with inner-city black groups and implementing a business program that could project a socially conscious corporate image, while also reassuring white licensees who did not want blacks in their market and who were not at all convinced that the company was moving in a good direction. Moreover, he bore the brunt of criticism from rank-and-file white employees who resented blacks' getting what they felt were "special favors." Finally, top management withheld full-fledged support because of what they perceived to be a lowering of program standards to bring more black licensees into the system.

In addition to these paradoxes, the job had a second, and inherently race-linked, set of difficulties. The company had financial criteria for franchising stores that whites could meet relatively easily but that most black licensing candidates could not. Furthermore, white banks would not lend money to black business, even though the company had a successful track record in starting up fledgling stores. Finally, blacks who had money to invest could not afford to quit their jobs and were often rejected by the company because policy stipulated the owner must also be the day-to-day operator on the premises. The executive said, "People that were declined would go to various groups, such as the NAACP, and say, 'Listen, they turned me down.'" Each of these quandaries, naturally, created its own tensions.

This manager, therefore, was saddled with what he called "an impossible task." And each of these race-linked conflicts made his career more vulnerable. In this job, the stakes were high, as was the likelihood of failure. The additional, if unintended, benefit of having a black fill such a job was the creation of a scapegoat. If the company effort failed, the black executive would take the blame from both top management and black community groups.

Affirmative-action, urban affairs, and community relations functions in companies worked in dual ways. They were mechanisms that made social and economic resources more available to the aggregate black population. They also were a system of occupations that helped to minimize change and maintain the status quo, for they alleviated political pressures on companies by defending them and deflecting attacks made on racial grounds. As a corporate tool, black executives' value was commensurate with their skill in abating political pressures, protecting profits, and not rocking the boat—not in consolidating black power and changing the makeup of an institution's power brokers.

It is easy to view these functionaries as gatekeepers and stumbling blocks to institutionalizing equality. Yet what were the alternatives? Did they, for example, hinder change more than white managers would have in similar positions? There is scant reason to believe that employers would have taken more progressive steps for blacks if these executives had not filled such roles.

Chapter 7

Blacks on the Bubble

> The company was . . . eliminating a lot of positions, you
> know, not just mine, and [I was told the company]
> would not be someplace that I wanted to be. To say that
> I was surprised is an understatement. I guess I just didn't
> think I would be affected because, number one, I was
> the only black senior manager in the company. And I
> had high visibility. [I] built all kinds of goodwill for the
> company . . . to the point that my name was almost
> associated with the corporation name when I would
> come to a place. So, you know, I overestimated my
> value to the company, I guess you could say that.
>
> —Director of community relations for a consumer
> goods company

✦ In the 1980s, the federal government's twenty-year commitment to policies and practices that helped blacks compete economically underwent a dramatic reversal. Throughout the decade the White House opposed race-based policies and protections, particularly the policy of affirmative action (Hudson and Broadnax 1982), and Supreme Court appointments and decisions signaled a retreat from race-based remedies to overcome historic discrimination (Wilson et al. 1991).[1]

The social and political mood of the decade was foreshadowed in the 1978 legal challenge to affirmative action by Allan Bakke, a white male who brought suit against the University of California Regents, claiming that affirmative action is "reverse discrimination" against better qualified white men and alleging that he had been discriminated against when he was denied admission to the university's medical program (Dreyfuss and Lawrence 1979; Sitkoff 1981).

The Supreme Court, in the Bakke decision, rejected the use of racial quotas to overcome the cumulative effects of past discrimination and found the special-admissions program for racial minorities unconstitutional. Although the Court subsequently ruled that race could be taken narrowly into account to remedy proven discrimination, the Bakke decision served as a prelude to the political viewpoints that would shape the 1980s. The Reagan administration took a stronger position, opposing "court-ordered and court-sanctioned racial preferences for non-victims of discrimination" (Hudson and Broadnax 1982) and actively arguing against the use of preferential treatment and quotas in employment practices (Reynolds 1983). Although the Court did twice endorse affirmative-action policies during the 1980s, most of its decisions clearly indicated that the use of preferential hiring policies was on soft ground.[2]

The attack on race-specific programs sent a strong signal that the role of the federal government as a strong advocate of African American employment was being reduced. Within this social and legal context, the question arises of whether substantive characteristics of African Americans' jobs make them particular targets to be downsized or cut from companies. I believe that the race-specific focus of these jobs makes them vulnerable to changes in the administrative structures of white corporations, which themselves fluctuate with political conditions.

Status of Jobs

In 1986, I asked the managers I interviewed if their jobs had changed in title, scope of responsibilities, budget, or functions since 1980. For example, had any of their responsibilities been increased, reassigned to other managers, or dissolved? Had departmental budgets and staff increased or decreased?

Racialized jobs in corporations were the most recent employment for about two-fifths (thirty of the seventy-six) of the managers I interviewed. Table 3 compares the reports of forty-five mainstream managers with these thirty managers in racialized jobs. (One chief

Table 3
Reports of Job Changes by Job Type

Job Changes	Racialized Jobs (N=30)	Mainstream Jobs (N=45)
Duties Downsized/Cut	50%	11%
Duties Increased/Unchanged	50%	89%

executive officer was excluded from this summary since a CEO might be fired by a corporate board but the job itself would not be increased, downsized, or eliminated.) The table conforms to what my hypothesis would predict. African American executives in racialized jobs, much more than African American executives in mainstream jobs (50 percent versus 11 percent), reported that the company had eliminated, reduced, or redistributed their functions to managers in other areas. The relevance of these reports may be seen as dependent upon comparing them to those of white managers, but my analysis links fragility to the characteristics of jobs rather than to the characteristics of job holders. Thus, I compare structures of occupations while keeping race constant. Hypothetically, whites concentrated in racialized jobs would confront the same economic fate. Comparing the status of managers by race, per se, is less pertinent to my argument than illustrating a specific type of occupational fragility.

The vulnerability associated with racialized jobs is evident in the career of a vice-president of urban affairs who started out in the financial industry in the early 1970s in affirmative action. The intensity of racial pressures in Chicago caused his job to evolve into a community liaison position that focused on placating blacks. He summarizes his job this way: "I kept my hands on the pulse of the [black] community—I sold the bank's story out to the community. I conducted the social audit, finding out where the bank was deficient, where they could come up to speed." This manager makes explicit the difference between the futures of the two tracks in public relations, mainstream and racialized (i.e., related to the problems of African Americans). The mainstream functions are responsible for maintaining traditional (white) civic contributions. The racialized

function for which this manager was responsible was essentially an
appeasement role. By the late 1970s, the rationale for appeasement
no longer existed in Chicago; pressure from blacks was absent. The
community affairs function oriented to African Americans was cut
back, as the manager explained.

> They kept talking, they had this term, "the long hot summer,"
> and up to about 1976 [or] 1977, they were talking about the
> long hot summer [but] nothing happened. [So] they just cut the
> money. I mean, their traditional lines of support . . . for the [Chi-
> cago] symphony, and the [Chicago] Art Institute and those
> kinds of things, they were still maintaining a level. But no
> [money for] community groups, grass-roots types. There was
> just a withdrawal.

African American executives in mainstream as well as in racialized jobs reported reductions in responsibility that suggest career
vulnerability (see Table 3). Such vulnerability in the climate of the
1980s, then, cannot be viewed merely as a product of racial inequality in the job structure. The job fragility of white-collar workers, including middle managers and financial industry employees, is a feature of macroeconomic changes such as greatly diminished growth
in service sector jobs (Hertz 1990; Starobin 1993).[3]

An example among the executives I interviewed is a vice-president
of sales for a publishing company who indicated that his company
changed dramatically when the breakup of phone companies made
the business climate in the mid-1980s "far more competitive and
risky than ever before." I asked if he felt his job might be threatened
as a result of increased competition, which is a market-mediated
source of job fragility. He responded, "Well, yes, if our new competitors get a disproportionate share of the work that we used to
get. No, if all those contracts come back to us." Indeed, in the late
eighties era of corporate mergers and economic restructuring, both
white and African American managers were victims of job loss in
major corporations, according to a 4 January 1987 *New York Times*
article, "The Ax Falls on Equal Opportunity," and a 1987 piece in

U.S. News and World Report, "You're Fired." Job insecurity con-
fronts all individuals in high-status occupations, not blacks per se.
We can still ask, however, whether there are substantive differences
within this strata of jobs that make the fragile position of African
Americans in management a somewhat different phenomenon.

The experiences of the executives I interviewed illustrate how
politically useful jobs in white companies can become economically
expendable, particularly in a context of corporate buyouts and economic reorganization. A director of community affairs and public
affairs who managed the African American component of public
relations for a major retail firm in Chicago had since 1972 been
charged with "keeping [up] the image" of the company in the African American community and representing the company at conven-
tions, on community boards, and on the committees of African
American community organizations. Between 1981 and 1982 the
company began streamlining the work force to maximize profits;
similar to the case just recounted, when pressure waned, the company image among African American consumers became less important. During this period, the director reported, his job in community
affairs "just wasn't important to them — they just didn't want to
spend money on that any more." He said that "they wanted to cut
the job, they just didn't want to cut me" when the component of
community affairs oriented to African Americans was dismantled
completely. Unable to shift successfully into a new role, he took an
early retirement.

A second individual, also in community affairs, observed that his
employer had become "a very different company" than the one he
entered. He went on to say that "there were certainly some things I
had to unlearn," because "everything [became] tied to [profit]. And
[my job] could not be tied to . . . revenue, at least in an immediate
sense. For over fifteen years I needed to be tuned in to what blacks
in the grass-roots community were thinking. But Chicago is changing. Now, all of a sudden, I [was not] getting any of the support I
needed."

Other managers in racialized jobs told similar stories. For instance, a twenty-year manager of corporate contributions and com-

munity relations for a steel company told me he was "looking for opportunities elsewhere." He reported that the programs he ran were funded by company reserves set aside during the 1960s and 1970s following rioting by African Americans in Chicago. These once large reserves diminished as company profits fell in the 1980s. In that comparatively calm racial atmosphere the incentive to tap limited reserves cooled. He said that his department "will have a totally different look. . . . [It] will honor the commitments we made for this year and phase out. That includes me and part of my staff."

Even managers who have not experienced immediate reductions say that racialized functions are easy targets because the political climate changed at the same time that corporations reorganized. Among them is a manager who had designed a minority purchasing program for his company and was considering retirement. The company began streamlining in 1984, reducing purchasing requirements and its base of minority and majority suppliers.

> They haven't tried to cut back my program; my budget grows every year. But they're trying to eliminate the supplier base. . . . They reduced it by half between last year and this, and they want to reduce it again by half next year. [Question: Do you think the company will fill your job once you leave?] No. When I leave I'm not even grooming anybody to take my job. As far as I know I don't think the company is [grooming anyone] either.

The highly specialized, once flourishing job was created to assist African Americans and other minority businesspeople during the 1970s gain company contracts through a minority set-aside program, typical of those developed by companies in response to mass demonstrations and consumer boycotts by African Americans. The program's mainstream, less specialized counterpart in the area of purchasing, although vulnerable in the 1980s, remained in place in the 1990s. But by the early 1990s the manager's part of the purchasing function, oriented toward African Americans, was dismantled and allowed to dissolve. Interviews support the conclusion that racialized jobs were created when economic expansion and race-specific employment demands converged. In the 1980s these trends reversed. Political pressure placed on employers by government and the African American public weakened. At the same time, competition for market share intensified. Racialized functions, therefore, have greatly reduced value.

The restrained federal approach to racial issues and the relative political quiescence of African Americans, in conjunction with economic transitions, undermined these jobs and made them rational targets for companies to let go. It is the intersection of deracialization and instability in the economy that creates a different (i.e., race-based) fragility to these managers' positions.

Job Security

I have predicted that the stability of racialized jobs would vary according to conditions, which suggests that a race-based structure of job opportunities actually works both ways. Where pressure declines, jobs become unstable; but where pressure is stable, so are these jobs. Race would make African Americans in these jobs vulnerable, but these jobs can also protect African Americans from job loss. This notion of a two-way street is supported by those managers with racialized responsibilities who did not express concern about their future in a company. These were that half of the managers (fifteen of thirty) who reported that their jobs had not been downgraded or cut (refer to Table 3). Yet these managers do not undermine my thesis that fragility is a component in racialized careers. Five executives identified current sources of political pressure on their employers.

Table 4 summarizes job changes among the thirty executives working in race relations. The jobs are sorted into two fields: affirmative action and community relations. I classified managers with EEO, personnel, and staff-training functions as mostly engaged in affirmative action, and managers with urban affairs, corporate contributions, marketing, and public relations functions as doing community relations work. Their experience showed that people who

Table 4

Reports of Job Changes by Racialized Job Type

Job Changes	Affirmative Action Jobs (N = 13)	Community Relation Jobs (N = 17)
Duties Downsized/Cut	67%	35%
Duties Increased/Unchanged	33%	65%

filled affirmative-action jobs faced fewer cuts than those filling community affairs jobs. About one-third of the managers (four of thirteen) in affirmative action reported job cuts compared to two-thirds (eleven of seventeen) in community relations.

Affirmative-action jobs stem from government requirements, a less volatile source of political pressure than community relations jobs, and they have been less vulnerable to cutbacks because the federal regulatory apparatus has not been dismantled. For instance, two affirmative-action managers who reported that their responsibilities had increased also reported that their current employer was under a consent decree. The rationale that created these jobs still remains, in spite of some attenuation in attention or enforcement. Even if a company wished to cut affirmative-action slots, it would still have government regulations and visits from compliance officers to contend with.

In contrast, African American jobs in community relations have proven more vulnerable, consistent with my notion of a politically mediated class position, for they were created in response to community pressures and company desire for visibility among African American constituencies. The greater variability in community-based political pressures compared to governmental legislation makes decline of community relations jobs more volatile.

A leading consumer goods company in Chicago illustrates these points. In 1989, the company was targeted for protest by a group of community activists because of its poor minority-hiring record. To avert a bad, and potentially embarrassing, public relations image, a mainstream company job in corporate giving was transformed into a community affairs job, and a new community affairs (African American) manager hired to negotiate with a small, but vocal, group of activists. In 1993, the company underwent corporate restructuring; in view of black community indifference at the time, the community affairs functions of this job declined. In 1994, the continued quiescence of blacks enabled the company to let the midlevel African American community relations manager go. When last I heard, he had moved to a lower-paid job in another midwestern city.

The case of Rebuild L.A. is a second illustration of how much more variable the cycles of community relations jobs are relative to affirmative-action jobs. Rebuild L.A. was created in 1992 as a clearinghouse for channeling more social and economic resources into the riot-torn areas of Los Angeles. A community activist and union organizer reported that African Americans and Hispanics were hired by that organization into highly visible roles similar to those of community affairs managers. The activist read their presence at meetings with union and community representatives as a way to deflect sensitive issues and protect the organization's credibility. In my model of class development, these professional jobs for nonwhite workers were created directly by urban turmoil; as the fear of new riots and community turmoil abates, these jobs — even, perhaps the organization — will no longer be necessary.

Overall, the pattern of job security or job vulnerability found among managers I interviewed in either affirmative action or community relations is grounded in the residuals of political pressure. Political conditions cause racialized jobs to erode, just as political conditions create these jobs and protect them. Job security is anchored to the governmental and community pressures from African Americans that remain in place today. It is reasonable to suggest that, as pressures from the government abate (as have pressures from the African American community), affirmative-action jobs will go the way of community relations jobs and become increasingly expendable in companies. Should that be the case, affirmative-action managers' reports of job stability would support the thesis of political dependency. When pressures for antibias legislation abate, these jobs would be eliminated.

My thesis also predicts that executives employed in consumer companies would report greater job security than those in capital goods companies, because consumer companies are more visible to the public, and therefore more vulnerable to direct social pressure. Moreover, they would be relatively more vulnerable to public action than capital goods companies because the public is their buyer, not other businesses. Indeed, during the 1960s and 1970s, boycotts by African Americans were an effective tool for gaining concessions from consumer companies because they directly threatened sales in inner-city markets. Manning Marable (1983:158) notes that "between 1960 and 1973, the estimated amount of goods and services purchased by African-Americans increased from $30 billion to almost $70 billion annually. By 1978 the African-American consumer market was the ninth largest in the world." Increased competition for market share during the 1980s may have given special markets greater economic significance.

I categorized respondents according to the industry of their private sector employer in 1986, coding companies such as McDonalds Corporation and Kraft as consumer/retail companies and companies such as Container Corporation of America and Brunswick as capital goods/manufacturing companies. The twenty-one managers employed in capital goods companies were more likely to report post-1980 job cuts than managers in consumer goods companies. Eight (38 percent) reported their jobs were cut or downsized, but seven of the nine managers (78 percent) employed in consumer goods reported this change. Whether this relationship holds when other factors are controlled was not part of my investigation, but it is at least consistent with what my thesis would predict. And, indeed, the potential for black consumer pressure during the 1980s remained just below the surface. PUSH orchestrated a boycott and negative publicity campaign against Nike, a sports goods manufacturer, and Anheuser-Busch to increase the representation of African Americans on their boards of directors and in managerial positions. Both companies overcame this threat to their profits. I suspect having an African American vice-president meet the press and defend the company's hiring record helped Anheuser-Busch deflect black criticism. As an African American director for public relations for a $6 billion consumer goods company in Chicago noted, "You can't be . . . based in Chicago and not fear . . . I shouldn't say fear . . . be aware of Jesse [Jackson]."

Attempts at Job Enhancement

During post-1980 cutbacks some managers tried to break out of racialized slots and into the mainstream of a company. At issue here is how much their relative success hinged on their previous work experience, and whether racialized human capital was a factor that limited their perceived value in mainstream corporate functions.

I categorized careers that incorporated some, but not a majority, of racialized jobs in 1986 as "mixed," and those composed of mostly racialized jobs in 1986 as "segregated." (I dropped executives who had held no racialized jobs from this portion of the analysis.) Although only illustrative, the following comparisons of segregated and mixed careers suggest that career segregation makes African American managers dependent on racial politics because they lack requisite experience in core corporate areas. The history of a former community relations manager for a major electronics corporation illustrates the idea that skills once in demand became a contributing factor to these managers' vulnerability. This man reported that his company's commitment to urban affairs began to decrease; observing the "handwriting on the wall," as he put it, he made multiple attempts to get out of urban affairs.

> I was just not able to make that break. I talked to [people] in various divisions that I was interested in, and I got the lip service that they would keep [me] in mind if something opened up. As it happened, that just did not develop. I can never remember being approached by anyone. Nothing [happened] . . . that I can really hang [onto] as an offer. People would ask, "Have you ever run a profit-and-loss operation?"

Finally, he describes himself as taking "hat in hand" and approaching senior management in 1982 to request duties he knew to be available in a general administrative area.

> Frankly, this was an attempt to seize an opportunity. This time I went and I asked for a [new assignment]. We had some retirement within the company and some reorganization. I saw an opportunity to help myself. The urban affairs was shrinking. A number of jobs we created [in urban affairs] were completely eliminated. It just happened that the opportunity [to pick up administrative services] was there. It had a significant dollar budget and profit-and-loss opportunity. . . . It was concrete and useful. So I asked for it.

He successfully diversified by combining urban affairs with the more stable functions of administrative services, but this strategy bought him only about two more years in the company before his job was eliminated entirely.

> You know . . . [whites] tend to stereotype black managers and say you can only do one thing. But they will take a white manager and they will allow him to try many different things, and I can think of somebody right away. But typically a black manager is pigeonholed. And he doesn't have the luxury of making a mistake either. A black manager can only make one mistake and he's branded forever.

When this man reentered the job market, the same skills that had created vulnerability in his old company were the ones for which he was in demand. Although he avoided applying for affirmative-action jobs and concentrated on administrative services, affirmative-action and related positions were the only offers that came his way.

An urban affairs manager who tried a move to warehouse distribution in a retail company was similarly constrained. He had previously constructed a successful career, but the trade-off for rising in a company in race-oriented jobs was being cut off from mainstream areas. He failed in his attempt to shift from what he termed the "money-using" to the "money-producing" part of the business. "I was too old to do what you had to do to compete. . . . I was competing with twenty-one and twenty-two year olds to get into the system. They couldn't charge [my salary] to a store and have me doing the same thing the others [were] doing [for much less money]. You need the ground-level experience. When I should have gotten it, I was busy running an affirmative-action department."

I explored with him possibilities for placement in other areas of the company. Why didn't he expand his job into mainstream public relations, an area he was apparently more qualified to pursue? He responded, "I thought about it very seriously. I wondered where I was going with the system. It came up quite often. I talked about it when I first accepted this job. And at the end. They told me, 'We don't know. We'll have to get back to you.' They never did." That his superiors never got back to him may have been because the organization needed him precisely where he was. Or it may have resulted from senior management's perception that he lacked the necessary skills to compete with younger mainstream managers who had moved up through that field.

The latter possibility is supported by the comments of a manager who was offered, and took, a job in compensation and benefits. However, he failed in his new job precisely because his past concentration in affirmative action underqualified him for it. "I moved over . . . as director," he said. "Now, mind you, I'm going from a corporate [affirmative-action] job . . . to . . . compensation and benefits. I told the chairman of the company I didn't have any experience in that field. I might not be his man."

These three men identified two routes in their attempts to buffer their position in a company, moving laterally either into an entirely different corporate area associated with mainstream planning, production, or administration or into the mainstream component of the racialized area. But the failure of their attempts show that these racialized jobs have walls; executives who specialized in affirmative action and community relations were stymied in both routes. In exchange for establishing expertise in racialized functions, these managers reduced their value in other areas.

One personnel specialist noted that moving creditable line managers into affirmative action legitimates the role in the eyes of other executives. On the other hand, he observed, individuals ought to be in such jobs only about three years, or they were lost to the larger system. And, indeed, the longer the executives studied were racialized, the greater their chances of staying racialized in a field. Because of circumscribed skills, their exclusion from in-house power networks, and their "black-only" track records, people who were concentrated in racialized roles were perceived to lack the experience to compete in mainstream company areas. The white power structure that one anointed these managers now perceives the skills that gave them value as outmoded. A director of affirmative action talked about this dead end. "Nobody ever told me . . . that if you stay in [this] job you'd be in [this] job forever. You don't move to vice-president of personnel from manager of EEO."

In sum, when economic and political challenges to corporate hiring policies abated during the 1980s, the value of racialized slots also abated. Either incumbents tried but failed to move into the corporate mainstream jobs, or they did not attempt a move because they failed to perceive another niche for themselves in their company. I asked an executive secretary of corporate contributions if, after twenty years with the company, there were other departments that could use him. He replied, "Apparently not. That's what they told me." When I asked a community affairs director if she had sought out other areas of employment in a company when her job was cut, she responded, "I didn't think there was any place in the company that I could fit."

People with mixed careers, in contrast, could enlarge their roles within core areas in the company. For example, one manager working for a major Chicago retailer had a nineteen-year career that had by 1986 alternated between personnel and labor relations and urban affairs and affirmative action. In 1985 he was appointed to replace an exiting African American vice-president of community affairs.

His new position in community affairs was a downgraded version of the old job; his title was director of community affairs. In 1986, the community affairs staff and budget once again were reduced. He explained how he aggressively enlarged his role, which had become a meaningless position.

> I . . . went into my boss and told him I could do it with one hand tied behind my back. I had a director title for something that took one day a week to do. I told him that he had to give me some more responsibilities in personnel. So that's how I got that. [The commitment to affirmative action had] gotten so bad, the firm moved its headquarters from O'Hare to Salt Lake City. I guess that's one way of getting the monkey off your back.

The fact that in the 1970s the company headquarters moved out of Chicago, with its highly politicized African American population, to a much less confrontational and predominantly white environment may indeed be part of the reason for reducing the budget for community affairs. To protect his future in the company this manager asked for, and received, more responsibilities in mainstream personnel, becoming director of community affairs and area personnel manager. Without critical experience in personnel functions, it is likely that continuing cutbacks in community affairs would have placed him in the ranks of vulnerable managers. By 1993 he had managed to shed all remnants of the racialized components of his job and acquire the title of area personnel manager.

Exiting Executives: 1986 and 1993

The theory of politically mediated opportunity structures posits that when social upheaval among African Americans abated the fragility of black professional and managerial advancement would be revealed. In 1986, eight of thirty managers in racialized jobs were leaving or had left the company with which they had been identified. None of the mainstream managers were leaving or had left.

Table 5
Characteristics of Exiting Executives

Job Title	Last Field	Exit Year	Job Status
Director	Community Relations	1982	Cut
Vice-President	Community Relations	1983	Cut
Manager	Affirmative Action	1984	Downgraded
Director	Affirmative Action	1984	Downgraded
Manager	Community Relations	1985	Cut
Vice-President	Community Relations	1986	Downgraded
Manager	Affirmative Action	1986	Downgraded

Table 5 records the status of civil rights jobs held by executives I interviewed who had left a corporation by 1986. All were either cut or downgraded. The suggestion might be made that these executives weren't talented and lacked the critical skills to compete in corporations and so were placed in racialized jobs unimportant to companies. But a review of their titles suggests that the opposite is true. Of eight exiting executives in 1986, two were functional vice-presidents. In comparison, in the same year only 4 of the 1,362 executives with a title of functional vice-president or above working in a Fortune 500 or Fortune Service 250 company (Korn/Ferry 1986) were African American. In the group of exiting managers, moreover, two were the highest-ranking African American managers in their respective companies. Another was cited in a major publication in 1982 as being among the top black managers nationwide in leading white corporations. Two managers were the first black persons to reach the level of director in their respective companies, and a third was the first to reach the level of full vice-president. One manager was among only three African American managers to succeed to the rank of midlevel manager in a company. This evidence is consistent with my point that racialized jobs were valuable functions in companies, and that in the 1980s their value unraveled.

By 1993, seven years later, thirty-six of the seventy-six people in my original study had left their companies,[4] collateral evidence that their racialized jobs — once useful for restoring peace in urban centers — are intrinsically fragile. When I compared managers in mixed and mainstream careers with those in racialized careers, I found that the latter had left their original employer at almost twice the rate of their mainstreamed counterparts (68 percent versus 35 percent). Thus, an executive's exodus depended on work experience. Theoretically, people in nonracialized careers in 1986 would fare relatively better over the decade than those in racialized careers because of the generalized nature of the functions they performed.

Executives in affirmative-action, community relations, and other black-related fields who had left companies by 1993 found that their value in the open market had eroded. Seven of the thirteen sought another position in a white corporation. The three who found jobs either moved down in job level or went to work for smaller firms in other midwestern cities. The other ten executives left the white private sector and returned to the niche that has historically supported the African American middle class — government jobs, entrepreneurship, and black self-help agencies.

As job competition intensifies, the ability of middle-class African Americans to protect and maintain their position in the broader labor market may erode. Layoff rates in managerial and professional specialty occupations almost doubled between 1981 and 1992 (Gardner 1995). One racialized executive recalled that when he was laid off in 1987, "a lot of people had not experienced it. Six years later, a hell of a lot of people have . . . — both white and black." By 1993 he had stopped worrying about his upward mobility and was looking for any stable work that would allow him to break even financially. He said, "I'm making about the same as I made in 1987, [but] I feel fortunate that I have at least landed on my feet to some extent and have a job. I've been level for all that time, but I mean, you know, what's the alternative?"

Indications are that the economic cleavage between the haves and have nots, wider than it was a decade ago, may get even wider (Starobin 1993). Under these conditions civil disorder among African Americans at the base of the economic ladder also may erupt again. At the same time, African Americans at every level feel that social justice is limited and that leaders, laws, and policies may not help them. They interpret the Los Angeles riots as an event caused by increased feelings of helplessness similar to the sense of deprivation that set off urban explosions in the 1960s. If government efforts abate and opportunities for good jobs decline, black community protest and social disruption may reach 1960s levels. Although white society is much better prepared to oppress racial uprising, the dilemma of African Americans may reemerge near the top of the national agenda. If so, roles that can explain and help calm black disruptive elements will regain prominence, and African Americans willing to fill these roles will once again be in demand.

Chapter 8

A Rash of Pessimism

> Question: When we talked in 1986, you were [more] optimistic about your future here. What changed?
>
> Answer: Moves . . . happening all around [me]. Individuals moving [up] who were at your level, who you know are not as competent and have not done as much, then you wonder. And you see others who have done a tremendous amount, either not being promoted or [being] terminated in the organization. You begin to be concerned. And I guess now I'm . . . I don't see myself as cynical, now I'm just much more realistic.
>
> —A mainstream director in a
> consumer goods company

✦ Racialized jobs are an obvious, but not the only, vulnerable facet of African Americans' private sector gains. In 1986, for example, a high-ranking retail executive was reputed to be "the one," a likely candidate for the first African American to head a Fortune 500 company. But his prospects for advancement in the company suddenly and dramatically altered, causing the African American business cognoscenti to radically change their assessment of him. They pointed to serious flaws in his management style. Some described him as a star in management who became too much of a company man to be a leader; others described him as too passive. Their judgment can be summarized in the words of an executive who suggested that "he gave it a good shot," but "he just didn't have it."

Beneath such sentiments lies the hope that lack of merit rather than racial discrimination produced this unsettling and unexpected career outcome. In 1986, the idea of one of their number reaching the "mahogany office" (the chief executive's office) in five or so years seemed highly possible to African American men and women in Chicago corporations. But in the 1990s, they witnessed repeated failure among African Americans as they approached positions to which they aspired. An optimistic belief in "the system" that once prevailed among nonracialized executives turned to a pessimism infused with scaled-down career expectations. In the 1980s, a forty-year-old department director boldly predicted she would one day become an officer of the company. In the 1990s the same woman, at forty-seven still a department director, planned her future in terms of "how much longer I'll be with the company, as opposed to how far up [in the company] I'll go." Executives who once buoyantly approached or occupied higher-level jobs now soberly analyze their careers and talk about their fight against downward mobility.

Face-saving explanations varied according to gender. Women tended to stress the importance of family and community over career. The forty-seven-year-old department director, for instance, pointed out that she was doing all she could to move to a higher level, "but if it doesn't work out, it doesn't matter to me anyway. I have a daughter . . . I'm active in the church . . . I'm not willing to give up my whole life to this." Men stressed the importance of their role in the company and itemized their responsibilities.

Although the self-protective stances differ, all reveal anxiety provoked by new challenges to these executives' occupational standing. A man in his midfifties offers a particularly poignant example. In 1986 he was the highest-ranking black manager in his company and the protegé of a powerful mentor. Thereafter his company downsized, his mentor left, and he held three jobs in different companies.

> You know, my life was . . . well grooved. I had everything planned to the T [and] I controlled the situation. [But] all that ended. And what happens is, you have to start all over. You . . . reorganize your lifestyle, your way of doing things, your way of thinking. All the time you . . . try to preserve some dignity. But you always go back and self-search and [ask], what did I do

wrong? What could I have done differently? Hopefully you come up with some good answers and you don't [do] too much inner destruction.

The new pessimism about career advancement among these African American executives is most apparent when the attitudes of nonracialized interviewees are compared to those reported by interviewees in racialized careers in 1986, who were about five times more likely to have had their jobs eliminated, trimmed, or dispersed. Such changes contributed both to the exit of racialized executives and to their perception that they had reached a dead end. In comparison, executives with principally mainstream experiences reported less (market-mediated) job fragility and consequently expressed much more optimism about their corporate futures. Three-quarters of mainstream compared to one-third of racialized individuals believed they had a good or excellent chance for promotion or for making a lateral move leading to promotion within five years. They also were much more likely to predict that they could replace a lost job with one at the same level or better (70 percent of mainstream executives versus 30 percent of racialized). But by the nineties, these executives on the fast track perceived changes in the organizational climate as detrimental for African Americans.

The emergence of a global economy, the continued trend in corporate downsizing, and rapid changes in workplace technology have swept aside layers of middle managers, African American and white alike (*Fortune* 1993). The old corporate norms of lifetime employment and regular promotions appear to be dissolving just when the first full generation of middle-class African Americans to benefit from civil rights–related policies arrived on the scene. In the 1990s a cohort effect has intertwined with a racial effect to short circuit black executives' careers. Most of them are now "on the beach," as one man termed it, that is, they have experienced little or no mobility since the mid-1980s. They have chosen either to hang on until they can retire, or to leave the white private sector altogether. The stars in this study are falling; the nine executives who were in the first and second executive tiers in 1986 (i.e., chief officers and senior vice-presidents) have either left their 1986 employer or simply are trying to survive in their jobs.

Informants indicate that job fluidity inside many companies has veered from historical norms. People no longer stay in their jobs over a long period of time but are moved from assignment to assignment. Nevertheless, vertical mobility that nonracialized executives once earned based on tenure and performance has become unpredictable. Since the 1986 study a sizable proportion of such individuals have stagnated inside companies; thirty-one of fifty were with the same company in 1993 that employed them in 1986.[1] Just seven of them (23 percent) — four in support positions and three in line positions — had received promotions designated by a new job title. Even individuals receiving salary increases or different sets of responsibilities since 1986 did not view these changes as promotions. Typical is one African American midlevel manager in a multinational service corporation who occupied a series of different slots of equivalent status. She believed that she might never, as she put it, "move to a higher level." Conversely, twenty-one of the thirty-one (68 percent) remained at their 1986 title or grade level or moved down. The remainder (three people) made lateral moves out of operations and into support jobs that signaled that, for them, the contest for upward mobility and power was over. All of these executives are on the beach; they have been taken out of the game. A look at the characteristics associated with executives in nonracialized careers shows that both market forces and race are operating to shape the economic future of African American executives.

Placement in Companies

Nearly half the managers in nonracialized careers in 1986 were in personnel-related, public relations, and other support jobs. This pattern is consistent with other surveys, which show that African Americans with executive titles in Chicago's white corporations were not in the profit-driven planning and production jobs that lead to power within organizations (Chicago Urban League 1977). Although support jobs are not racialized, neither do they lie within the corporate loop of power or the mainstream work arena (Kanter 1977). They are peripheral functions with no responsibilities for profits or loss, out of the mainstream route for upward mobility. Only five of twenty-nine managers in this study who made it to the level of vice-president or above specialized in personnel. The remainder were in production, operations, and sales.

White graduates from the country's top business schools (Kellogg, Wharton, Harvard, and Stanford, for example) apparently avoid such support jobs on their road to the top of major companies. Of 1,362 executives with an MBA degree responding to a survey of senior level executives in Fortune 500 Industrial companies and Fortune Service 500 companies only 4 percent started in personnel and just 6 percent were in personnel when the study was conducted (Korn/Ferry 1986). In comparison, in my study 20 percent of non-racialized executives with a graduate degree in business started in personnel, and 30 percent were in personnel when they were interviewed. If racialized careers were included in this profile, the disparity in the occupational outcomes of black and white executives would be even larger. In sum, personnel jobs in this study, but not in studies of successful white executives, represent a sizable proportion of the opportunity structure filled by black men and women deemed to be successful managers. In 1980, African Americans employed in management-related occupations were almost twice as likely as whites to be in a personnel, training, and labor relations job (28 percent versus 15 percent), a proportion that had increased by 1990.

Corporate support areas such as personnel and public relations jobs mirror racialized jobs in a company. Black visibility in personnel and public relations positions is symbolically meaningful to black publics and useful for projecting an image of commitment to racial equality and sensitivity. In the 1980s it was hard for a major employer to justify the absence of black managers in a personnel department, given the overall high concentration of blacks in personnel areas. Individuals in these jobs, however, typically have no budget and few, if any, direct staff reporting to them; they are politically and operationally useful but not critical to an organization, and typically are the least prestigious administrative team members. Operations people, for example, viewed support managers as holding dead-end jobs, that is, out of the running for the top positions. Jackall (1988) notes managers' tendency to characterize the myriad of people in their occupational world, and the phrase "personnel types" was used repeatedly by operations people in my study to characterize personnel workers. "Personnel types have no power" over line managers, noted one, because they tend to lack operating experience. "Personnel types don't want to get their hands dirty," said another line manager, "and basically they're lazy." It is no surprise, then, that recommendations from personnel executives have little credibility with the central core of a company's directors (Kanter 1977). The positions' deficits in corporate power and prestige, in turn, force incumbents to beg to sit at the table where hiring decisions are made.

In sum, the typical African American executive career path, racialized or not, converged in corporate arenas that neutralized their power to change the culture of companies. Although both racialized and nonracialized jobs pushed them a certain distance up the corporate ladder, the jobs offered the least chances to wield influence, control resources, or sustain upward career mobility. White executives view personnel as one of the worst routes to top jobs in a company (Korn/Ferry 1990) — the crumbs at the corporate table.

Where the advancement of African Americans is tied to support functions, as in personnel, further advancement becomes differentially problematic in a rapidly changing economy with periods of job shortages. First, the mobility of African American managers is relatively more limited in corporate settings because of the nature of the jobs in which they are concentrated. Second, when corporate mergers and buyouts displace workers, personnel executives will fare relatively poorly compared to well-educated workers who perform the essential operations at the heart of a company, positions central to a firm's purpose. Workers employed in jobs less central to core operations are easier to discard. Third, support staff may confront a job market whose demand for their talent has diminished. One estimate from out-placement specialists is that displaced managers and professionals in human resources, information resources, and finance and accounting functions take about eight to ten months to find new jobs, compared to about 6.4 months for those in marketing and sales (Whittingham-Barnes 1993). For these reasons, perhaps, placement specialists urge African American executives to move out of support areas, as Jonathan Hicks reported in "Blacks Refashion Their Careers" in the *Wall Street Journal,* 11 November 1985. Knowledgeable insiders view these jobs in the business world as having been hit especially hard over the last several years.

The direct experience of executives I interviewed offers some support for this perspective. One reported that executive heads of support staff were the first positions to go after a company acquisition, while line officers were transferred to the parent company because they had an essential expertise. Another executive experienced a corporate buyout in which most of the officers in support jobs were treated in "quasi-patronage" terms and replaced by the new CEO's own team; line officers were retained. The exception was this African American chief financial officer, who believed his technical expertise (and his likability) prompted the new owner and CEO of the parent company to retain him.

Thus, within the "nonracialized" structure of managerial jobs the division of labor is racially differentiated, with mixed implications for African Americans in the 1990s. On the positive side, because support positions serve a permanent function for employers, unlike the transient role of racialized positions, blacks would be no more likely than whites to absorb the cost of private sector restructuring. On the negative side, support positions, outside the mainstream of companies, are soft roles in a hard environment where the pressure is on individuals to demonstrate a tangible contribution to the corporate bottom line. Performance standards in support jobs like personnel management are tied to subjective rather than to quantified measures, such as profit, sales, or production figures. The profit-generating capacity of personnel jobs would be hard to quantify. In this scenario, the "soft" positions of African American managers in the corporate division of labor would make them more superfluous and gains in these areas relatively more fragile and vulnerable to erosion. Equally negative, African American managers in these jobs are sidetracked from routes to positions of power. Even in the market-spawned (rather than politically spawned) sector of jobs in corporations, they are concentrated in useful but powerless positions.

Downsizing, Flattening, and Affirmative Action

African Americans attribute their career stagnation to processes they associate with corporate downsizing and restructuring, such as the trimming of functions and excess personnel. Some black executive careers are casualties of corporate attempts to meet new competition from domestic and international businesses. For instance, a company once considered the leader in its industry went through a period of flat sales brought on by stiffer competition, followed by several phases of staff reductions, departmental reorganizations, and new standards of profit accountability from middle management. In this process the size of a black executive's staff had been reduced from three people reporting to him in 1986 to one in 1992. Other individuals stagnated because of a relatively new business trend in which hierarchical, vertically integrated corporations are "flattening." In Chicago, for example, one nationally known professional service firm implemented a team-leader strategy, dismantling the hierarchical framework that more typically defines promotions in corporate structures. Such organizational changes make it harder to move up in a company and easier to fail. One executive who had worked for the company for twelve years said, "I don't think I'll make twenty." Nineteen of the fifty nonracialized executives in this study (38 percent) left companies altogether between my first and second interviews and went out on their own.

African American sentiment mirrors that of nonblack managers. The displacement of executives, administrators, and managerial professionals due to cutbacks or plant closings jumped by 50 percent between 1987 and 1992 (U.S. Bureau of Labor Statistics, unpublished tables). In the theoretical framework linking blacks' attainments to political pressure, however, one may plausibly identify the brunt of such changes as inequitably borne by African American executives. The most powerful weapon against bias in administrative hiring and promotion — the OFCC's ability to affect profits based on affirmative-action compliance — was undermined during the 1980s (Leonard 1988), as the agency's use of sanctions, such as back-pay awards and show-cause notices declined sharply, and compliance audits (which doubled between 1979 and 1985) became more perfunctory than critical (Leonard 1984, 1988; Orfield and Ashkinaze 1991).

Slack federal enforcement is compounded by macroeconomic changes to create an environment more lenient about implementing problack employment efforts. With such efforts aimed at now quiescent constituencies, it has become easier to neglect the quest for a more equalitarian society. In a survey of business leaders, for instance, affirmative action appeared far down on the list of business priorities, ranked twenty-third of the top twenty-five human resource management issues (Dingle 1988). Similarly, Hanigan Consulting of New York City found that recruitment of minority college graduates by Fortune 500 companies fell after 1989. Although overall hiring at U.S. campuses was down for the same period, one might have predicted that a greater percentage of minorities would be hired because minority enrollment on college campuses is growing.

Even where affirmative-action goals are implemented, they can be reoriented to recruit nonblack, nonracial, and more politically active minorities. Since the dual political pressures that underpinned blacks' mobility — federal sanctions and black collective action — have receded, blacks predictably would lose ground to other segments of labor. A retired affirmative-action manager noted that his old employer was "once characterized as . . . doing . . . things to help stabilize African American communities, [but now] initiatives are appropriated by other groups or are going away." In a similar vein, in a 14 September 1993 *Wall Street Journal* article, "Losing Ground," Rochelle Sharpe raises the possibility that corporate restructuring of major employers during the 1990–1991 recession disproportionately displaced working-class African Americans, while whites, Hispanics, and Asians gained thousands of jobs. However, Sharpe reported that African Americans' share of managerial, professional, and technical jobs in large corporations, although very small, did not decline.[2]

Between 1980 and 1990 nonracial minorities, such as Hispanics and women, entered managerial occupations at a much faster rate than did African Americans; in the previous decade the reverse trend was true for blacks vis-à-vis other ethnic groups protected by Title VII (Anderton, Barrett, and Bogue forthcoming). One explanation is that 1970s labor markets were shaped by race-conscious activism on the part of blacks and the federal government. After 1980 blacks became much less vocal and politically active, while other groups adopted the political model for status attainment that had served blacks so well, as evidenced by a *Chicago Sun-Times* 12 December 1994 article, "Hispanics Condemn Lack of School Jobs" and other sources (Ramos 1994). Although blacks still made gains in managerial jobs, this trend would explain why, relative to other minorities, their gains are rapidly winding down.

Relative changes in employment distribution are most striking in the progress made by women. Between 1968 and about 1971, the proportion of black men managers only briefly equaled that of white women; from about 1972 onward the proportion of white women managers shot up and rapidly surpassed the black male proportion (Bureau of Labor Statistics n.d.; U.S. Equal Employment Opportunity Commission n.d.). A vice-president of personnel in the financial service industry observed, "If you look at the numbers, there are more white women moving along . . . at a faster rate, than there are African American males. I . . . guess that white women, . . . Hispanics, . . . [and] Asians are closer to him [i.e., white men] in culture . . . [even if] just the texture of hair."

There is no way to know whether hiring, firing, or promotional decisions discriminate against these black executives. Motives that shape job allocations always look ambiguous to people outside the decision-making process. Ultimately, the extent to which employers view fair employment goals as "quotas" and African American workers as "tokens" predicts the shape of blacks' professional and managerial job opportunities in the future. With no race-based legislation or framework for community activism in place, race may once again interact with class to produce the more traditional and discriminatory results.

Gentlemanly Quotas and Glass Ceilings

A group of "heavy hitters" in private industry stood out from the typical pattern because of their positions as corporate officers.[3] Three of the executives I interviewed were chief officers, five were senior vice-presidents in operations, two were vice-presidents in corporate finance, and one was a chief finance officer (CFO). Finance carries exceptionally high status, although it is a support position, because the occupant typically plays a powerful role in a company.

Between 1986 and 1994 the number of this chosen few, already very small, eroded. People once near or at the peak of the job pyramid had ceded their line power in the company or had become self-employed. Each of the three chief officers — the CEO, CFO, and chief operating officer (COO) — were self-employed. Of five senior vice-presidents, two were self-employed, and two had moved into social policy–related areas from core line functions. Further down the pyramid, four of the twelve vice-presidents were self-employed, and two others had moved into support areas.

It was rare for executives in these higher echelons to experience lateral movement or promotions by changing companies. The market is soft for high-level executives, so few who left a company found equivalent or better jobs in the white private sector. Ultimately they started their own businesses.

Reasons for this may be racial in nature. The early 1990s was an employer's market, and the recession of 1990–1991 hit older and white-collar workers. Conventional wisdom holds that corporations seek to hire younger people, a state of affairs that would handicap higher-ranking African American and whites due to age. On the other hand, the market may be soft because the boardrooms of major companies remain all-white bastions of power (Heidrick and

Struggles 1979a, 1979b, 1984). In 1995 there was still not one African American CEO among the Fortune 500 industrial corporations. The only African American, Reginald Lewis, who ever attained that position became the CEO of a billion-dollar firm not by moving up the ladder, anointed by whites, but by buying the company.

Members of a national association of African American executives at officer or equivalent levels in Fortune 1,000 companies across the country reported that, with some exceptions, fellow members seldom leave a company in midstream. Conversely, they perceive that a market exists for similarly situated whites, who achieve promotions by jumping from company to company. Accurate or not, their self-reports reflect the mobility pattern of black Chicago executives. Most with titles of vice-president or higher had reached that level with one employer.[4] Only about 18 percent of 698 white executives working for major employers reported in 1990 that they had worked for only one firm (Korn/Ferry 1990).

Given these top black executives' reemployment records, one questions their chances of capturing equivalent jobs should they fail in one company. An intriguing case among my interviewees is that of the first African American ever to be appointed COO in a Fortune 500 company. In 1986 he indicated no plans either to leave or to start his own business. On the contrary, a popular national business magazine featured him as a major corporate star likely to become the first African American CEO of a Fortune 500 company. Reportedly, however, his eventual exit was connected at least partially to the publicity generated by this article. He left the company, expecting to move laterally to another Fortune 500 company, but reported that he did not receive an acceptable offer. After an extended job search, he left the white private sector altogether.

As the restructuring of the labor market interacts with the fragmentation of blacks' power, their chances of entering the corporate sanctum today monopolized by whites appears even more problematic. Some evidence indicates that blacks did best in companies where we can infer race-conscious intent operating in the background. In 1979 the largest proportion of black senior executives were employed in billion-dollar companies, in firms headquartered in the East, in organizations with the largest work forces, and in consumer products companies (Heidrick and Struggles 1979b). Sizable companies and consumer product companies face a trio of race-related considerations: federal regulatory requirements, public relations, and labor relations. Yet even under relatively supportive political conditions and less uncertainty in the economy, the corporate world enforced a gentlemanly quota on blacks in senior level executive positions. Excluding chief executive officers and presidents, nonwhites constitute less than 1 percent of senior level executives (Korn/Ferry 1986, 1990). White corporations, both nationwide and in Chicago, consistently fail to include senior blacks at these levels beyond token numbers, despite legislative efforts (Chicago Urban League 1977; Heidrick and Struggles 1979a; Korn/Ferry 1986, 1990; Theodore and Taylor 1991).

Corporate Politics and the Comfort Criterion

Political games are played at every level of a company, and "at the top of the pyramid the stakes just get higher," noted one executive, particularly in political battles over who will take the helm of a rapidly changing company. Over time, ascension in the corporate structure means that executives who play well — whether white or black — will move ahead. Those who do not will be moved out or moved aside.

Yet black executives cannot be viewed simply as political players like any others in a competitive corporate culture, for an abundance of research indicates comfort and compatibility as important considerations when incumbents select people to work with and to succeed them (Fernandez 1981; Kanter 1977). The comfort criterion interacts with race and creates another perspective for understanding black executives' standing. An illustrative case concerns one of the most senior executives in the study.

In 1986 this man noted that he didn't want to "seem too boastful" when I asked where he saw himself in the next five years, but he thought he had an excellent chance ("95 percent sure") of being promoted to the next executive level. When a major internal shake-up occurred because profits were falling, the senior executive level was reorganized, some members were reshuffled, and new operating positions were put in place to steer a rapidly changing organization into new national and international markets. Only one person jumped to the next executive level, and that was not the executive I interviewed, who by 1993, seven years after our first talk, appeared more guarded about his future, an executive "on the beach," although this was an image he clearly wished to avoid presenting. In the year following our second interview, he was forced out of operations and moved into diversity, an incongruous shift considering his skills.[5] Apparently a decision had been made to move him aside but not release him from the company.

Selection for advancement in the upper regions of the corporate world rarely hinges solely on the ability to do a job but is connected as well to subjective criteria and the extent work groups believe that each job contender "fits in" (Jackall 1988; Kanter 1977). This subjectivity makes African American people — through color, culture, and sometimes political consciousness — obvious targets of bias in the selection process. A senior executive who left a company to start his own business rather than be passed over continually for promotion said,

> As I went higher in the organization . . . other factors [overshadowed my] performance. I don't think that I was particularly successful in negotiating those factors to my favor. [Question: Other factors?] Well it couldn't have been my performance. I not only met, but I exceeded, my performance goals every year, and many times I was the only person to do so. Your race. Your attitudes. I wasn't politically conservative, and I wasn't too good at hiding that. And on more than one occasion, I developed a feeling that I wasn't being selected for further promotion.

Having gotten "the message," as he put it, he left the company by 1994. At least in his own eyes the talents and skills that had until then sustained his mobility became irrelevant when he stood one job away from the company's inner circle. Objective performance goals were overshadowed by subjective dynamics that help people to feel at ease. He left the company when he realized he would never be "picked by someone and brought to their table."

Neither differences in appearance nor a wide range of other differences are easily tolerated in the ranks of managers (Jackall 1988; Kanter 1977). African Americans contending against whites for power are both isolates and tokens. Race-related stereotypes and dissimilarities heightened by tokenism are likely to emerge (Kanter 1977). Race becomes a wild card easily played against them, particularly in the competition for jobs involving uncertainty in which stereotypes and tradition work against exotic workers (Kanter 1977; Pfeffer 1982). Corporate slots rarely held by African Americans have been relinquished by them—and filled by whites.

Apart from race, tokenism of any kind hinders one's ability to form the political coalitions necessary to win out against contenders for top jobs. Blacks must be adept at forming coalitions across racial lines, which is exceedingly hard to do. Furthermore, the dominant group becomes alert to token individuals, a heightened awareness that accentuates differences (Kanter 1977). Historically, the exaggeration of racial differences in particular has caused those differences to become devalued and inequality rationalized. I know of only one person who has successfully transcended race on his journey to the top of a white-dominated bureaucracy, the former head of the Joint Chiefs of Staff, Colin Powell.

Marginal Men

At the same time executives must be selected by others to move up, they must adapt themselves to fit in. For black executives, successful adaptation undermines racial cohesion and racial group solidarity, creating marginal men and women with one foot in each racial culture who belong to neither.

African American managers were outsiders in all-white institutions in the 1970s, unwanted participants in a socially and racially homogeneous occupation, who survived and thrived more by consciously conforming and submerging racial differences than by aggressive individualism. The beached executive who held the highest rank of those I interviewed, for example, succeeded in a white culture where loyalty and commitment to the company, and total adherence to company norms, are highly valued. He was reminiscent of the "other-directed" corporate man of the 1950s who followed a safe path, looked good, and made no waves (Fromm 1955; Harrington 1959; Mills 1951; Riesman 1950; Whyte 1956). Repeatedly he mentioned that he would do what it took to "serve the company." He used the word "we" (referring to himself and the company) continually, creating and reinforcing my perception that he was a company man, a team player, an image that had surely helped him get ahead.

The role conformity played in black executives getting ahead becomes more clear in the vocabularies of people with titles of vice-presidents and above compared to those of lower-ranked executives. Higher-ranked executives were generally less likely to distinguish themselves in racial terms during our interview. They had, for example, a greater tendency to speak of the company using terms such as "family," and to use inclusive terms such as "we" and "us" rather than "they" and "them." Fewer of their references to white peers and superiors or to career routes could be construed as criticism. Overall, they were much less likely to frame their understanding of their careers in racial terms. The most senior executives in this study had internalized a norm that called for demonstrating loyalty, treating even the privacy of our interview as a chance to demonstrate color-blindness, company commitment, and social conformity.

As blacks in a white world their survival and success relied, at least to some degree, on being "nonblack" to win white acceptance. In order to succeed, that is, this vanguard of blacks minimized conflict, didn't make waves, and avoided controversy. For African Americans, but not for whites, fitting into the bureaucratic culture means shedding the racial self and submerging racial history and differences. The corporate world is, after all, an environment where one gains advantages from interpersonal relationships. Put another way, these managers' success in the conservative world of large corporations requires that they gloss over potentially volatile differences where race matters.

The marginal run the risk of being perceived as outsiders — that is, culturally and politically suspect — by generalized others in the black community. One very high-ranking executive was characterized by lower-ranking blacks as the company's "fair-haired boy" who symbolized a degree of co-optation and communal betrayal. His strategies for moving ahead were provocative to these blacks; they viewed him as isolated from other African Americans in the company and as attempting to "fit in with whites" and to garner white approval. This executive also was summed up as a person who had put himself in "an awfully lonesome spot [vis-à-vis other blacks in the company] for an awfully long time." Other blacks in the company demeaned him because, I was told, "he became someone who didn't remember where he came from."

Executives who didn't have, develop, or keep a black constituency that would come forward when their jobs were threatened decreased their power base. In the 1960s, pressure from black constituencies created new opportunities that these executives exploited. In the 1990s, the lack of a constituency means these executives can be treated like any other member of the team, but with vastly different group consequences. The "fair-haired boy" just described represented a dramatic measure of black gain. Between our first and second interview he was promoted from vice-president of operations to company president. But he had no black constituency to call on later to protest his firing from that position. In 1995 this once high-powered man was without a job. If blacks' racial group solidarity is splintered by the process of upward mobility, the capacity of the group is undermined.

Overall, it is doubtful that the federally mandated race-based employment programs that began in the 1960s ever were intended to be a permanent part of the legislative landscape. The hope instead, it would seem, was that the proximity, the influence, and the achievements of African Americans would deracialize the cultures of employment and achievement. The movement of employers toward a color-blind posture in hiring and promotion would take on a life of its own. Thus, from the point of progress, it was — and still is — good news that most African American executives who participated in this study in 1986 did not occupy race-specified roles in Chicago corporations. Finding them employed outside the corporate niche of politically mediated jobs was a positive forecast for the continued integration of African Americans throughout the private sector. Perhaps race-based job allocation in the postindustrial economy *can* give way to race-blind fair employment policies. Perhaps this competitive African American executive elite *can* protect and reproduce its position via human capital and business networks, without federal protections.

However, labor market restructuring increases competitiveness along lines drawn by race, gender, ethnicity, and ideology inside the culture. At the same time, the significance of net gains or net loss in the job market and downward mobility for each race and ethnic group are hardly equal, a concept essential for recognizing discriminatory processes in a labor market context.[6] That is, even if equal numbers of African Americans and whites lose jobs, the impact on the aggregate status of African Americans would be relatively much greater. The progress made since the 1960s could be seriously undermined.

Chapter 9

Bursting the Bubble: The Failure of Black Progress

✦ Most scholars agree that dramatic progress was made in blacks' access to white-collar occupations over the past three decades. One explanation points to the forces of an impersonal labor market that rewarded improvements in black skills and education with occupational mobility in a growing service economy (see Smith and Welch 1978b). And, indeed, both aggregate data and individual cases show that college-educated blacks improved their position when the economy was on the upswing and the need for skilled labor was expanding.

Yet the timing of blacks' attainment as revealed by my study, in conjunction with other research (such as Freeman 1976a; Leonard 1984), strongly links this attainment to political pressures on employers exerted by government and by the black community. Since black executives in Chicago entered the job market in an environment dominated by black political activism and governmental intervention, I suggested that the elaboration of the black middle class that has been attributed to their entry into higher-paying white-collar jobs is not grounded solely in their educational attainments and in economic trends. The growing demand for blacks in higher-paying jobs is also a function of a shift in hiring to conform to blacks' demands for increased access to economic resources and to government regulations. I do not mean that education, skills, and related factors such as motivation and hard work are trivial to blacks' gains and greater competitiveness. I mean, rather, that blacks' objective qualifications are still strongly intertwined with racialized processes that continue to operate in the labor market.

Before the 1960s the market economy that spawned the black middle class was dominated by artificial and race-based barriers, not free and impersonal exchange. It was an economy shaped by restrictions that limited interaction between the races and that allowed black business and professional opportunities only in the segregated environment of black ghettos. This economy created opportunities in a narrow range of service areas for a very small number of blacks, such as doctors, morticians, and lawyers. The opportunity structure was broadened somewhat when blacks were able to enter the public sector as teachers or other professionals. But even this somewhat broadened economic system was structured along racial lines: Black professionals distributed services to other blacks, as opposed to the total (i.e., predominantly white) community.

These race-linked occupational restrictions are reflected in executives' descriptions of their labor market experiences. Most college-trained blacks entering the labor market before 1965 went to work for the government. Only the most highly educated broke into the professions in the white private sector. Even in these cases, characteristics ascribed to a free market were not in play; they worked almost exclusively in industries that depended heavily on government contracts for survival.

From the mid-1960s onward, the labor market for educated blacks underwent a transformation. College-educated blacks entered the economic mainstream in jobs similar to those of their white counterparts that better reflected their training and abilities. It appeared that artificial racial barriers had lifted as college-educated blacks captured the incomes and occupations to support the middle-class life-styles previously reserved for whites. For the first time in history, the structure of opportunity for college-educated blacks shifted significantly to include higher-paying white-collar jobs in the central economy.

My study documents this shift in executives' reports of job search experiences, which showed a much broader spectrum of jobs available to them after the mid-1960s than before. After 1965 their rate for going into sales jobs tied to unstable black consumer sectors and into low-paying clerical positions decreased by two-thirds. At the same time, at the other end of the white-collar job spectrum, twice the proportion had access to professional and managerial jobs in the white private sector. Overall, these changes in patterns of employment are consistent with other research showing that the socioeconomic status of blacks jumped dramatically when college-educated blacks entered professional-managerial jobs inside government and private industry.

My study reflects the increase in compliance pressures on the part of the EEOC and the OFCCP at the same time that managerial and professional job opportunities in white organizations increased. Executives entering business-related fields credited the activism of black organizations and civil disturbances with helping to create their job opportunities; two-thirds of them who entered business-related fields after 1965 knew of the vulnerability of their company to black consumer boycotts and the vulnerability of company property to urban upheaval, and three-quarters of this group also knew that their employment was a result of a company effort to hire blacks.

Further evidence of a connection between political variables and black employment opportunities appears in employment patterns where blacks fill jobs mediating black demands for white institutions. That is, the alchemy of market and political forces not only influenced new behavior on the part of employers, it also increased demand for blacks by increasing social service and manpower development programs in government and in private industry. Although blacks clearly made gains in a variety of jobs, the new black middle class is grounded in professional and business roles created or reoriented to nullify pressure from black people.

A sizable majority of African American managers in the upper echelons of Chicago's white corporations were channeled into an occupational structure that evolved from the pressures of the civil rights period. Two-thirds of those I interviewed had held at least one job that was oriented toward blacks over the course of their private sector career. One-third of the managers were concentrated in race-oriented jobs throughout their career. This nascent business elite moved, in particular, into personnel areas of corporations to administer affirmative-action policies and into public relations areas to respond to turbulent black communities. One-third of the managers I interviewed who entered the white private sector after 1965 were recruited to fill jobs in one of these two areas; others were enticed to transfer into these jobs by salary increases, better job titles, and promises of future rewards. Even those with incompatible backgrounds and highly technical skills — such as accountants, engineers, and chemists — were tracked into affirmative-action and public relations areas. The forces expanding blacks' economic opportunities were protest related, and the new black middle class is at least in part a politically mediated phenomenon.

My portrayal of the black middle class is both consistent with and different from the dominant theoretical perspectives on race and labor markets. While I agree that unprecedented advancement occurred within the black middle class, I do not agree that attainment among blacks is evidence of the deracialization of labor markets (cf. Wilson 1978). Middle-class attainments among blacks reflect a dependence on employment practices that are sensitive to race. Better job opportunities for blacks are connected to federal legislation and to the expansion of social service bureaucracy and other administrative apparatuses to implement social policies designed to appease disruptive black constituencies.

My observations of the characteristics of jobs that blacks hold contradict the notion that spontaneous market demands in tandem with education and skill lifted the employment barriers faced by talented blacks. Such qualifications are necessary, but they have never been sufficient to ensure the relative success of blacks in the economy. Governmental mandates — not the forces of free markets — are critical to expand and stabilize the black middle class.

Finally, my findings go beyond research that highlights the effects of affirmative-action kinds of government policy on the black middle class to indicate that the intent of such policy is not assimilated by the marketplace. An analysis of the careers of highly successful black executives with great potential as competitors inside the mainstream labor market showed that the economy opened up in only a distinctive and marginalized way. Private employers channeled a group of people with a variety of talent into racialized careers during the 1960s and 1970s. Even the majority of mainstream, that is, non-racialized, careers found in this group have marginalized features. Executives in this study are concentrated in support areas where institutional objectives reflect policy attempts to nullify blacks' potential for disruption, not in the planning and operations functions oriented toward profit that lead to power in an organization. My view of the black middle class is consistent with a conflict model of class relations rather than with traditional status attainment models. Blacks' advancement is a function of protest, and this protest was not resolved by true deracialization in the labor market. Rather, the role of race has been reconfigured in the modern economy and continues to have an impact on blacks' access to middle-class positions.

Implications for Racial Equality

What do a politically mediated model and racialized jobs have to say about blacks' chances for economic equality? The short answer is that they are not likely to eradicate inequality.

In the scenario of this book, demand is mediated by an interrelationship among economic upswings, political pressures, and a labor supply of qualified workers. The intersection of these three factors created the conditions for some blacks to rise and compete for higher-paying jobs against whites. Yet the processes of attainment reported here show that blacks' new socioeconomic status does not necessarily indicate racial equality in institutions or in labor markets. First, some types of racial segregation and economic inequality are maintained and even facilitated by the mechanisms associated with black middle-class mobility. Second, if the black middle class results from special political and legal conditions, then it can be argued that it occupies a fragile economic position.

Maintaining Inequality

Since the mid-1960s, the black middle class no longer has been relegated strictly to the lower ends of occupation and income hierarchies or restricted by geography. However, a new structure of inequality was created by a system of employment opportunities that channeled some blacks into racialized functions. On the positive side, this system of jobs afforded some blacks a chance to succeed economically and garner unprecedented, albeit temporary, status in white institutions. The negative half of the equation is that this system could not solve the problem of blacks in the long run. Affirmative-action managers, for instance, wrote hiring plans and were in charge of their implementation. But affirmative action was implemented with an eye toward appeasing governmental and public relations requirements, not changing the color of power brokers in white institutions. Data show that black men in particular are underrepresented — even after thirty years of affirmative-action efforts — in managerial jobs and in almost every business-related profession.

Progress was limited because the incumbents of racialized jobs simply did not have the power to change institutional practices. The ability to get things done required access to and persuasiveness with the CEO and other top management. Executives I interviewed who survived in these roles were not individuals likely to risk their jobs by "pushing the envelope" and disrupting the equilibrium of their employing institution. It is logical to presume that they survived in these jobs because of their ability to accommodate whites, not embarrass the company, and not cause trouble.

Their jobs helped companies conform to federal regulations, and programmatic allocations (such as technical assistance, corporate funding of community-based projects and job training) both quelled urban pressures and undermined claims of racism, creating a progressive, more socially conscious, corporate image. But these allocations were too small and were viewed by companies as a short-term atonement for past grievances, rather than as a long-term commitment to justice. Racialized jobs, therefore, were instrumental in negotiating the needs of the black community and in distributing corporate resources, but they ultimately maintained inequality by temporarily abating black pressures and meanwhile marginalizing the incumbent.

At the same time, this study makes it clear that racialized jobs effectively kept their incumbents from traveling conventional routes up the corporate ladder. When individuals in this study entered race-oriented and staff positions in white companies, they assumed career tracks that typically do not lead to line power in a company. Once in these jobs, many of them were constrained by the perception or the reality that they lacked the necessary skills to contribute in a mainstream function. Consequently, gains they made over the last three decades did not—and will not—blossom into meaningful numbers of executives heading production and planning areas. The tracks their careers took in the 1960s and 1970s diminished the pool of blacks in Chicago corporations who could compete to manage mainstream production units in the 1980s and beyond.

In sum, managers in this study are part of a black middle class that has occupied a useful but nonadversarial position in white companies. Ultimately this means that racialized jobs were a factor in reducing competition for power in organizations along racial lines. And, since they were unable to succeed in policy- and decision-making positions in meaningful numbers, it is doubtful that the makeup or resource allocation of organizations will change dramatically. Even current policy decisions to continue affirmative-action programs are contested outside any arena in which blacks exercise power.

What I have found, therefore, is a structure of achievement that preserved inequality while it carried out its role in reinstating social order, and that established a class position with obsolete features built in. Executives in this study were desirable candidates for affirmative-action and public relations jobs at a time of intense social upheaval; when pressure from blacks abated, the status of many of these executives tumbled. Specifically, as racial pressures were ameliorated in Chicago, racialized jobs lost their value. Given the relationship between these jobs and the political pressures faced by white corporations during the 1960s and 1970s, one could take the position that affirmative-action and urban affairs managers did their jobs for companies too well. In relieving the pressure on companies, they not only helped shut the window of opportunity through which other blacks could follow but they undermined the very element that had produced their own positions.

African American Middle-Class Fragility

Although job opportunities for college-educated blacks and whites have converged since the 1960s, class mobility for these two groups stems from different factors. In my view of black attainment, the opportunity to earn income and maintain middle-class life-styles depended as heavily on blacks' broad social status and the activity of the state apparatus as on general economic trends. First, African Americans were viewed by major employers as desirable candidates for professional and managerial jobs because of governmental sanctions. Second, meeting the needs of, or solving, the problems of blacks was near the top of the public policy agenda.

Extending this view leads to the conclusion that different factors insure the economic viability of blacks and whites. Thus, if the federal government dismantles strong race-specific programs and affirmative-action mandates, employer effort to create and maintain equal employment opportunities may shrink accordingly. Indeed, this study suggests that the institutional mechanisms for protecting black attainments in the white private sector — affirmative-action and community relations departments — *are* shrinking and becoming watered down. Weakening these areas weakens the race-conscious influence on employment decisions that in the past protected blacks. Moreover, departures from the liberal social thinking that dominated Congress in the 1960s make likely a radically different political agenda and an alteration in the basis for federal policy and spending decisions.

The federal government's incentives and sanctions in the 1960s and the 1970s were not meant to change the intrinsic nature of the economy but only to get employers to respond differently. Two outcomes are possible if these incentives and sanctions are dismantled. First, despite race-conscious political supports, blacks' qualifications and the greater acceptance of blacks in white-dominated settings may enable the same or a larger relative proportion of blacks to move further up the economic ladder. Or, second, the proportion of employed blacks in the next cohort to enter higher-paying white-collar occupations may erode. In the first outcome we would see significant further progress in the black middle class in capturing positions of power within white institutions. In the second outcome, we would see stagnation or vastly diminished rates of change.

Of these two options, the second, I believe, is the more plausible. The link between black class mobility and political pressure predicts the fragility of the middle-class position. Although black protest and government intervention theoretically occur outside the market-place, they are two of the three ingredients necessary for blacks' economic progress. That is, the status of the economy, the level of black activism, and the public policy agenda all assisted blacks to rise. I believe that gains made in the black middle class will dissolve in the next generation of labor force participants for reasons related to all three of these ingredients.

Changes in Public Policy

The race-based legislation and spending that assisted blacks to rise is being challenged and dismantled. As a general proposition, as government funding dwindles from social service areas and federal efforts abate in employment legislation, the ability of blacks to maintain and to continue their gains will also erode. If federal antibias employment policy and government spending in social service arenas created the conditions that opened up nontraditional white-collar jobs to blacks, it follows that cutbacks in affirmative-action pro-grams in major companies and cutbacks in government social ser-vices would decrease the demand for blacks in specified occupations.

In the national debate, some social critics see blacks' dependence on government as a negative outcome of federal protections. How-ever, I see government dependency as an unavoidable partner in blacks' progress. Because the federal government did not change the intrinsic nature of the economy or of employers, equal employment and color-blind hiring are not institutionalized in the labor market, and fair employment practices would not continue in the govern-ment's absence. We must think of affirmative action not only as reparation for past discrimination but also as an instrument neces-sary to prevent present acts of employment discrimination. And there is plenty of evidence to indicate that racial discrimination, both economic and social, still exists (Cose 1993; Feagin and Sikes 1994; Jones 1986; Kirschenman and Neckerman 1991; Massey and Denton 1993; U.S. Department of Housing and Urban Development 1991).

Black Power Versus Passivity

Within a politically mediated model of middle-class ascendancy, black gains may be fragile for another reason. The rise of the new black middle class is tied to black collective action, but splintered interests and large economic divisions now exist within the black community. In the 1950s, class divisions existed, but segregation forged a racial group consciousness that, in the face of white terrorism, transcended factionalism based on occupational and income differences.[1] In the 1990s, however, the gap caused by the assimilation of a skilled and highly educated black middle class and the dislocation and exclusion of a black underclass may mean that blacks have less ability to form alliances and harness the power of collective action.

The emergence of the new black middle class and blacks' new forms of institutional participation, in part, occurred by siphoning activists and future leaders out of the black community. This siphoning process mirrors that which William Wilson (1987) describes when arguing that desegregation spurred the exodus of middle-class blacks from inner-city neighborhoods. Chicago alderman Bobby Rush, former head of the Black Panther Party, and Chief Justice Thurgood Marshall, the NAACP's chief legal strategist for ending segregation in the South, are prominent individual examples. Much less prominent but still relevant examples are managers in this study who, during the 1960s, were community activists with networks that made them useful recruits for manpower training and development jobs in white corporations. They left behind a dispirited black community increasingly beset by social and economic problems.

At the same time, a new type of black accommodationism and white paternalism emerged. As the new black middle class became dependent on whites for jobs, the raised arms with clenched fists associated with "black power" and solidarity that dominated the 1960s were dropped and sleeved among the middle class in the grey flannel associated with being a "company man" and "fitting in." Integration of this black middle-class vanguard into white corporations set it upon a path of achievement and upward mobility that, having separated it from the black collective, also divests it of both the license and the constituency to argue group claims. Nathan Hare once observed that a vanguard detached from the mass becomes an elite. This reasoning, perhaps, allows whites and black neoconservatives disingenuously to dismiss, as a means to exploit race for its own thinly veiled interests, black middle-class complaints about racial barriers.

The problem blacks confront is whether to attempt to go all the way into the system or to go back to their roots. Assimilation has given rise to a new set of community problems without a corresponding rise in economic power. Put another way, the dilemma is whether to continue to pursue the assimilationist goals embodied by traditional civil rights organizations such as PUSH, the Urban League, and the NAACP or to embrace the separatist ideology exemplified by the Black Muslims.

I predict one of two outcomes. In a politically mediated middle class, when racial pressures on white employers emerged, the system of racialized jobs enabled white bureaucracies to mediate the goals and options open to black communities. If black solidarity, filtered through white bureaucracies, grows weaker, then the power of protest blacks exercised three decades ago is weakened also. African Americans would then find it hard to organize protest strategies on a level impressive enough to protect and provoke further economic gains. Under these conditions advancement among middle-class blacks will erode in the future.

In an alternate scenario, having been exposed to education and the subjective philosophy of merit, the black middle class may become more race conscious. Black executives who played by the rules of the game and now confront glass ceilings may become an angry, alienated middle class that gives birth to new and radicalized black leadership. The broad coalition of disparate organizations that supported the 1995 March on Washington and the antipathy that cuts across class lines against denouncing Louis Farrakhan may cue future coalitions. Limitations to black power inside the system may radicalize the black middle class and spur race-based challenges to institutional practices.

The Status of the Economy

Finally, blacks gains will erode because of macroeconomic factors. As companies continue to prune functions and excess personnel to be more competitive, and as a global economy and technological advancements sweep away layers of middle managers, the economic positions of both whites and blacks grow more fragile.

Even in the best of times, when the economy is expanding and good jobs are plentiful, blacks make gains but continue to trail their white counterparts. For example, despite the advancement of black men in managerial and business-related professions, they remain greatly underrepresented in jobs at the core of white corporate America. Forecasts of the status of blacks become even gloomier when one considers the predictable effects of economic recessions. That is, in a soft economy, the competition for jobs that secure middle-class lifestyles stiffens. Indeed, this may be a reason that race-based employment legislation finds little support among whites, as the *Pollwatcher Letter* columnist suggested in the *New York Times* on 12 May 1992. Thus, while both races are affected by restructuring in the marketplace, the risk, the meaning, and the impact differ considerably for blacks and whites.

In the final analysis, using Christian morality and street militancy to provoke white guilt and embarrassment no longer works. First, these strategies depended in large part on shock value, the contrast between American ideals of justice and equality and the reality of blacks' daily existence, particularly in the South. That contrast alone was powerful enough to give black leadership the moral authority to challenge the economic status quo. Now the degree of poverty among African Americans is no longer shocking, in part because what once was hidden beneath the image of America as a middle-class society now is obvious in urban ghettos. Nor is there national sympathy for the claims of blacks. When race-based demonstrations and appeals arise, they are more likely to provoke disdain in whites, to raise the question, What do those people want from us anyway?

Second, the onus of guilt over what went wrong has shifted. Blame once borne by the national conscience has been placed by economic and cultural conservatives on the poor and "undeserving" blacks doomed by white presumptions of inherent black immorality. Dooming them also is the political expediency of using blacks as scapegoats in an era of economic uncertainty. What remnants of guilt remain have been cleansed away by the rise of a black middle class, and by whites' sense of cultural sacrifice and racial beneficence for implementing thirty years of social welfare programs and affirmative action.

Thus, the controversy once tempered by compassion that began with the passage of the Civil Rights Act in 1964 and with affirmative action has reemerged full blown. Widespread public sentiment among whites increasingly rejects the notion of government intervention, according to a *New York Times* Pollwatcher Letter on 12 May 1992. Given the results of the 1995 congressional elections, it also appears that the white electorate now rejects leaders who have a tradition of supporting black causes. This public resentment appears to bring with it a different political climate for the 1990s and perhaps beyond in which whites' political pressure demands that protective policies oriented toward blacks be retracted and allowed to erode.

As a society we seem to disdain history. We ignore the intransigence, the meaning, and the magnitude of racial inequality, thereby making it easier for social critics to trivialize the need for affirmative action. Ahistorical assessments of the impact of antibias employment legislation, moreover, fail to recognize the potential for future racial conflict in the United States. If affirmative action and other race-specific legislation are dismantled, no mechanisms exist to re-place them. Affirmative action has always been a hotly contested and controversial policy. It was implemented as a last step in a series of escalating antibias policies because racial strife threatened the fabric of the country and because the mechanisms for racial separation in the core economy are tough and enduring.

NOTES

Chapter 1

1. The term *underclass* designates a heterogeneous grouping at the very bottom of the economic class hierarchy that includes low-paid workers whose income falls below the poverty level, the long-term unemployed, workers who have dropped out of the labor market, and permanent welfare recipients. The heads of households in the underclass are primarily women. Men are primarily unattached and transient. In this definition I am borrowing heavily from the work of William Wilson (1981:21).

2. See, for example, Leonard (1984) and Heckman and Payner (1989) for research on the effects of federal contract compliance programs.

3. One person declined to be interviewed; ten others were not interviewed because of logistical reasons or because they did not meet my criteria.

4. I was able to update information on all the original executives in the study, locating seventy-four of seventy-six and interviewing fifty-nine. Of the seventeen not interviewed, six had moved out of state, two had left a company and could not be located, seven failed to respond to my request for a follow-up interview, and two had died.

Chapter 2

1. Smith and Welch (1986) examined the occupational attainments of the black middle class and argue that upward mobility among blacks preceded the civil-rights legislation of the 1960s. In addition, earlier discussions of federal antibias regulations suggest that the legislation had minimal effects (Adams 1972; Flanagan 1976; Ornati and Pisano 1972; Wolkinson 1973).

2. For comprehensive discussions of affirmative-action and equal employment opportunity programs see Benokraitis and Feagin (1978), Hausman et al. (1977), and Nathan (1969).

3. No data on the percent of federal government procurement going to minority businesses have been published since about 1982.

4. See Brown and Erie (1981) for discussion of the number and source of jobs created in the public sector between 1960 and 1976; see Betsey (1982) on the growth of the public sector between 1939 and 1981.

Chapter 3

1. See Isaac and Kelly (1981) for an analysis of the riot-welfare relationship.

Chapter 4

1. Due to strict segregation in the educational system, black college students were then heavily concentrated in black colleges.

2. *Griggs v. Duke Power* required that if employment tests were shown to have an adverse impact on protected groups, the firm must demonstrate that the test is job related; employers have found it difficult to do so (Burstein and Pitchford 1990; Kirschenman and Neckerman 1991).

3. In identifying the stages, groups, and activities of the civil rights movement I borrow heavily from earlier research and scholarship: Bloom (1987), Sitkoff (1981), Broderick and Meier (1971), and the Civil Rights Education Project of the Southern Poverty Law Center.

4. In one week in June 1963, the Justice Department cited several hundred demonstrations taking place not only in southern cities such as Tallahassee, Savannah, and Jackson, but also in the North, in Providence, Columbus, New York City, and Los Angeles (White 1964).

5. Brand names are deleted because these products are identified with producer and thus might break the executive's anonymity.

6. In 1964, riots erupted first in Harlem, then in Brooklyn, Bedford-Stuyvesant, Rochester, and New York City, and later in Jersey City and Philadelphia. In the summer of 1965, riots broke out in the black ghetto of Watts and in Chicago and San Diego. In 1966, more than two dozen cities were struck by riots; in 1967, Detroit experienced the largest and the most destructive of the series of urban upheavals; and in 1968, waves of rioting followed the assassination of Martin Luther King, Jr.

Chapter 6

1. Set up by the Department of Labor, PICs consisted of local businesses and minority entrepreneurs working with the private and public sector

to identify ways in which private enterprise could take advantage of existing Department of Labor initiatives to increase employment in targeted areas.

2. Under the bonus plan in this company, an employee earned a percentage of the net profit: 1 percent of the first $50,000, 2 percent of the second $50,000, 3 percent of the next, and 4.5 percent of anything over $150,000 in profits. This man reported being "one of the few managers who got in that fourth category."

Chapter 7

1. *Memphis Firefighters v. Stotts* (1984), *Wygant v. Jackson Board of Education* (1986) and *J. A. Croson Company v. City of Richmond* (1989) are examples of key decisions that undermined the principle of racial preferences in employment.

2. In 1986 the Supreme Court in *Local 93 v. City of Cleveland* held that local courts can approve settlements that involve the preferential hiring of blacks. In the same year, the decision in *Local 28 v. Equal Employment Opportunity Commission* approved a lower-court order requiring a union local to hire a fixed quota of blacks.

3. Information conflicts about what in particular happened to managerial jobs during the 1980s. Contrary evidence suggests that black managers are the least likely among workers to lose their jobs in companies during periods of economic distress and that black managers increasingly have been victims of job loss in major corporations.

4. In the 1990s, I found twenty-six executives (100 percent) relocated in segregated careers and interviewed nineteen. I also found forty-nine of the fifty executives (98 percent) in nonracialized careers and interviewed thirty-nine.

Chapter 8

1. Large corporate mergers have somewhat complicated the concept of changing employers, as corporations sometimes become operating divisions of the buying unit. For instance, a vice-president of finance in 1986 worked for a firm that by 1993 was acquired by a larger company, itself a division of a multinational electronics corporation. If individuals were employed in some part of either the old or new entities in this exchange, I viewed them as remaining with their employers. Conversely, I viewed individuals no longer

with their original employers, in whatever form those entities now exist, as leaving their employers.

2. Sharpe's analysis is based on a matched sample of private companies reporting to the EEOC. The U.S. Bureau of Labor Statistics study of 1990–1991 employment data, focused on aggregate level data, found that blacks and whites both suffered a net job loss in this period.

3. Individuals at the corporate officer level are elected by a company's board of directors and generally include the chief executive officer (CEO), chief operating officer, chief financial officer, chief information officer, and the executive vice-presidents of manufacturing, sales, marketing, legal, and human resources. Other executives, appointed at the discretion of the CEO, generally include division presidents and vice-presidents. Consequently, an enlightened CEO can have a significant impact on the racial makeup of senior corporate positions. However, the full board of directors ultimately determines who sits in the corporation's inner circle.

4. Three of twenty-nine vice-presidents or higher entered the first or second tier of a corporate hierarchy by changing companies, recruited in the pre-Reagan years when my theory of a politically mediated labor market would predict a greater demand for African Americans at the executive level.

5. I characterize the move as forced because it was not consistent with what this manager had planned for career advancement. Moreover, he was fully aware that affirmative-action and related jobs are, as he put it, "dead-end jobs [with] no power."

6. There are two ways to prove discrimination in the labor market. The first is by showing disparate *treatment* and the second is by showing disparate *impact*. See Lazaar (1991) for a straightforward account of what constitutes labor market discrimination.

Chapter 9

1. See Bloom (1987) for a perceptive analysis of the emergence of new civil rights leadership in the South.

REFERENCES

Adams, Arvil V. 1972. *Toward Fair Employment and the E.E.O.C.: A Study of Compliance under Title VII of the Civil Rights Act of 1969.* Washington, D.C.: Equal Employment Opportunity Commission.

Allen, Robert L. 1970. *Black Awakening in Capitalist America: An Analytic History.* Garden City, N.Y.: Doubleday.

Althauser, Robert P. 1975. *Unequal Elites.* New York: Wiley.

Ames, Charles B., and James D. Hlavacek. 1989. *Market Driven Management: Prescriptions for Survival in a Turbulent World.* Homewood, N.J.: Dow Jones–Irwin.

Anderton, Douglas L., Richard E. Barrett, and Donald J. Bogue. Forthcoming. *The Population of the United States.*

Ashenfelter, Orley, and James J. Heckman. 1976. "Measuring the Effect of an Anti-discrimination program." In *Evaluating the Labor Market Effects of Social Programs,* ed. O. Ashenfelter and J. Blum, 46–89. Princeton: Industrial Relations Section, Princeton University.

Becker, Brian, and Stephen Hills. 1979. "Today's Teenage Unemployed–Tomorrow's Working Poor?" *Monthly Labor Review* 102 (January): 69–71.

Becker, Gary. 1981. *A Treatise on the Family.* Cambridge: Harvard University Press.

Belohlav, James A., and Eugene Ayton. 1982. "Equal Opportunity Law: Some Common Problems." *Personnel Journal* 61: 282–285.

Benokraitis, Nijole V., and Joe Feagin. 1978. *Affirmative Action and Equal Opportunity.* Boulder: Westview.

Betsey, Charles. 1982. *Minority Participation in the Public Sector.* Washington, D.C.: Urban Institute Press.

Bloom, Jack M. 1987. *Class, Race, and the Civil Rights Movement.* Bloomington: Indiana University Press.

Blumberg, Paul. 1980. *Inequality in the Age of Decline.* New York: Oxford University Press.

Branch, Shelly. 1993. "America's Most Powerful Black Executives." *Black Enterprise,* February, 79–134.

Brimmer, Andrew. 1976. "The Economic Position of Black Americans." *Special Report to the National Commission for Manpower Policy.* No. 9. Washington, D.C.: Commission for Manpower.

Broderick, Francis L., and August Meier. 1971. "Black Protest Thought in the Twentieth Century." Ed. August Meier, Elliott Rudwick, and Francis L. Broderick. 2d. ed. Indianapolis: Bobbs-Merrill.

Brown, Michael K., and Steven P. Erie. 1981. "Blacks and the Legacy of the Great Society: The Economic and Political Impact of Federal Social Policy." *Public Policy* 29 (Summer): 299–330.

Burstein, Paul. 1985. *Discrimination, Jobs, and Politics: The Struggle for Equal Employment Opportunity in the United States Since the New Deal.* Chicago: University of Chicago Press.

Burstein, Paul, and Susan Pitchford. 1990. "Social-Scientific and Legal Challenges to Education and Test Requirements in Employment." *Social Problems* 37 (May): 243–257.

Chicago United. 1980. *Chicago United Compendium of Minority Professional Service Firms.* Chicago: Chicago United.

Chicago Urban League. 1977. *Blacks in Policy-Making Positions in Chicago.* Chicago: Chicago Urban League.

Clark, Kenneth B. 1965. *Dark Ghetto: Dilemmas of Social Power.* New York: Harper and Row.

Cohn, Jules. 1975. "Is Business Meeting the Challenge of Urban Affairs?" *Equal Opportunity in Business.* Harvard Review Reprint Series, no. 21132.

Collins, Sharon M. 1983. "The Making of the Black Middle Class." *Social Problems* 30 (April): 369–382.

Colton, Elizabeth O. 1989. *The Jackson Phenomenon: The Man, the Power, the Message.* New York: Doubleday.

Cose, Ellis. 1993. *The Rage of a Privileged Class: Why Are Middle-Class Blacks Angry? Why Should America Care?* New York: HarperCollins.

Davis, George, and Glegg Watson. 1982. *Black Life in Corporate America: Swimming in the Mainstream.* New York: Doubleday.

Dingle, Derek. 1988. "Will Black Managers Survive Corporate Downsizing?" *Black Enterprise,* March, 51.

Donovan, J. C. 1967. *The Politics of Poverty.* New York: Pegasus.

Drake, St. Clair, and Horace R. Cayton. 1962. *Black Metropolis.* Vol. 2. New York: Harper and Row.

Dreyfuss, Joel, and Charles Lawrence III. 1979. *The Bakke Case: The Politics of Inequality.* New York: Harcourt, Brace, Jovanovich.

Eccles, Mary. 1975. "Race, Sex, and Government Jobs: A Study of Affirmative Action Programs in Federal Agencies." Ph.D. diss., Harvard University.

Farley, Reynolds. 1977. "Trends in Racial Inequalities: Have the Gains of the 1960s Disappeared in the 1970s?" *American Sociological Review* 42 (April): 189–208.

———. 1984. *Blacks and Whites: Narrowing the Gap?* Cambridge: Harvard University Press.

Farley, Reynolds, and Walter R. Allen. 1987. *The Color Line and the Quality of Life in America.* New York: Russell Sage.

Farley, Reynolds, and Suzanne M. Bianchi. 1983. "The Growing Gap between Blacks." *American Demographics,* July, 15–18.

Feagin, Joe R., and Melvin P. Sikes. 1994. *Living with Racism: The Black Middle-Class Experience.* Boston: Beacon.

Featherman, David, and Robert Hauser. 1976. "Changes in the Socioeconomic Stratification of Races, 1962–1973." *American Journal of Sociology* 82 (November): 621–651.

Fernandez, John. 1981. *Racism and Sexism in Corporate Life: Changing Values in American Business.* Lexington, Mass.: Lexington Books.

Flanagan, Robert J. 1976. "Actual versus Potential Impact of Government Anti-Discriminating Programs." *Industrial and Labor Relations Review* 29 (July): 486–507.

Fortune. 1968. "The Editor's Desk." January, 127–128.

Franklin, Raymond S., and Solomon Resnik. 1973. *The Political Economy of Racism.* New York: Holt.

Frazier, E. Franklin. 1957. *The Black Bourgeoisie: The Rise of a New Middle Class.* New York: Free Press.

Freeman, Richard. 1973. "Changes in the Labor Market for Black Americans, 1968–1972." *Brookings Papers on Economic Activity 1* (Summer): 57–120.

———. 1976a. *The Black Elite.* New York: McGraw-Hill.

———. 1976b. *The Over-Educated American.* New York. Academic Press.

———. 1981. "Black Economic Progress after 1964: Who Has Gained and Why." In *Studies in Labor Markets,* ed. S. Rosen, 247–295. Chicago: University of Chicago Press.

Fromm, Erich. 1955. *The Sane Society.* New York: Rinehart, Mills.

Gershman, Carl, and Kenneth Clark. 1980. "A Matter of Class." *New York Times Magazine,* October 5, 22.

Gibson, Parke D. 1978. *$70 Billion in the Black: America's Black Consumers.* New York: Macmillan.

Glasgow, Douglas G. 1980. *The Black Underclass: Poverty, Unemployment, and Entrapment of Ghetto Youth.* San Francisco: Jossey-Bass.

Hampton, Robert E. 1977. "The Response of Governments and the Civil Service to Antidiscrimination Efforts." In *Equal Rights and Industrial Relations,* ed. L. Hausman, O. Ashenfelter, B. Rustin, R. F. Schubert, and D. Slaiman. Madison, Wis.: Industrial Relations Research Association.

Harrington, Alan. 1959. *Life in the Crystal Palace.* New York: Knopf.

Harrington, Michael. 1984. *The New American Poverty.* New York: Holt, Rinehart, and Winston.

Harris, Ron. 1994. "The Riots Helped No One—Except the Well-Placed Few." *Los Angeles Times,* 1 May, sec. B, 1.

Hauser, Robert, and David Featherman. 1974. "White/Non-White Differentials in Occupational Mobility among Men in the United States, 1962–1972." *Demography* 11: 247–266.

Hausman, Leonard, Orley Ashenfelter, Bayard Rustin, Richard F. Schubert, and Donald Slaiman, eds. 1977. *Equal Rights and Industrial Relations.* Madison, Wis.: Industrial Relations Research Association.

Haworth, Joan, James Gwartney, and Charles Haworth. 1975. "Earnings, Productivity, and Changes in Employment Discrimination during the 1960s." *American Economic Review* 65 (March): 158–168.

Haynes, Ulric, Jr. 1968. "Equal Job Opportunity: The Credibility Gap." *Harvard Business Review,* May–June, 113–120.

Heckman, James J. 1976. "Simultaneous Equation Models with and without Structural Shifts in Equations." In *Studies in Non-linear Estimation,* ed. Stephen Goldfeld and Richard Quandt, 235–272. Cambridge, Mass.: Ballinger.

Heckman, James J., and Brook S. Payner. 1989. "Determining the Impact of Federal Antidiscrimination Policy on the Economic Status of Blacks: A Study of South Carolina." *American Economic Review* 79: 138–177.

Heckman, James J., and Kenneth Wolpin. 1976. "Does the Contract Compliance Program Work? An Analysis of Chicago Data." *Industrial and Labor Relations Review* 29 (July).

Heidrick and Struggles, Inc. 1979a. *Chief Personnel Executives Look at Blacks in Business.* Chicago: Heidrick and Struggles.

———. 1979b. *Profile of a Black Executive.* Chicago: Heidrick and Struggles.

Henderson, Hazel. 1968. "Should Business Tackle Society's Problem?" *Harvard Business Review,* July–August, 77–82.

Herring, Cedric, and Sharon Collins. 1995. "Retreat from Equal Opportunity? The Case of Affirmative Action." In *The Bubbling Cauldron,* ed. J. Feagin. Minneapolis: University of Minnesota Press.

Hertz, Diane. 1990. "Worker Displacement in a Period of Rapid Job Expansion: 1983–87." *Monthly Labor Review* (Bureau of Labor Statistics), May, 21–33.

Hill, Herbert. 1977. "The Equal Employment Acts of 1964 and 1972: A Critical Analysis of the Legislative History and Administration of the Law." *Industrial Relations Law Journal* 2, 1 (Spring): 1–98.

Hout, Michael. 1984. Occupational Mobility of Black Men: 1962–1973. American Sociological Review 49: 308–322.

Hudson, William, and Walter Broadnax. 1982. "Equal Employment Opportunity: A Public Policy." *Public Personnel Management* 11: 268–276.

Isaac, Larry, and William R. Kelly. 1981. "Racial Insurgency, the State, and Welfare Expansion: Local and National Level Evidence from the Postwar United States. *American Journal of Society* 6 (May): 1348–1385.

Ismail, Sherille. 1985. "Despite Gains of the 1960s and 1970s Blacks' Progress Lags Behind Whites'." Pp. 10–12 in *Point of View* 20 (Spring/Summer). Washington, D.C.: Congressional Black Caucus Foundation Inc.

Jackall, Robert. 1988. *Moral Mazes: The World of Corporate Managers.* Oxford: Oxford University Press.

Jackson, Jesse L. 1979. *Straight from the Heart.* Philadelphia: Fortress Press.

Jaynes, Gerald David, and Robin M. Williams, Jr. 1989. *A Common Destiny: Blacks and American Society.* Washington, D.C.: National Academy Press.

Jones, Edward W. 1986. "Black Managers: The Dream Deferred." *Harvard Business Review,* May–June, 84–89.

Kanter, Rosabeth. 1977. *Men and Women of the Corporation.* New York: Basic Books.

Kasarda, John D. 1980. "The Implications of Contemporary Redistribution Trends for National Policy." *Social Science Quarterly* 61: 373–400.

———. 1986. "The Regional and Urban Redistribution of People and Jobs in the U.S." Working paper prepared for the National Research Council Committee on National Urban Policy, National Academy of Sciences.

Kirschenman, Jolene, and Kathryn M. Neckerman. 1991. " 'We'd Love to

Hire Them, But . . .': The Meaning of Race for Employers." In *The Urban Underclass,* ed. C. Jencks and P. Peterson, 203–232. Washington, D.C.: Brookings.

Korn/Ferry. 1986. *Korn/Ferry International's Executive Profile: A Survey of Corporate Leaders in the Eighties.* New York: Korn/Ferry. Pamphlet.

———. 1990. *Korn/Ferry International's Executive Profile: A Decade of Changes in Corporate Leadership.* New York: Korn/Ferry. Pamphlet.

Landry, Bart. 1987. *The New Black Middle Class.* Berkeley: University of California Press.

Lazaar, Edward. 1991. "Discrimination in Labor Markets." In *Essays on the Economics of Discrimination,* ed. E. Hoffman. Mich.: Upjohn Institute for Employment Research.

Leonard, Jonathan S. 1982. "The Impact of Affirmative Action on Minority and Female Employment." Working paper, School of Business Administration, University of California, Berkeley.

———. 1984. "The Impact of Affirmative Action on Employment." *Journal of Labor Economics* 2: 439–64.

———. 1988. "Women and Affirmative Action in the 1980s." Paper presented at the American Economic Association Annual Meeting, October.

Levitan, Sar. 1969. *The Great Society's Poor Law: A New Approach to Poverty.* Baltimore: Johns Hopkins University Press.

Levitan, Sar, William B. Johnson, and Robert Taggart. 1975. *Still a Dream: The Changing Status of Blacks since 1960.* Cambridge: Harvard University Press.

Loury, Glenn C. 1985. "The Moral Quandary of the Black Community." *Public Interest* 79 (Spring): 9–22.

Lydenberg, Steven D., Alice Tepper Marlin, Sean O'Brien Strub, and the Council on Economic Priorities. 1986. *Rating America's Corporate Conscience: A Provocative Guide to the Companies behind the Products You Buy Every Day.* Reading, Mass.: Addison-Wesley.

Maccoby, Michael. 1976. *The Gamesman.* New York: Simon and Schuster.

Marable, Manning. 1983. *How Capitalism Underdeveloped Black America: Problems of Race, Political Economy, and Society.* Boston: South End Press.

Mare, Robert D., and Christopher Winship. 1980. "Family Background, Race, and Youth Unemployment, 1968–1978: Evidence for a Black Underclass." Paper presented at the meetings of the American Sociological Association, New York, August.

References ✦ **179**

Massey, Douglas S., and Nancy A. Denton. 1993. *American Apartheid: Segregation and the Making of the Underclass.* Cambridge: Harvard University Press.

McKersie, Robert. n.d. *Minority Employment Patterns in an Urban Labor Market: The Chicago Experience.* Report Commission of the Equal Employment Opportunity Commission. Washington, D.C.: Equal Employment Opportunity Commission.

Meier, August. 1967. "Civil Rights Strategies for Negro Employment." In *Employment, Race, and Poverty,* ed. A. M. Ross and Herbert Hill, 186–188. New York: Harcourt, Brace, and World.

Mills, Charles. 1951. *White Collar: The American Middle Classes.* New York: Oxford University Press.

Murray, Charles. 1984. *Losing Ground: American Social Policy, 1950–1980.* New York: Basic Books.

Nathan, Richard P. 1969. *Jobs and Civil Rights: The Role of the Federal Government in Promoting Equal Opportunity in Employment and Training.* Prepared for the U.S. Commission on Civil Rights. Washington, D.C.: Brookings.

Newman, Dorothy K., Nancy J. Amidei, Barbara L. Carter, Dawn Day, William J. Kruvant, and Jack S. Russell. 1978. *Protest, Politics, and Prosperity: Black Americans and White Institutions, 1940–1975.* New York: Pantheon.

Orfield, Gary, and Carol Ashkinaze. 1991. *The Closing Door: Conservative Policy and Black Opportunity.* Chicago: University of Chicago Press.

Ornati, Oscar A., and Anthony Pisano. 1972. "Affirmative Action: Why It Isn't Working." *Personnel Administration* 35 (September): 50–52.

Parsons, Donald O. 1980. "Racial Trends in Male Labor Force Participation." *American Economic Review* 70 (December): 911–920.

Peters, Tom. 1988. *Thriving on Chaos: Handbook for a Management Revolution.* New York: Knopf.

Pfeffer, Jeffrey. 1982. *Power in Organizations.* Boston: Pitman.

Piven, Frances Fox, and R. A. Cloward. 1971. *Regulating the Poor: The Functions of Public Welfare.* New York: Pantheon.

———. 1977. *Poor People's Movements: Why They Succeed, How They Fail.* New York: Pantheon.

Poole, Isiah. 1981. "Uncle Sam's Pink Slip." *Black Enterprise* 12 (December): 52.

Purcell, Theodore V. 1977. "Management and Affirmative Action in the Late Seventies." In *Equal Rights and Industrial Relations,* ed. L. Hausman, O. Ashenfelter, B. Rustin, R. F. Schubert, and D. Slaiman, 71–103. Madison, Wis.: Industrial Relations Research Association.

Ramos, Dante. 1994. "White Minorities." *New Republic,* 17 October, 24.

Reynolds, William B. 1983. "The Justice Department's Enforcement of Title VII." *Labor Law Journal* 34: 259–265.

Riesman, David, Nathan Glazer, and Reuel Denney. 1953. *The Lonely Crowd: A Study of the Changing American Character.* New York: Doubleday.

Ross, Heather L., and Isabel Sawhill. 1975. *Time of Transition: The Growth of Families Headed by Women.* Washington, D.C.: Urban Institute Press.

Sheppard, Harold L., and Herbert E. Stringer. 1966. *Civil Rights, Employment, and the Social Status of American Negroes.* Based on a report for the U.S. Commission on Civil Rights (contract no. CCR-66-5). Kalamazoo, Mich.: Upjohn Institute for Employment Research.

Siegel, Paul M. 1965. "On the Cost of Being Negro." *Sociological Inquiry* 35 (Winter): 41–57.

Sitkoff, Harvard. 1978. *A New Deal for Blacks: The Emergence of Civil Rights as a National Issue.* New York: Oxford University Press.

———. 1981. *The Struggle for Black Equality: 1954–1980.* New York: Hill and Wang.

Smith, James P., and Finis R. Welch. 1977. "Black-White Male Wage Ratios, 1960–1970." *American Economic Review* 67 (June): 323–338.

———. 1978a. *The Convergence to Racial Equality in Women's Wages.* Santa Monica, Calif.: Rand Corporation.

———. 1978b. *Race Differences in Earnings: A Survey and New Evidence.* Santa Monica, Calif.: Rand Corporation.

———. 1983. "Longer Trends in Black/White Economic Status and Recent Effects of Affirmative Action." Paper presented at the Social Science Research Council Conference, Chicago.

———. 1984. "Affirmative Action and Labor Markets." *Journal of Labor Economics* 2: 269–299.

———. 1986. *Closing the Gap: Forty Years of Economic Progress for Blacks.* Santa Monica, Calif.: Rand Corporation.

Sowell, Thomas. 1983. "The Economics and Politics of Race." *Firing Line,* November. Transcript.

Starobin, Paul. 1993. "Unequal Shoes." *National Journal,* 11 September, 2176–2179.

Theodore, Nikolas C., and D. Garth Taylor. 1991. *The Geography of Op-*

portunity: The Status of African Americans in the Chicago Area Economy. Chicago: Chicago Urban League.

Thurow, Lester. 1969. *Poverty and Discrimination.* Washington, D.C.: Brookings.

Urban League. 1961. *Equal Rights — Greater Responsibility: The Challenge to Community Leadership in 1961.* Chicago: Urban League.

U.S. Bureau of the Census. 1963. *Occupational Characteristics.* Series PC(2)-7A. Washington, D.C.: Bureau of the Census.

———. 1973. *Occupational Characteristics.* Series PC(2)-7A. Washington, D.C.: Bureau of the Census.

———. 1979a. *Social and Economic Status of the Black Population in the United States, 1790–1978: A Historical View.* Series P-23, no. 80. Washington, D.C.: Bureau of the Census.

———. 1979b. *1977 Survey of Minority-Owned Business Enterprises: Black.* Series MB77-1. Washington, D.C.: Social and Economic Statistics Administration.

———. 1980a. *Characteristics of the Population: Detailed Population Characteristics.* pt. 1, sec. A. Series PC-80-1-D1-A. Washington, D.C.: Bureau of the Census.

———. 1980b. *Current Population Reports: Money and Income of Persons in 1979.* Series P-60, no. 129. Washington, D.C.: Bureau of the Census.

———. 1982. *1982 Survey of Minority-Owned Business Enterprises: Black.* Series MB82-1. Washington, D.C.: Bureau of the Census.

U.S. Bureau of Labor Statistics. 1982. *Current Population Survey: 1983 Annual Averages. Basic Table.* Washington, D.C.: Bureau of Labor Statistics.

———. n.d. Office of Employment and Statistics tables. Washington, D.C.: Bureau of Labor Statistics.

U.S. Commission on Civil Rights. 1969. *Staff Memorandum.* February 4. Washington, D.C.: U.S. Government Printing Office.

U.S. Department of Commerce. 1979. *A Strategy for Minority Business Enterprise Development.* Washington, D.C.: Minority Business Development Agency.

———. 1980. *1978 Status of Minorities and Women in State and Local Governments.* Washington, D.C.: Equal Employment Opportunity Commission.

———. 1981a. *Minority Business Development in the Eighties.* Washington, D.C.: Minority Business Development Agency.

———. 1981b. *Performance for Minority Business Development, Fiscal Year 1980.* Washington, D.C.: Minority Business Development Agency.

U.S. Department of Housing and Urban Development. 1991. *Housing Discrimination Study.* Washington, D.C.: U.S. Government Printing Office.

U.S. Equal Employment Opportunity Commission. 1978. *Ninth Annual Report.* Washington, D.C.: U.S. Government Printing Office.

———. 1980a. *1978 Status of Minorities and Women in State and Local Governments.* Washington, D.C.: Equal Employment Opportunity Commission.

———. 1980b. *Federal Civilian Work Force Statistics.* AR-80-21 (November). Washington, D.C.: Office of Personnel Management.

———. 1982. *Job Patterns for Minorities and Women in Private Industry.* Vol. 1. Washington, D.C.: U.S. Government Printing Office.

———. n.d. Report. Washington, D.C.: Office of Program Research, Survey Division.

U.S. Kerner Commission. 1968. *Report of the National Advisory Commission on Civil Disorders.* New York: Bantam.

U.S. Senate Committee on Labor and Public Welfare, Subcommittee on Civil Rights. 1954. *Antidiscrimination in Employment: Hearings 5.692.* 23–25 February, 1–3 March.

Vroman, Wane. 1974. "Changes in Black Workers' Relative Earnings: Evidence from the 1960s." In *Patterns of Racial Discrimination,* ed. George M. Furstenbery, Bennett Harrison, and Ann Horowitz. Lexington, Mass.: Heath.

Wall Street Journal. 1984. "Taking a Chance: Many Blacks Jump off the Corporate Ladder to be Entrepreneurs." 2 August, sec. 1, 1.

———. 1988. "Labor Letter: A Special News Report on People and Their Jobs in Offices, Fields, and Factories." 9 February, 1.

White, Theodore. 1964. *The Making of the President.* New York: Atheneum.

Whittingham-Barnes, Donna. 1993. "Workforce Trends: Is There Life after Unemployment?" *Black Enterprise,* February, 181–186.

Whyte, William H., Jr. 1956. *The Organization Man.* New York: Simon and Schuster.

Williams, Walter. 1982. "Rethinking the Black Agenda." In *Proceedings of the Black Alternatives Conference,* 16–31. San Diego, Calif.: New Coalition for Economic and Social Change.

Wilson, Cynthia A., James H. Lewis, and Cedric Herring. 1991. *The 1991 Civil Rights Act: Restoring Our Basic Protections.* Chicago: Chicago

Urban League and Chicago Lawyers' Committee for Civil Rights under Law.

Wilson, William J. 1978. *The Declining Significance of Race*. Chicago: University of Chicago Press.

———. 1981. "The Black Community in the 1980s: Questions of Race, Class, and Public Policy." *Annals of the American Academy of Political and Social Sciences* 454: 26–41.

———. 1984. *Race, Economics, and Corporate America*. Wilmington, Del.: Scholarly Resources.

———. 1987. *The Truly Disadvantaged: The Inner City, the Underclass, and Public Policy*. Chicago: University of Chicago Press.

Wolkinson, Benjamin W. 1973. *Black Unions and the EEOC*. Lexington, Mass.: Heath.

Yarmolinsky, A. 1969. "The Beginning of OEO." In *On Fighting Poverty: Perspectives from Experience,* ed. James L. Sundquist, 34–51. New York: Basic Books.

Zweigenhaft, Richard L., and G. William Domhoff. 1991. *Blacks in the White Establishment? A Study of Race and Class in America*. New Haven: Yale University Press.

INDEX

accounting: affirmative action programs in, 61; blacks underrepresented in, 50; racialization of jobs in, 38–42

advertising firms, black-owned, racialization of jobs in, 38–42

affirmative action: black CEOs in, 73–75, 147–151, 172n.5; black economic opportunity and, 17–27, 118, 158–159; black groups' pressure for, 107–111; career-enhancing strategies and, 81–84; challenges to, xi–xii, 17, 169n.1 (Chapter 2); compliance with anti-discrimination legislation through, 100–107; downsizing and flattening of corporations and, 144–147; economic conditions and, 166–168; education levels and, 9–10; expansion of, 18–19; federal contract compliance and, 20–21, 60–63; fragility of African American middle class and, 162–168; individual and group activism of black executives and, 90–93, 104–105; job security issues facing, 126–129; legal challenges to, 119–120, 170nn.1–2; mainstreaming of black executives and, 73–75, 85–89; as mobility trap for black executives, 77–80, 102–107; racial equality and, 159–161; racialized jobs and, 77–80, 157–159

African American business elite. *See* black corporate executives

African American–owned business sector, racialized roles in, 37–42

Aid to Families with Dependent Children, 24; racialized jobs for African Americans in, 31

Allen, Walter R., 17

Althauser, Robert, 25

American Can Company, 59

American Telephone and Telegraph (AT&T), EEOC charges against, 19

Anderton, Douglas L., 146

Anheuser-Busch, 128–129

Arthur Andersen & Company, 61

Ashenfelter, Orley, 20–21, 60

Ashkinaze, Carol, 145

Bakke, Allan, 119–120

Barrett, Richard E., 146

Bay Area Rapid Transit system (BART), 60

"beached" executives, 150, 152

Becker, Brian, 6–7

black corporate executives: career-enhancing strategies of, 81–89, 129–133; defined, 13; downsizing and flattening of job market for, 144–147; elite status of, xi–xiii; exiting executives, characteristics of, 133–136; external relations with black groups, 107–111; golden handcuffs for, 93–96; history of hiring practices for, 45–48, 116–118; individual and group activism of, 90–93, 102–107;

sex discrimination, black business elite and, 14

Sharpe, Rochelle, 145–146, 172n.2

Siegel, Paul M., 48

Sikes, Melvin P., 6, 164

Sitkoff, Harvard, 63, 66, 119

Smith, James P., 3–4, 6, 47, 155

social conditions, African American economic mobility and, 15–16, 162–163

social policy. *See* public policy

social science occupations, black men employed in, 30

social services occupations: black employment opportunities and expansion of, 23–26, 30; black men employed in, 30; dominance of federal funding for, 34–35. *See also* public sector

South Carolina textile industry, federal contract compliance and minority hiring in, 21

Southern Christian Leadership Conference (SCLC), 65–68

Sowell, Thomas, 9–10

Starobin, Paul, 122

Student Nonviolent Coordinating Committee (SNCC), 64

Students for a Democratic Society (SDS), 64

Supreme Court (U.S.), retreat of, from affirmative action, 119–120, 170nn.1–2 (Chapter 7)

"tan" territories, black consumers and sales employees confined to, 53–54

Taylor, D. Garth, 149

Theodore, Nikolas C., 149

Thurow, Lester, 48

Touche Ross, 61

Truman, Harry S (President), 17

underclass, defined, 169n.1; deterioration in, 5–6; growth of, despite affirmative action, 1; militancy versus passivity in, 164–166; pacification of, through racialized jobs, 31

underemployment patterns, pre-1960s job market, 50–57, 155–159

unskilled labor market, jobs for African Americans in, 48–50

upward mobility for African Americans: blue-collar employment and, 8; cultural factors in, 7–9; fragility of African American middle class and, 162–168; mobility trap for black executives, 77–81; pessimism from black executives about, 140–153, 171n.1 (Chapter 8); placement patterns in corporate positions and, 141–144, 147–151, 172n.4; political conditions and, 15–16; racialization of jobs and, 12, 29–43, 73; sex-related differences in, 3–5

urban affairs programs, black CEOs in, 73–75, 157–158; corporations' use of, 107–111, 115–116, 118

urban crisis, expansion of black job opportunities and, 68–72, 157–158, 170n.6; racialized jobs as result of, 77–80

Urban League, 165

U.S. News and World Report, 123

Voluntary Organization of Blacks in Government, 26

Voting Rights Act of 1965, 2

Vroman, Wane, 60

Wall Street Journal, 143, 145

War on Poverty, 23–24